Molecular Basis of Neurological Disorders and their Treatment

Molecular Basis of Neurological Disorders and their Treatment

Edited by
J.W. Gorrod,
O. Albano,
E. Ferrari and
S. Papa

CHAPMAN & HALL
London · New York · Tokyo · Melbourne · Madras

UK	Chapman & Hall, 2–6 Boundary Row, London SE1 8HN
USA	Chapman & Hall, 29 West 35th Street, New York NY10001
JAPAN	Chapman & Hall Japan, Thomson Publishing Japan, Hirakawacho Nemoto Building, 7F, 1–7–11 Hirakawa-cho, Chiyoda-ku, Tokyo 102
AUSTRALIA	Chapman & Hall Australia, Thomas Nelson Australia, 102 Dodds Street, South Melbourne, Victoria 3205
INDIA	Chapman & Hall India, R. Seshadri, 32 Second Main Road, CIT East, Madras 600 035

First edition 1991

Typeset in 10/12pt Baskerville by Mews Photosetting, Kent
Printed in Great Britain by St. Edmundsbury Press,
Bury St Edmunds, Suffolk

ISBN 0 412 40410 9

British Library Cataloguing in Publication Data

Molecular basis of neurological disorders and
their treatment.
I. Gorrod, J.W.
616.8

ISBN 0-412-40410-9

Library of Congress Cataloging-in-Publication Data
Available

Contents

List of contributors

1 A. PULLMAN
Institut de Biologie Physio-chimique, Paris, France

2 S. PAPA, F. GUERRIERI AND F. ZANOTTI
Institute of Medical Biochemistry and Chemistry and Centre for the Study of Mitochondria and Energy Metabolism, University of Bari, Italy

3 M.N. GADALETA, V. PETRUZZELLA, M. RENIS, F. FRACASSO AND P. CANTATORE
Department of Biochemistry and Molecular Biology, University of Bari and CSMME, Bari, Italy

4 M.A. JOHNSON
School of Neurosciences, University of Newcastle upon Tyne, UK

5 P. RICCIO, A. BOBBA,[1] G.M. LIUZZI, T. ZACHEO[2] and E. QUAGLIARIELLO
Dipartimento di Biochimica e Biologia Molecolare, Università di Bari, Italy; [1]Centro Studio sui Mitocondri e Metabolismo Energetico, CNR, Bari, Italy; [2]Istituto di Nematologia Agraria, CNR, Bari, Italy

6 A. BRUNI,[1] L. MIETTO,[2] F. BELLINI,[2] D. PONZIN,[2] E. CASELLI[2] and G. TOFFANO[2]
[1]Department of Pharmacology, University of Padova, Italy; [2]Fidia Research Laboratories, Abano Terme, Italy

7 G. GENNARINI, F. VITIELLO, P. CORSI, G. CIBELLI, M. BUTTIGLIONE AND C. DI BENEDETTA
Institute of Human Physiology, University of Bari, Italy

8 C. ALAFACI, F.M. SALPIETRO AND F. TOMASELLO
Department of Neurosurgery, University of Messina, Italy

9 A. HIRANO AND R. WAKI
Division of Neuropathology, Department of Pathology, Montefiore Medical Center Albert Einstein College of Medicine, Bronx, New York, 10467, USA

10 W. HEINDEL, W. STEINBRICH, J. BUNKE AND G. FRIEDMANN
Department of Diagnostic Radiology, University of Cologne Medical School, Köln, Germany

11 G.L. LENZI, A PADOVANI, P. PANTANO, M. IACOBONI, M. RICCI, P. FRANCO AND V. DI PIERO
Neurolological Clinic, Department of Neurological Sciences, University 'La Sapienza', Rome, Italy

12 H. SOMER AND R.O. ROINE
Department of Neurology, University of Helsinki, Finland

13 D.F. CONDORELLI AND A.M. GIUFFRIDA STELLA
Institute of Biochemistry, Faculty of Medicine, University of Catania, Viale A. Doria 6, 95125 Catania, Italy

14 GREGORY J. DEL ZOPPO
Department of Molecular and Experimental Medicine and Division of Hematology/Medical Oncology, Scripps Clinic and Research Foundation, La Jolla, California USA

15 S.M. MEDD,[1] J.E. WALKER,[1] I.M. FEARNLEY,[1] R.D. JOLLY[2] AND D.N. PALMER[2]
[1]Medical Research Council, Laboratory of Molecular Biology, Hills Road, Cambridge CB2 2QH, UK; [2]Department of Veterinary Pathology and Public Health, Massey University, Palmerston North, New Zealand

16 A. FEDERICO
Istituto di Scienze Neurologiche e Centro Per lo Studio delle Encefalo-Neuro-Miopatie Genetiche, Università degli Studi di Siena, Italy

17 T. OZAWA, M. TANAKA, W. SATO, K. OHNO, M. YONEDA AND T. YAMAMOTO
Department of Biomedical Chemistry, Faculty of Medicine, University of Nagoya, Japan

18 J. POULTON, M.E. DEADMAN, M. SOLYMAR, S. RAMCHARAN AND R.M. GARDINER
University of Oxford Department of Paediatrics, John Radcliffe Hospital, Headington, Oxford OX3 9DU, UK

19 B. KADENBACH,[1] P. SEIBEL,[1] M.A. JOHNSON[2] AND D. TURNBULL[2]
[1]Fachbereich Chemie, Philipps-Universität, Hans-Meerwein-Strasse, D-3550 Marburg, Germany; [2]Department of Neurology, University of Newcastle upon Tyne, UK

20 E.A. SCHON,[1,2] C.T. MORAES,[2] S. MITA,[1] H. NAKASE,[1] A. LOMBES,[1] S. SHANSKE,[1] E. ARNAUDO,[1] Y. KOGA,[1] M. ZEVIANI,[1] R. RIZZUTO,[1] A.F. MIRANDA,[1,3] E. BONILLA,[1] AND S. DiMAURO[1]
Departments of [1]Neurology, [2]Genetics and Development, and [3]Pathology, Columbia University, New York

21 C. MARSAC,[1] F. DEGOUL,[1] N.B. ROMERO,[2] J.L. VAYSSIERE,[4] P. LESTIENNE,[3] I. NELSON,[3] D. FRANCOIS[1] AND M. FARDEAU[2]
[1]INSERM U75, 156 rue de Vaugirard, 75015 Paris, France; [2]INSERM U153, 17 rue du Fer à Moulin, 75005 Paris, France; [3]INSERM U298, CHR, 49033 Angers, France; [4]Collège de France, Biologie Cellulaire, 11 place Marcelin-Berthelot, 75231 Paris Cedex 05 France

22 N. BRESOLIN, I. MORONI, E. CIAFALONI, M. MOGGIO, G. IMEOLA, A. GATTI, G. COMI AND G. SCARLATO
Istituto di Clinica Neurologica, Università degli Studi di Milano, Centro Dino Ferari, Via F. Sforza 35, 20122 Milan, Italy

23 A. OLDFORS,[1] E. HOLME,[2] B. KRISTIANSSON,[3] N.-G. LARSSON[2] AND M. TULINIUS[3]
[1]Department of Pathology, Gothenburg University, Sahlgren's Hospital, S-413 45 Gothenburg, Sweden; [2]Department of Clinical Chemistry, Gothenburg University, Sahlgren's Hospital, S-413 45 Gothenburg, Sweden; [3]Department of Paediatrics, Gothenburg University, East Hospital, S-416 85 Gothenburg, Sweden

24 C. ANGELINI, A. MARTINUZZI, M. FANIN, M. ROSA, R. CARROZZO AND L. VERGANI
Regional Neuromuscular Centre, Neurological Clinic, University of Padova, Italy

25 M.R. CARRATÙ and D. MITOLO-CHIEPPA
Institute of Pharmacology, Medical Faculty, University of Bari, I-70124 Bari, Italy

26 U. QUAST
Preclinical Research, Sandoz Ltd, CH 4002 Basel, Switzerland

27 Z. DRAHOTA,[1] J. MOUREK,[2] H. RAUCHOVÁ[1] AND S. TROJAN[2]
[1]Institute of Physiology, Czechoslovak Academy of Sciences, and [2]Institute of Physiology, Faculty of Medicine, Charles University, Prague, Czechoslovakia

28 D. DI MONTE, K.P. SCOTCHER AND E.Y. WU
California Parkinson's Foundation, San Jose, California, USA

29 E. DANIELE,[1] M.D. LOGRANO,[1] C. LOPEZ,[2] F. BATTAINI,[3] M. TRABUCCHI[3] AND S. GOVONI[1]
[1]Department of Pharmacobiology, University of Bari, Italy; [2]Institute of Pharmacology, University of Milan, Italy; [3]Chair of Toxicology, 2nd University of Rome, Italy

30 J.W. GORROD, M.S. FELDMAN, A. KLER AND A. ROSEN
Chelsea Department of Pharmacy, King's College London (University of London), Manresa Road, London SW3 6LX, UK

International Biomedical Institute

International Biomedical Institute (IBMI) was founded in Bari in May 1986 thanks to the efforts and the financial donations of an initial group of Founder Members. These were later joined by others to constitute the **College of Founder Members,** who promote, support and project the activities of IBMI.

IBMI is recognized as Ente Morale by Regione Puglia, and is a private non-profit organization. Its objectives are to promote and develop scientific, clinical and technical research in selected areas of Biological Sciences and medicine. IBMI will pursue its objectives by obtaining the necessary financial resources and utilizing them for:

building and financing international research institutes in Puglia;

development of biotechnologies and training of experts;

promoting and financing research programmes in its own laboratories or in other research institutes and clinics; also providing Fellowships;

organizing and conducting courses, scientific meetings and issuing technical-scientific publications.

In order to programme and evaluate the scientific activities of the Association, the College of Founder Members is assisted by a Scientific Committe, formed by scientists from all over the world.

The financial resources necessary for the activities of IBMI in addition to donations of Founder Members and contributions of Supporters will also derive from funds, grants and contracts from public and private institutions and companies.

Programming of scientific activities and research laboratories is developed by the administrators and scientific staff on the basis of the directions and appraisals of the College of Founder Members and the Scientific Committee.

IBMI has identified three general sectors of biomedical research in which significant progress of important and practical application could be made in the near future:

identification and mechanism of action of macromolecules involved or affecting cellular growth, differentiation, transformation and defence;

biomembranes: compartmentation, transfer of information, excitation phenomena;

mechanisms of action of molecules of pharmacological or toxicological interest.

Research in these areas will turn out to be essential to achieve scientific and clinical progress for aetiology, prevention, diagnosis and treatment of major pathologies like cardiovascular diseases, neuro-psychiatric diseases, neoplasia, hereditary diseases and ageing disfunctions, viral diseases and disfunctions of the immune system.

IBMI intends to promote through meetings, courses, publications and fellowships biomedical knowledge and to invest financial resources in research in these areas. The present volume is part of this effort.

International Biomedical Institute

Advisory Scientific Committee

College of Founder Members

Acknowledgements

The Symposium of Molecular Basis of Neurological Disorders and their Treatment Fasano, Italy, 11–14 September 1989, was generously supported by UNESCO and IUB.

The organizers wish to express their appreciation for the invaluable assistance of Dr Helena Kirk of the International Biomedical Institute and Dr Carla Del Pesce of the Institute of Medical Biochemistry and Chemistry, Bari.

The organizers gratefully acknowledge the generous support of the following pharmaceutical companies to the Meeting:

Bayer Italia (Milano)
Beckman Analytical Instruments (Milano)
Bioresearch (Milano)
Ciba-Geigy (Basel)
Farmitalia Carlo Erba (Milano)
Fondazione Sigma-Tau (Ponezia-Roma)
Italfarmaco (Milano)
Kontron Instruments (Milano)
Lederle (Milano)
Master Pharma (Parma)
Neopharmed (Milano)
Zambeletti (Milano)

The Meeting was also supported by the National Research Council, Italy, the Commune of Fasano, the Province of Brindisi and the Chamber of Commerce, Bari.

PART ONE

Biochemical Function, Probes and Imaging

1 Towards a molecular model of the acetylcholine receptor channel

A. PULLMAN
Institut de Biologie Physio-chimique, Paris, France

Due to the difficulty (Kühlbrandt, 1988) in obtaining appropriately diffracting crystals of the acetylcholine receptor (AChR), there is not yet precise information on the three-dimensional structure of this membrane protein at atomic resolution. Thus an understanding at the molecular level of the functioning of this channel-making protein relies heavily on the development of an appropriate model incorporating as well as possible the structural information deduced from a variety of experimental investigations. Such a model of the ion-channel part of the AChR has been developed recently in our laboratory (Furois-Corbin and Pullman, 1988, 1989a, 1989b; Pullman and Furois-Corbin, 1990), using the presently available structural data, together with theoretical calculations and computer-aided molecular modelling. An outline of this model is given below with a brief discussion of some of its implications.

1.1 THE PREMISES

The basic features concerning the AChR which served as a starting point for building the model (for exhaustive source references and discussion of the evidence quoted see Changeux, in press; Numa, 1989) are the following:

1. The receptor protein possesses four subunits α, β, γ, δ, of known sequence, arranged pseudopentagonally, in the stoichiometry $\alpha^2\beta\gamma\delta$, around a central axis, in the order αβαγδ seen from the synaptic side (Fig. 1.1).
2. Each subunit possesses four hydrophobic segments, noted MI to MIV, which are supposed to cross the membrane as α-helices.
3. The N- and C-terminals of the subunits are both on the synaptic side

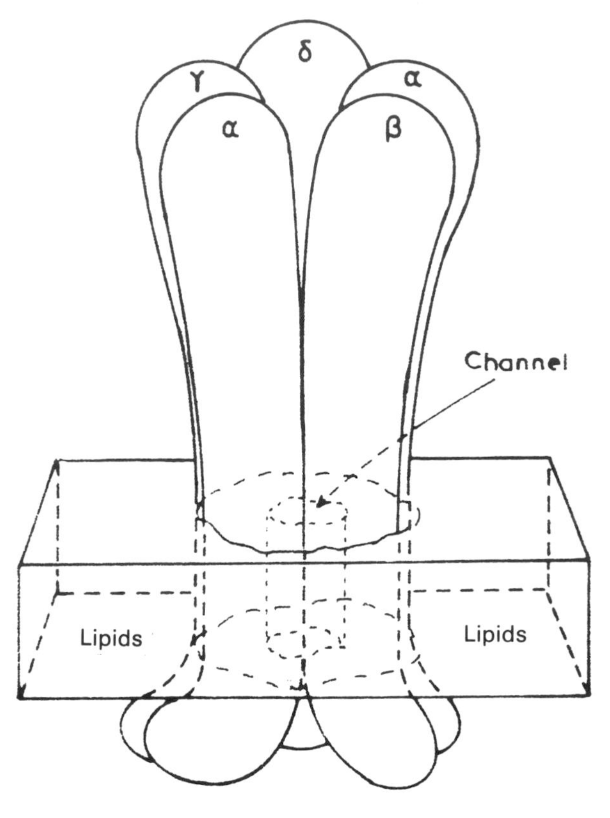

Fig. 1.1 Schematic disposition of the subunits of the AChR.

of the postsynaptic membrane; as a result the successive orientations (from the N- to the C-terminal) of the segments MI to MIV are as viewed schematically in Fig. 1.2.

4. Helix MII in each subunit participates in the inner wall of the channel: evidence based on the converging indications of conductance measurements on chimaera of different species, affinity labelling by non-competitive blockers (NCB) and more recently the probing of the binding site by QX222 in appropriately chosen mutants of the mouse AChR (a very careful chronological relation of the development of this evidence can be found in Changeux (1990), p. 86.
5. The labelling of homologous serines (248, 254, 257 and 262 in

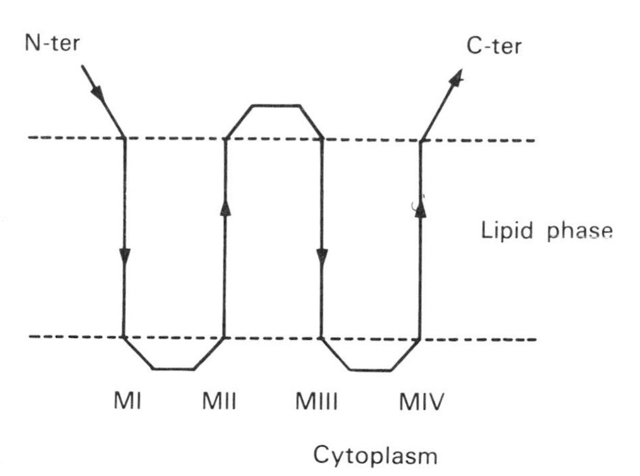

Fig. 1.2 Sequential order and orientation of the segments MI to MIV (lengths of the loops not respected: MIII to MIV is much longer than the three others).

α, β, γ, δ respectively) on the MII segments by irradiated tritiated chlorpromazine (CPZ) imposes that:

(a) these serines essentially face the centre of the pore (Giraudat *et al.*, 1986);
(b) the CPZ-blocking site is situated at, or near, the level of the labelled serines (Changeux and Revah, 1987).

1.2 THE MODEL OF THE INNER WALL

A model of the inner wall of the open channel has been developed on the basis of the preceding propositions. Its characteristics are the following:

1. The five MII helices are disposed with exact pentagonal symmetry around a central axis.
2. Their sequences are aligned in such a way that the labelled homologous serines are at the same level.
3. The helices are orientated so that the α-carbons of the labelled serines point exactly towards the central axis of symmetry.
4. The upper (synaptic, C-terminal) limit of the helical part of the MII hydrophobic segments is fixed at the amino acid situated four residues below the conserved prolines (265, 271, 274, 279, in α, β, γ, δ respectively (see Fig. 1.3)).

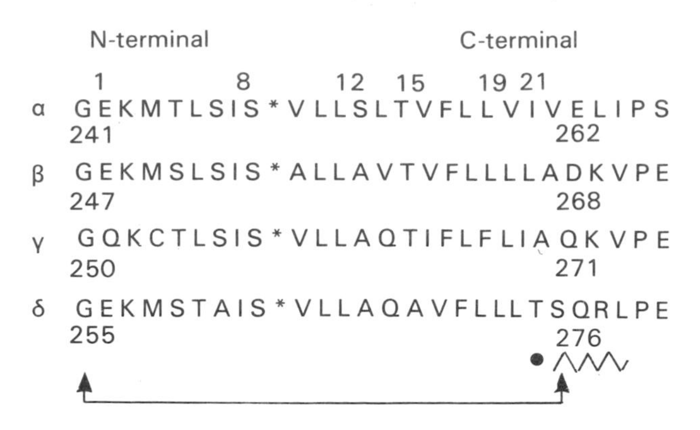

Fig. 1.3 The sequences of the MII segments and the limits adopted for the helices (*Torpedo*).

5. The lower limit (N-terminal, cytoplasmic) of the helices is taken to be at the negatively charged glutamate (glutamine in δ) numbered *one* in the simplified numbering of Fig. 1.3.
6. The helical structure, together with conditions 1–3, fix the orientation of all the α-carbons relative to the interior of the pore (2, 3), *imposing in particular that the glutamates point towards the interior and the adjacent lysines towards the exterior, while bulky side chains (e.g. 15 and 19) crowd the upper internal part of the helices* (see Fig. 1.4).
7. The minimal dimensions of the pentagonal prism formed by the axes of the five MIIs are *calculated* using two conditions deduced from these properties:
   (a) The closest approach of two adjacent MIIs at the level of the labelled serines is determined so that the space enclosed by the five helices is just sufficiently narrow to block CPZ at this level.
   (b) The closest approach of the helices in the upper region of the channel is determined so that CPZ can diffuse freely to its blocking site through this part. Due to the presence of bulky side chains this imposes a tilt of the helices away from the central axis by 7°, a value sufficiently small for an easy obturation of the gap by a neighbouring helix. Note that this tilt *is imposed by the two conditions above* and *not* (Furois-Corbin and Pullman, 1989b) by the 'funnel' shape of the extramembranar part of the receptor seen in the electron microscopic images (Toyoshima and Unwin, 1988). The tilt of the present model is compatible with the funnel (which is situated rather above it) but not imposed by it.
8. Owing to their location within the helical stretch and their resulting orientation, the N-terminal glutamates can form salt bridges with

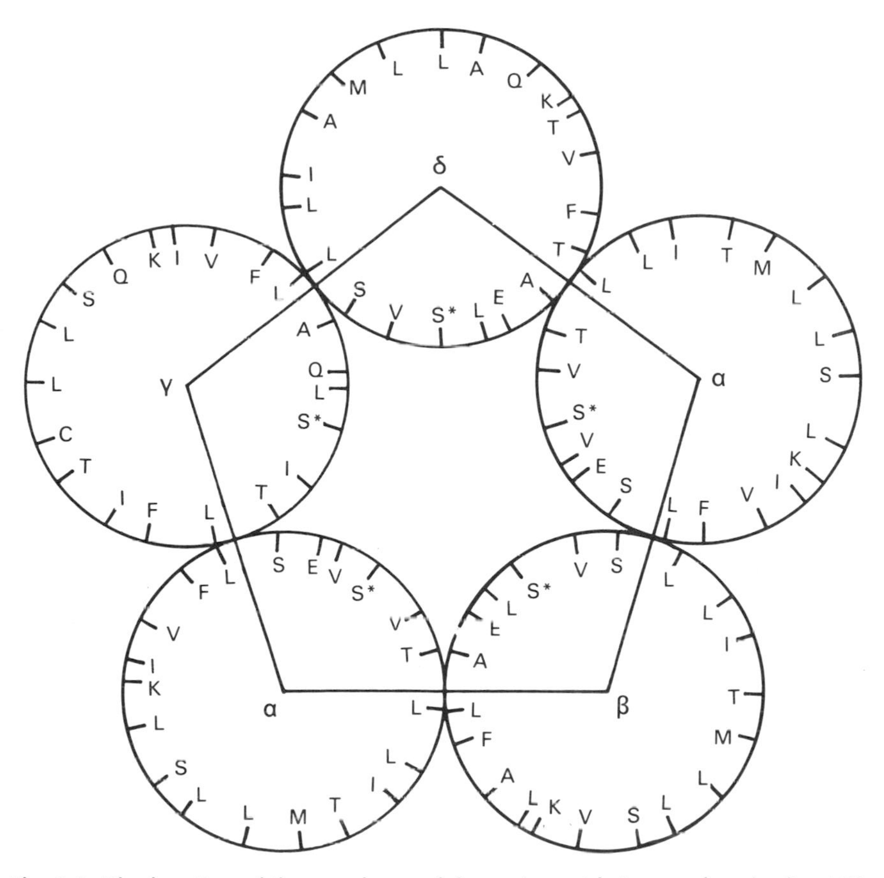

Fig. 1.4 The location of the α-carbons of the amino acids imposed to the five MIIs by the model. (E or Q in the oval and K in the rectangle are the 1st and 2nd residues respectively; the labelled serines carry an asterisk.)

the adjacent lysines (Furois-Corbin and Pullman, 1988) (with Gln–Lys in δ forming a corresponding hydrogen bond), a conformation which leaves an open space at the bottom of the pore sufficient (Fig. 1.5) to accommodate the largest permeant ion DMDEA (Furois-Corbin and Pullman, 1988, 1989a). Note that the bottom dimensions of the inner pore are not imposed by the dimension of the largest permeant ion but are compatible with it.

The role played by the Glu–Lys residues at the bottom of the channel is a key feature of the model: indeed, calculations of the energy profile felt by a cation showed that when these charged residues are not included in the helices, the profile presents a strong energy barrier before the

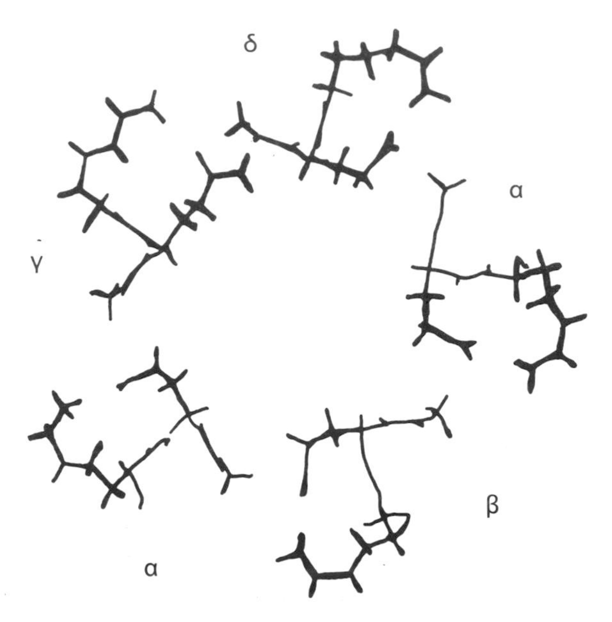

Fig. 1.5 The opening left at the bottom of the helices by the conformations of the EK and QK residues.

exit (Furois-Corbin and Pullman, 1988, 1989b); this barrier disappears when the two charged residues are included in the helices and properly disposed in the above-described bridges. Note that such a disposition is warranted only by the helical structure: otherwise, only the location of the C_α of Lys 2 is imposed, so that nearly any random disposition of its end charges and of those of the Glu and Gln with respect to the pore is possible. The decisive role of these residues inferred from the calculations (Furois-Corbin and Pullman, 1988) has been fully confirmed since (Imoto *et al.*, 1988) by the finding that the conductance of the channel was strongly affected by mutations directed at the negatively charged residues concerned, while mutations of the negative charges near the C-terminal or in the MI–MII and MII–MIII linkers have much less influence. Also, the negligible effect observed (Imoto *et al.*, 1988) when mutating the lysines number 2 confirms their location in the model. Still more striking confirmations have been brought about by calculations of energy profiles in different mutants (see section 1.4).

1.3 INSUFFICIENCY OF A PURE MII MODEL

The calculation of energy profiles in fact allowed one to go one step

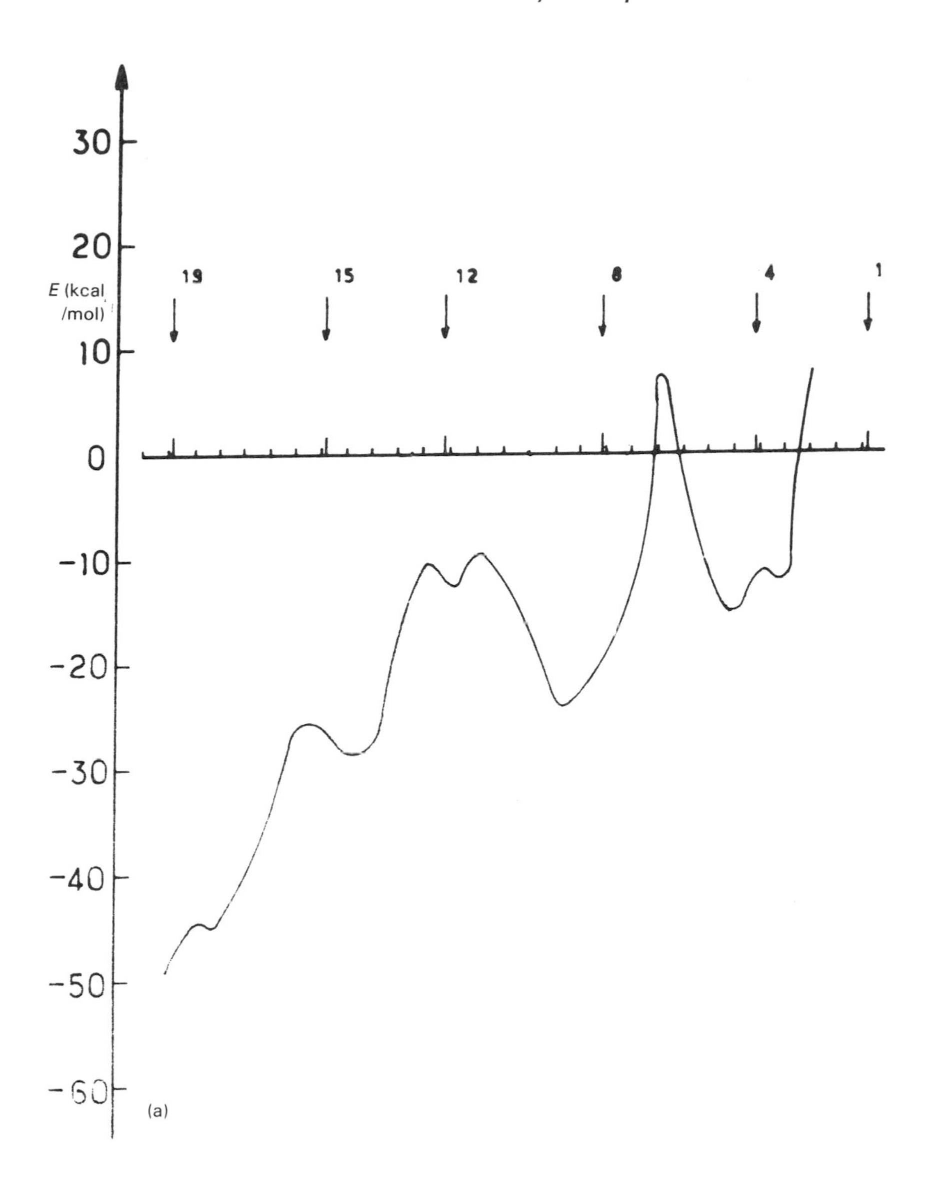

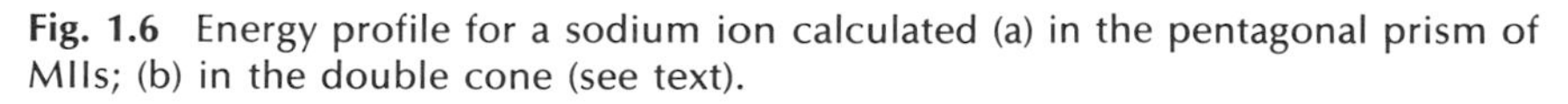
Fig. 1.6 Energy profile for a sodium ion calculated (a) in the pentagonal prism of MIIs; (b) in the double cone (see text).

further in the refinement of the model (Furois-Corbin and Pullman, 1989b).

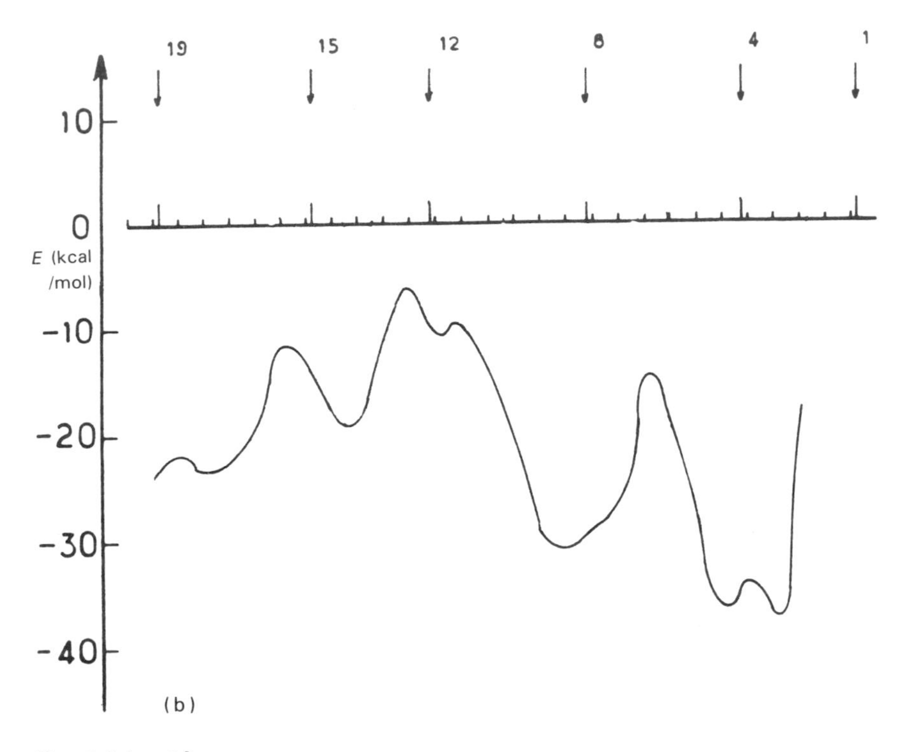

Fig. 1.6 (contd)

The energy profile for a sodium cation generated by the pentagonal prism of MIIs described above is shown in Fig. 1.6(a). It is seen that, although the interaction energy between the ion and the channel is negative, thus favourable everywhere, the overall shape of the profile is clearly against a favourable transfer of a positive ion towards the cytoplasmic side. An analysis of the components of the energy (Furois-Corbin and Pullman, 1989b) indicates that it is dominated by the evolution of its electrostatic component, more and more unfavourable towards the exit of the pore. This feature is clearly a result of the fact that the five nearly parallel MIIs are orientated in such a way that their rather strong dipole moments (about 70 debye units) each with their positive ends at the N-terminal add up their effects, disfavouring strongly the transit towards this terminal. If, however, it is remembered that each MII is part of a subunit comprising three other helices, the artefactual character of the energy result appears immediately: due to the sequential disposition of the MI–MIII segments and to the shortness of the

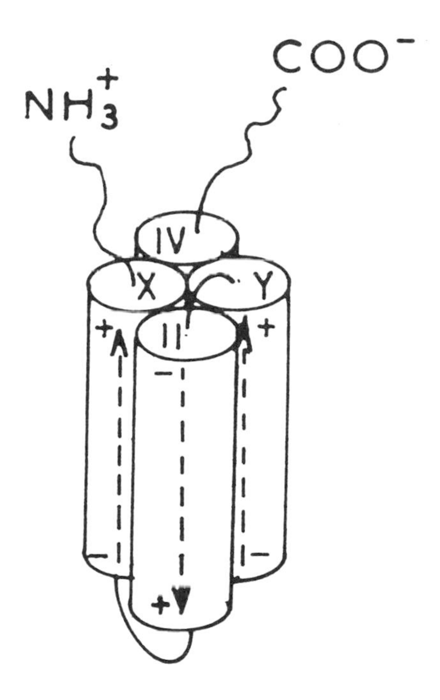

Fig 1.7 The relative orientation and location of the dipole moments of the other helices with respect to that of MII in each subunit (X and Y stand for MI or MIII; MIV may be further out) (see Pullman and Furois-Corbin, 1990).

MI–MII and MII–MIII linkers, the two segments MI and MIII (which have dipole moments opposed to that of MII) are necessarily close to the MIIs, somewhat behind them (see the scheme of Fig. 1.7). It was shown that the combined effect of their values and approximate locations (mimicked by placing five polyglycine helices appropriately oriented in a second cone surrounding the cone of MIIs (Furois-Corbin and Pullman, 1989b)) cancels the handicap produced by the dipole moments of the MII helices. The profile for Na^+ (Fig. 1.6(b)) recalculated in the double cone is seen now to present appropriate characteristics for ion transfer to occur.

1.4 FURTHER CONFIRMATIONS: ENERGY PROFILES FOR MUTANTS

We mention at the end of section 1.2 how the decisive role of the N-terminal glutamates suggested by energy calculations were globally confirmed by conductance measurements (Imoto *et al.*, 1988) on a series of mutants implying these and other charged residues. More detailed confirmations have been obtained (Pullman *et al.*, 1990; Furois-Corbin *et al.*, in preparation) by explicit calculation of energy profiles for Na^+ in different mutants and their comparison to that obtained in the wild-type. This is exemplified below for two cases: one involving mutation of glutamate to aspartate in α, the other mutation of glutamate to glutamine in δ. The two profiles (calculated in the double cone described in section 1.3 with appropriately modified salt-bridge and hydrogen-bond conformations for the mutated residues) are shown in Fig. 1.8, in comparison to that obtained in the wild-type channel: it appears clearly that, on the basis of these energy results, the transit of the ion should

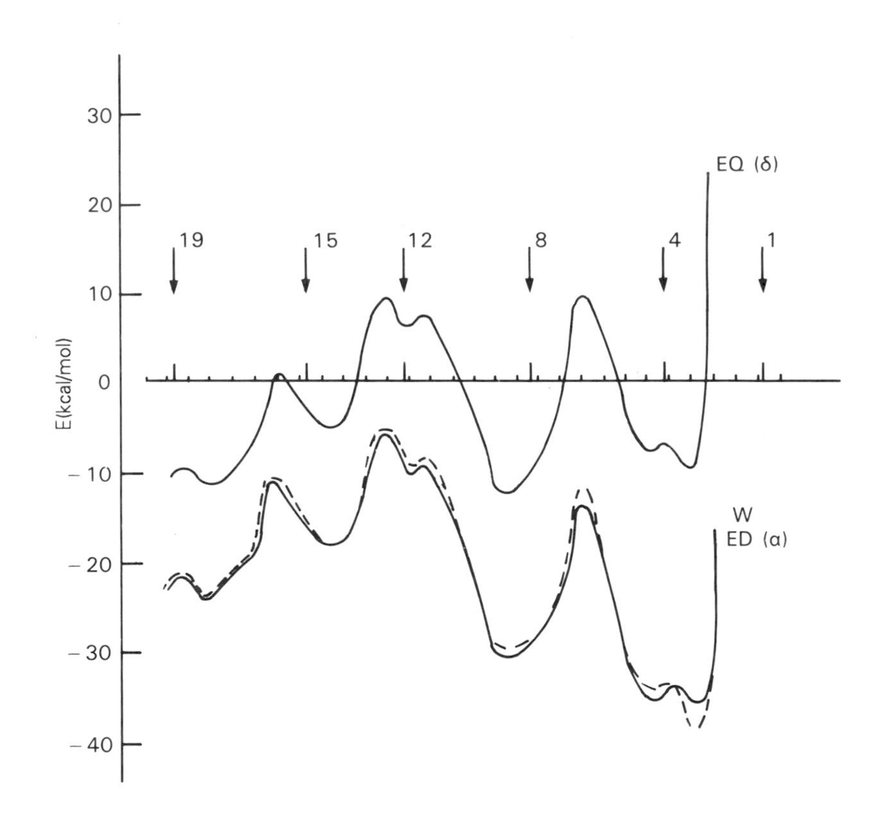

Fig. 1.8 Energy profiles for Na^+ in the double cone (see text) of the two mutants (EQ in δ and ED in α as indicated) compared to the wild-type W. (The numbered arrows locate the levels of the corresponding residues in the pore.)

be appreciably less favourable in the Eδ → Q mutant than in the wild-type, whereas it should be about as favourable in the Eα → D mutant as in the wild-type. This is in remarkable agreement with the observations (Imoto *et al.*, 1988) which show a lowering of the conductance by about one half in the first case, and very little modification of the conductance in the second case.

The detailed analysis (Pullman *et al.*, 1990; Toyoshima and Unwin, 1988; Furois-Corbin *et al.*, 1991) of the results confirms and specifies the role proposed (Furois-Corbin and Pullman, 1988) for the negatively charged residues considered, and the fact that the conductance results agree with the profiles calculated within the model and can be taken as a confirmation of its *essential* structural characteristics.

1.5 CONCLUDING REMARKS

Decisive elements in the development of the model have been the location of the labelled serines on the MII segments *and* their global orientation towards the centre of the pore, together with the assumptions that the MIIs are helical *and* that the N-terminal charged residues belong to the helices. The exact symmetry of the pentagonal arrangement, the exact central location of the serines and the exactly regular helical structure are convenient working hypotheses which have made the constructions and calculations possible. Needless to say, these conditions can be released in further refinements of the model. In view of the overall confirmations mentioned earlier, it may be hoped, however, that the essential conclusions will remain.

REFERENCES

Changeux, J.P. (1990) Fidia Research Foundation Neuroscience Award Lectures, Vol. 4, pp. 21–168.
Changeux, J.P. and Revah, F. (1987), *TINS*, **10**, 245–9.
Furois-Corbin, S. and Pullman, A. (1988) in *Transport through Membranes, Carriers, Channels and Pumps* (21st Jerusalem Symposium) (eds A. Pullman *et al.*), Kluwer, London, pp. 337–57.
Furois-Corbin, S. and Pullman, A. (1989a) *Biochim. Biophys. Acta*, **984**, 339–50.
Furois-Corbin, S. and Pullman, A. (1989b) *FEBS Lett.*, **252**, 63–8.
Furois-Corbin, S. and Pullman, A. (1991) *Biophysical Chemistry*, **39**, 153–9.
Giraudat, J., Dennis, M., Heidmann, T. *et al.* (1986) *Proc. Nat. Acad. Sci, USA*, **83**, 2719–23.
Imoto, K., Busch, C., Sakmann, B. *et al. Nature*, **335**, 645–8.
Kühlbrandt, W. (1988) *Q. Rev. Biophys.* **21**, 429–77.
Numa, S. (1989) *The Harvey Lectures*, **83**, 127–65.

Pullman, A. and Furois-Corbin, S. (1990) *Sixth Conversation Biomol. Struct. Dyn.*, Vol. II, 11, 195–210.

Pullman, A., Furois-Corbin, S. and Andrade, A.M. (1990) in *Modelling of Molecular Structure and Properties* (proceedings of the 41st International Meeting on Physical Chemistry) (ed. J.L. Rivail) Elsevier, Amsterdam, **71**, 527–39.

Toyoshima, C. and Unwin, N. (1988) *Nature*, **336**, 347–60.

2 Proton-motive ATP synthase and energy transfer in the cell

S. PAPA, F. GUERRIERI AND F. ZANOTTI
Institute of Medical Biochemistry and Chemistry and Centre for the Study of Mitochondria and Energy Metabolism, University of Bari, Italy

The H^+-ATP synthase of mitochondria is the key enzyme of energy metabolism in eukaryotic cells (Senior, 1988). It is responsible for the synthesis of ATP supported by respiration, which covers, under normal conditions, more than 80% of the energy demand of mammalian cells. An adult man with a daily energy demand of some 3000 kcal turns over some 400 mol of ATP (200 kg). The brain is the organ with the highest demand for aerobic ATP, accounting, with only 2% of the body weight, for 20% of total oxygen consumption at rest (Erecinksa and Silver, 1989).

In the brain most of the ATP is expended by the Na^+-K^+ pump of the plasma membrane which converts the free energy made available by ATP hydrolysis into electrochemical ion gradients. In the brain, as well as in other tissues, there are a number of H^+-ATPases associated with intracellular organelles where ATP is utilized for the storage of neuromediators and other substances (Pedersen and Carafoli, 1987).

Under hypoxia, a condition which can be experienced by the brain in a variety of physiological and pathological conditions, production of ATP by glycolysis, which normally accounts for not more than 20% of the total ATP demand, becomes critical. In the absence of regulatory devices, ATP produced by glycolysis would be rapidly dissipated by the mitochondrial H^+-ATP synthase which, in the absence of respiratory proton-motive force, displays a high hydrolytic activity.

Whilst the prokaryotic enzyme functions under physiological conditions as ATP synthase or ATP hydrolase, the mammalian enzyme is apparently made to function as ATP synthase, thus being organized so as to prevent ATP dissipation (Senior, 1988).

The H^+-ATP synthase of coupling membranes is structurally and functionally made up of three parts: 1. the catalytic sector or F_1 (knob) universally consisting of five subunits; 2. the H^+–translocating or membrane-integral sector (F_0) (consisting of a variable number of subunits); 3. the stalk (coupling sector or gate) made up by some F_1 and F_0 components.

This tripartite structure was first revealed by electron microscopy of negatively stained vesicles of the inner mitochondrial membrane (Fernandez-Moran *et al.*, 1964). Recently this structure has also been obtained by image analysis of electron micrographs of rapidly frozen unstained samples of *Escherichia coli* F_0F_1 H^+–ATP synthase (Gogol *et al.*, 1987), thus indicating that the knob-of-stalk structure represents a general feature of the enzyme.

The F_1 catalytic sector has both in prokaryotes and eukaryotes an invariable composition of 3α, 3β, 1γ, 1δ, and 1ε subunits (Senior, 1988). The F_0 membrane-intrinsic factor, responsible for H^+ translocation, consists in *E. coli* of three subunits a, b and c (Senior, 1988).

In mitochondria more than three proteins appear to constitute the F_0 and stalk sectors (Papa, 1990; Papa *et al.*, 1988) which are involved in H^+ conduction and/or coupling of H^+ translocation to the catalytic process in the F_1 moiety (Table 2.1) (Papa, 1990; Papa *et al.*, 1988). Two of these are the products of the mitochondrial genoma: the

Table 2.1 Subunits of the F_0 and stalk moieties of bovine heart ATP synthase

Genes	*Calculated M_r(kDa)*	*N-terminus*	*Denomination*	*Function*
Nuclear gene	24.67	PVP (1 Cys)	PVP subunit	Binding of F_1 to F_0; H^+ conduction in F_0; involved in oligomycin and DCCD sensitivity
More nuclear pseudogenes	20.97	FAK (1 Cys)	OSCP	Functional connection of F_1 to F_0
Mitochondrial gene	24.82	f-SFI	ATPase 6	Proton conduction? Oligomycin sensitivity
Nuclear gene	18.60	Ac-AGR (1 Cys)	d	Unknown
Nuclear gene	7.96	NKE	F_6	Binding of F_1 to F_0
Mitochondrial gene	7.96	f-MPQ	A6L	Unknown
Two nuclear genes encoding different pre-sequences	8.00	DID (1 Cys)	c	Proton conduction
Nuclear gene	9.58	GSE	IF_1	Inhibits catalysis and H^+ translocation

ATPase 6, homologous to subunit *a* of *E. coli* and likely to be involved, as in this (Von Meyenburg *et al.*, 1986; Lightowlers *et al.*, 1988; Eya *et al.*, 1988; Paule and Fillingame, 1989), in H^+ conduction, and A6L in mammals (aap1 in yeast), whose function is as yet unknown (Papa, 1990). Five other proteins have been identified in the mammalian enzyme which are encoded by nuclear genes (Walker *et al.*, 1987a, 1987b). These include subunit *c* which is directly involved in H^+ conduction (Hoppe and Sebald, 1984; Kopecky *et al.*, 1983) and OSCP and F_6 which contribute to the stalk connecting F_1 to F_0. This represents, in our opinion, the sector of H^+-ATP synthase which has undergone the most extensive evolution, acquiring in high eukaryotes a number of additional subunits, which confer to the system a regulatory flexibility so that its activity can be adapted to the continuously changing energy demand of mammalian cells like heart and brain.

We shall describe here, in some detail, the structural and functional characteristics of subunits contributing to the stalk sector of mitochondrial H^+-ATPase.

Fig. 2.1 shows the sodium dodecyl sulphate–polyacrylamide gel electrophoresis (SDS–PAGE) of F_0 purified from F_1-depleted bovine heart submitochondrial particles (USMP) (Guerrieri *et al.*, 1989). The preparation exhibits the characteristic components of F_0 and the stalk listed in Table 2.1. The bands of interest were electroeluted in glycerol and sequenced or used for raising polyclonal antibodies in rabbits (Guerrieri *et al.*, 1989; Zanotti *et al.*, 1988). The band of apparent M_r 27 000 was found from the amino acid sequence (Zanotti *et al.*, 1988) to correspond to the nuclear encoded protein characterized by Walker *et al.* (1987a) and considered by them to be analogous to the *b* subunit of *E. coli*. We denominate this protein PVP from the first three N-terminal residues. The band of apparent M_r 25 000 consisted of a closely spaced doublet; both components reacted, like the PVP protein and subunit *c*, with the fluorescent thiol reagent *N*-(7-dimethylamino-4-methylcoumarinyl)-maleimide (DACM). The lower band of M_r 23 000 was not labelled by thiol reagents and may represent the product of the ATPase 6 gene, which has no codon for cysteine (Fearnley and Walker, 1986).

Removal of the ATPase inhibitor protein (Pullman and Monroy, 1963) from the F_1F_0 complex in submitochondrial particles by chromatography on Sephadex results in twofold stimulation of the hydrolytic activity as well as of oligomycin-sensitive passive H^+ conduction (Table 2.2) (Guerrieri *el al.*, 1987a, 1987b). Both activities were brought back to control values by adding the purified inhibitor protein to the depleted particles. The inhibitor protein exerted the

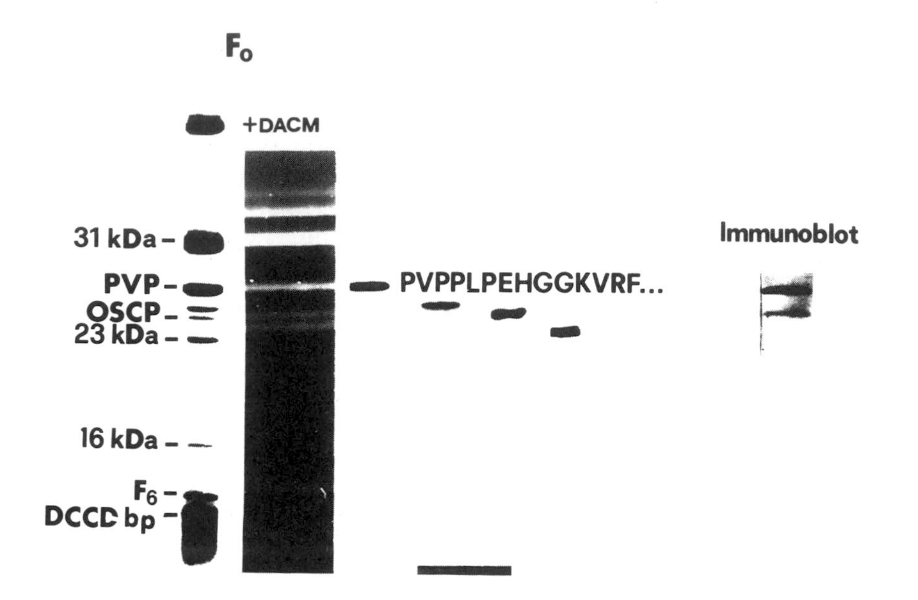

Fig 2.1 SDS-PAGE of F_0 and isolated F_0 polypeptides. Immunoblot with antisera against PVP protein and OSCP. F_0 was purified by extraction with CHAPS from F_1-depleted bovine heart submitochondrial particles (USMP) or from USMP treated with the fluorescent thiol reagent DACM (Papa *et al.*) as reported in Guerrieri *et al.* (1989). SDS-PAGE, immunoblot analysis and isolation of individual F_0 proteins were carried out as reported in Houstek *et al.* (1988). Amino acid sequence analysis was carried out as reported in Papa *et al.* (1989) and Zanotti *et al.* (1988).

same effect when added to F_1-depleted particles supplemented with purified F_1.

Fig. 2.2 shows a titration of the inhibitory effect exerted by addition of the purified inhibitor protein on the ATPase activity and proton conductivity of depleted particles. Treatment of the isolated inhibitor protein with ethoxyformic anhydride (EFA) under conditions resulting in modification of one out of the five histidine residues of this protein (Guerrieri *et al.*, 1987b) suppressed its inhibitory activity. Critical importance of a histidine residue is also indicated by the pH dependence of the inhibitory activity, which increased with acidification of the reaction mixture, with a pK_a around neutrality typical of histidine-imidazole.

It is known that the inhibitor protein binds to the ß-subunit of F_1, also after the removal of this sector from the membrane sector, and inhibits its hydrolytic activity. The inhibitor protein may represent

Table 2.2 Effect of F_1 inhibitor protein (IF_1) on ATPase activity and oligomycin-sensitive H^+ conduction in submitochondrial particles with various degrees of resolution of H^+-ATPase complex

	ATPase activity (μmol ATP hydrolysed min $^{-1}$ mg protein^{-1})		*Anaerobic H^+ release $1/t_{1/2}$ (S^{-1})*	
	Control	*+ IF_1*	*Control*	*+ IF_1*
ESMP	1.25	0.92	1.00	0.75
Sephadex-ESMP	2.88	0.90	2.00	0.91
Urea-ESMP	0.08	0.08	2.74	2.80
Urea-ESMP + F_1	1.00	0.50	1.67	1.11

Preparations of ESMP, Sephadex-treated ESMP, urea-ESMP, F_1 and IF_1 were carried out as described in Guerrieri *et al.* (1987b). F_1 was prepared as described in Guerrieri *et al.* (1989). ATPase activity and oligomycin-sensitive H^+ conductivity were determined as described in Guerrieri *et al.* (1987a, 1987b). Incubation particles with IF_1 (4 μg/mg protein) were carried out at 25 °C as described in Guerrieri *et al.* (1987a, 1987b).

an important component of the F_0/F_1 junction, where it may serve a critical regulatory role in preventing futile ATP hydrolysis as well as free escape of H^+ from the M mouth of the transmembrane H^+ channel (see also Hashimoto *et al.*, 1983, 1984; Yoshida *et al.*, 1984).

It appears that binding of 1 mol of inhibitor protein per ATPase complex is enough to exert full inhibitory activity (Klein *et al.*, 1980). However, since the protein can be dissociated from the complex under the influence of the proton-motive force, its function could be modulated at the expression level and different tissues may contain different amounts of inhibitor protein.

The content of the inhibitor protein can be estimated by immunological methods. We have produced polyclonal antibodies against the isolated inhibitor protein and developed a semiquantitative immunoblot procedure shown in Fig. 2.3.

By use of immunological titration it has, for example, been shown that certain experimental tumours have an enhanced content of inhibitor protein (Chernyak *et al.*, 1987). Furthermore, evidence has been presented indicating an enhanced level of inhibitor protein in ischaemic tissues (Rouslin and Pullman, 1987).

Other information on the role of the inhibitor protein and other possible components of the stalk or coupling sector of the ATP synthase are provided by the effects exerted by certain amino acid modifiers on proton conduction in submitochondrial particles. These are EFA, specific for histidine residues, and diamide and Cd^{2+}, which oxidize vicinal

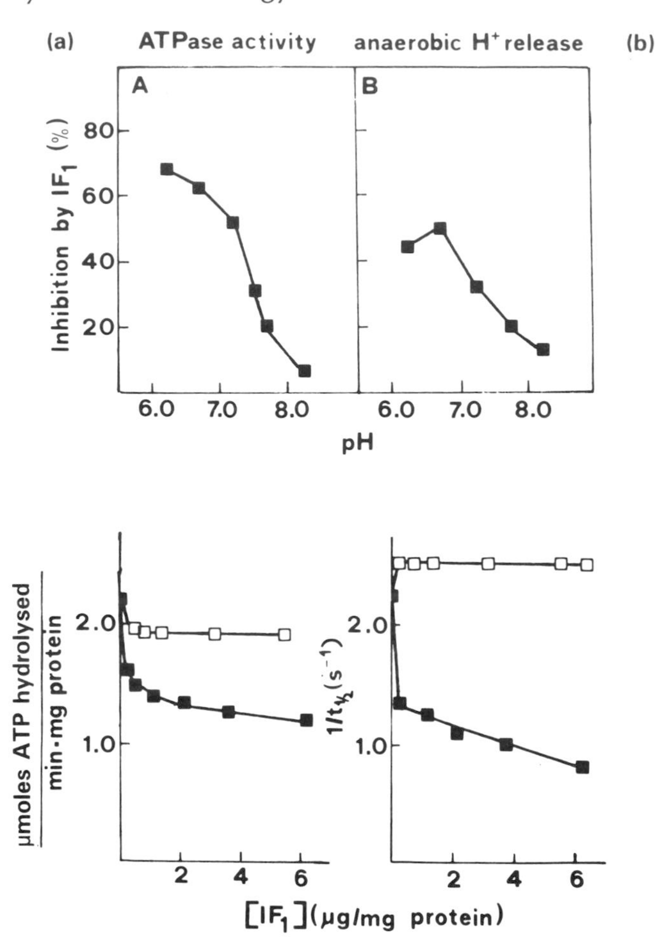

Fig 2.2 Titration of inhibition of purified inhibitor protein on ATPase activity (A and a) and on passive proton conductivity (B and b) of IF_1-depleted particles; effects of diethylpyrocarbonate (EFA) treatment of IF_1 and of pH. For preparations of Sephadex-EDTA particles, purification of IF_1, its treatment with EFA, determination of ATPase activity, measurement of oligomycin-sensitive proton conduction and pH dependence see Guerrieri *et al.* (1987b). ■—■, IF_1; □–□, IF_1 modified by treatment with 0.5 mM EFA.

dithiols, inducing formation of disulphide bridges. Treatment of EDTA treated submitochondrial particles (ESMP) with each one of these reagents (Fig. 2.4) induces a dramatic acceleration of passive H^+ conduction in the particles. Differently from the promotion of H^+ conduction induced by carbonyl cyanide p-trifluoromethoxyphenylhydrazone (FCCP), whose effect is oligomycin insensitive (the

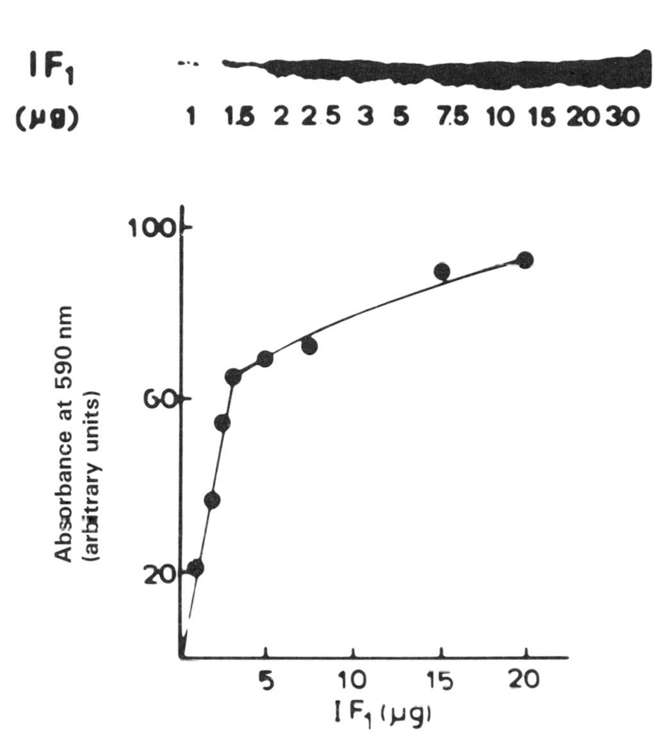

Fig. 2.3 Immunoblot of purified bovine IF_1: a semiquantitative analysis. IF_1 was purified as described in Guerrieri *et al.* (1987a, 1987b). After gel electrophoresis (Houstek *et al.*, 1988) the IF_1 was electroeluted and injected in a rabbit to obtain the polyclonal antibody (Guerrieri *et al.*, 1989; Houstek *et al.*, 1988). Immunoblot analysis of IF_1 (at the concentrations reported in the figure) was carried out as reported in Guerrieri *et al.* (1989), Papa *et al.* (1989), Houstek *et al.* (1988) and Zanotti *et al.* (1988) and semiquantitative densitometric analysis was carried out (Houstek *et al.*, 1988).

uncoupler promotes H^+ conduction in the phospholipid bilayer), the enhanced H^+ conductivity induced by the amino acid reagents is fully suppressed by oligomycin, this proving that it is, in fact, H^+ escape from the F_0 channel to be promoted by the reagents. It can be noted that the stimulatory effect of EFA, diamide and Cd^{2+} is followed, after maximal stimulation at a critical concentration, by inhibition at higher concentrations. Evidently the thiol reagents modify two classes of residues, differently exposed to the reaction medium. Modification of superficial thiols results in promotion of H^+ conduction.

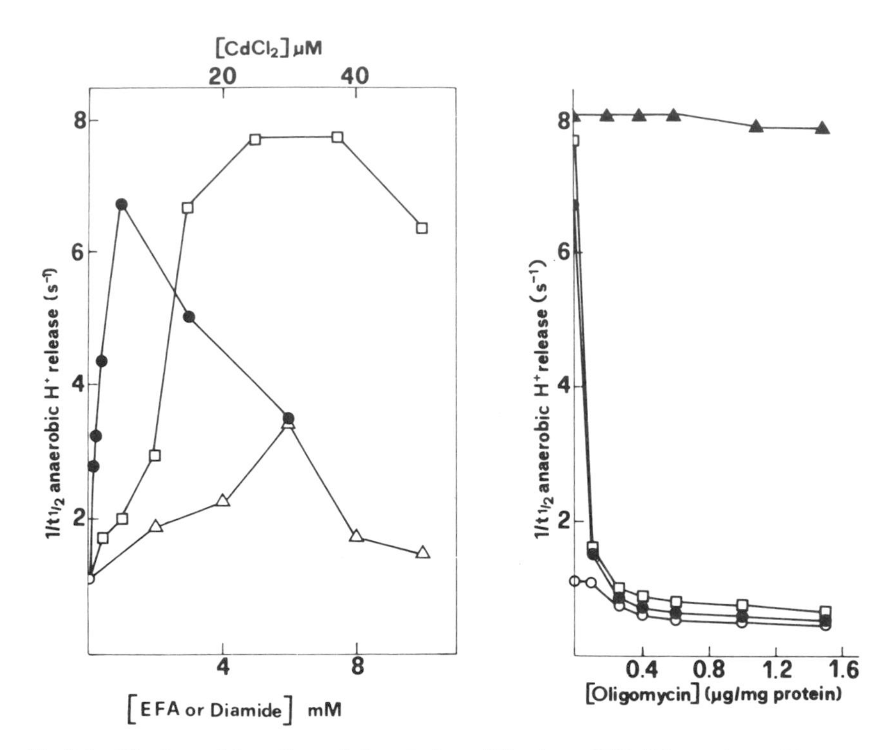

Fig 2.4 Titration of the effect of chemical modification of ESMP by amino acid reagents on oligomycin-sensitive H^+. ESMP were preincubated with EFA (●—●) at the concentrations reported in the figure (left-hand panel) as described in Guerrieri *et al.* (1987a) before oxygen induced proton translocations. ESMP were preincubated 2 min with diamide (□–□) or Cd^{2+} (Δ—Δ) at the concentrations reported in the figure. The oligomycin titration (right-hand panel) was carried out on control (o—o) or ESMP treated with 1 mM EFA (●—●); 5 mM diamide (□–□) or in the presence of 5 µM FCCP (▲—▲).

Modification, at higher concentrations, of thiols located more deeply in the membrane results in inhibition of a step in H^+ conduction which obscures the stimulatory effect exerted on more superficial steps of the process.

The stimulatory effect of EFA was lost in particles deprived of the inhibitor protein, thus confirming that the stimulatory effect exerted on H^+ conduction was specifically due to modification of this protein.

In isolated F_0 reconstituted in liposomes, diamide caused inhibition of proton conduction (Zanotti, *et al.*, 1985). It is conceivable that the stimulatory effect caused by diamide possibly results from oxidation and disulphide-bridging of thiol groups in neighbouring superficial subunits.

The inhibitory effect exerted by diamide in F_0-liposomes may be due to modification of those thiols in membrane integral proteins, whose modification by monofunctional thiol reagents (Guerrieri and Papa, 1982; Zanotti *et al.*, 1987) results in inhibition of H^+ conduction. Thiol groups in F_0 components are very limited in number; there is only 1 Cys residue each in the PVP protein, in the protein *d* of Fearnley and Walker (1986), in *c* subunit and in factor B, an additional putative component of F_0 (Sanadi, 1982).

These thiol groups, of which only one is present in the *E. coli* (*b* subunit) enzyme, may play critical roles in the mitochondrial system.

A component of the F_0 sector which seems to play an important role in the functional coupling of H^+ transport to hydro-anhydro catalysis in F_1 is the PVP protein. Interesting information on this subunit has been obtained by selective enzymatic digestion of membrane proteins and reconstitution with the native isolated components (Guerrieri *et al.*, 1989; Papa *et al.*, 1989; Houstek *et al.*, 1988; Zanotti *et al.*, 1988). Treatment of submitochondrial particles with trypsin results, only after removal of F_1, in digestion of the PVP protein and oligomycin sensitivity conferring protein (OSCP) (Fig. 2.5). Immunoblots with anti-PVP serum of F_0 extracted from trypsin-digestion of F_1-depleted bovine-heart submitochondrial particles (USMP) show that the PVP protein can be digested to an immunoreactive fragment some kilodaltons smaller. The fragment retains the N-terminal region as well as the only

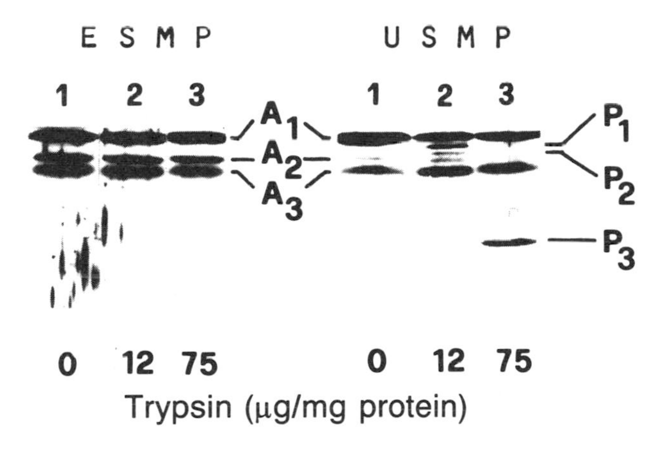

Fig. 2.5 Immunoblot analysis of trypsin digestion of PVP protein and OSCP in ESMP and USMP. Immunodecoration with antiserum against 25-27 kDa protein fractions of F_0 was carried out as described in Houstek *et al.* (1988). ESMP or USMP was prepared and treated with trypsin (at the concentrations reported in figure). A_1 identifies the PVP protein; A_2 OSCP and A_3 does not correspond to any F_0 protein. P_1, P_2 and P_3 are the products of tryptic digestion of PVP protein.

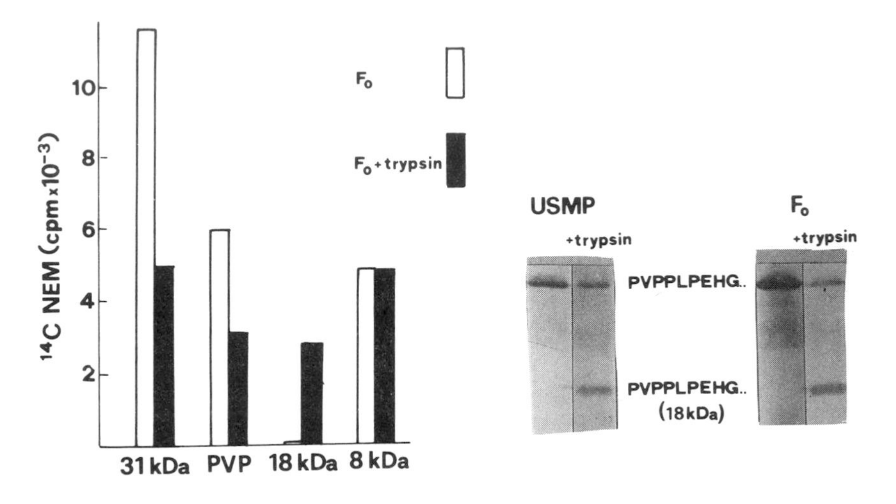

Fig 2.6 Immunoblot analysis of trypsin digestion of PVP protein in USMP and F_0, isolated from USMP, before and after trypsin digestion; analysis of binding of [^{14}C]NEM to F_0 proteins. USMP and F_0 were prepared as reported in Guerrieri *et al.* (1989). Trypsin (50 μg/mg protein) digestion was carried out for 20 min as described in Houstek *et al.* (1988), and immunoblot analysis of USMP (20 μg) or F_0 (20 μg) was carried out with specific anti-PVP protein serum as described in Guerrieri *et al.* (1989). The 18 kDa protein indicates the specific product of trypsin digestion. Left-hand panel reports the binding of [^{14}C]NEM to F_0 proteins. F_0 was isolated from [^{14}C]NEM-treated control and trypsin-treated USMP. Then 50 μg protein of F_0 were subjected to SDS-PAGE and the polypeptide isolated from the gel as reported in Houstek *et al.* (1988) and used for determination of radioactivity. On the abscissa the apparent molecular masses of isolated proteins are reported.

Cys residue in this protein (reactive with [^{14}C]N-ethyl maleimide (NEM)) at position 197 (Walker *et al.*, 1987a), 12 residues away from the C-terminal Met (Fig. 2.6). Thus trypsin cleaves off selectively the C-terminal region of PVP down to Lys 202 or Lys 206. It can be noted from the loss of [^{14}C]NEM binding that also the 31 kDa protein is digested by trypsin; subunit *c*, on the other hand, retains [^{14}C]NEM binding.

The experiment of Fig 2.7 shows that progressive digestion of PVP by trypsin is linearly associated to depression of proton conduction (Zanotti *et al.*, 1988). It should be noted that, after complete digestion of PVP, there still remains a substantial proton conductivity, which is, however, insensitive to oligomycin and N,N′dicyclohexylcarbodilimide (DCCD) (see below and Figs 2.8 and 2.9).

When F_0, extracted from trypsin-digested particles, was incorporated

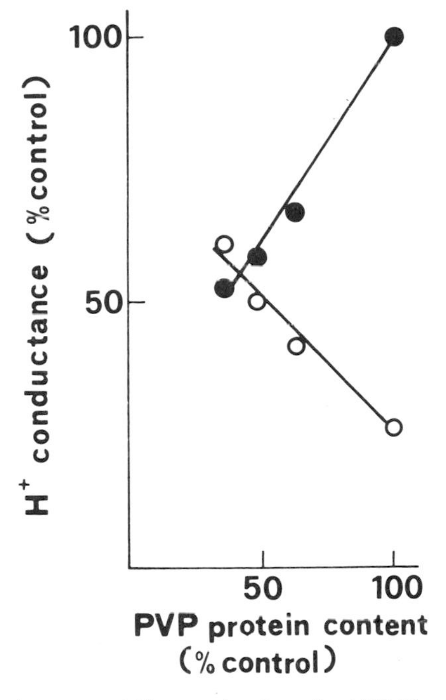

Fig. 2.7 Relationship between H^+ conduction in USMP and the content of PVP protein; effect of oligomycin. For USMP preparations, immunoblot analysis, trypsin treatment, measurement of oligomycin-sensitive H^+ conduction and evaluation of PVP protein content see Houstek *et al.* (1988) and Zanotti *et al.* (1988). Symbols: •—• , control; o—o, + 1 μg/mg particles protein oligomycin.

in liposomes it exhibited a markedly depressed, oligomycin-sensitive H^+ conductivity as compared with that of F_0 extracted from untreated particles (Fig. 2.8). Proton conduction in liposomes reconstituted with digested F_0 was progressively restored to the control values of undigested F_0 by the addition of increasing amounts of isolated PVP protein, which also restored oligomycin sensitivity (Zanotti *et al.*, 1988) (Fig. 2.8). These effects of the intact PVP protein were highly specific in that the truncated protein of apparent M_r 18 000 or the protein of M_r 31 000 were totally ineffective (Table 2.3). Furthermore it can be noted that modification of Cys 197 by DACM in the isolated PVP did not affect its capacity to restore oligomycin-sensitive H^+ conduction (Table 2.3). This is consistent with previous observations showing that the inhibition by NEM of proton conduction in F_0-liposomes was directly correlated with the binding of [^{14}C]NEM to subunit *c* (and/or subunit d) but not to the binding to the PVP protein (band of apparent M_r 27 000 in Houstek *et al.*, 1988).

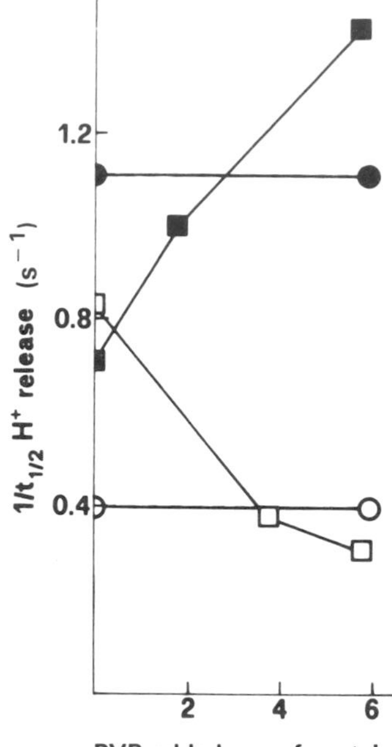

Fig. 2.8 Reconstitution of H^+ conduction and oligomycin sensitivity of F_0 extracted from trypsin USMP by addition of isolated PVP protein. For reconstitution conditions see Guerrieri *et al.* (1989) and Zanotti *et al.* (1988). Symbols: •—•, F_0 prepared from control USMP; o—o, F_0 prepared from control USMP + oligomycin (2 μg/mg F_0); ■—■, F_0 isolated from trypsin-treated USMP (50 μg/mg particle protein; □—□, F_0 isolated from trypsin-treated USMP + oligomycin (2 μg/mg F_0).

Table 2.3 Restoration of H^+ conduction and oligomycin sensitivity by addition of purified F_0 proteins to USMP treated with trypsin

	$1/t_{1/2}$ H^+ release (s^{-1})	
	Control	*+ Oligomycin* (2 μg/mg protein)
USMP	1.82	0.25
Trypsin-USMP	0.91	0.80
Tryspin-USMP + PVP	2.00	0.57
Trypsin-USMP + 18 kDa protein	1.00	0.90
Trypsin-USMP + M_r 31 kDa protein	1.00	0.91
Trypsin-USMP + PVP-DACM	2.00	0.59

Trypsin-USMP was preincubated for 15 min at 25 °C with purified PVP, 31 or 18 kDa proteins (6 μg/mg particle protein) before measurement of passive proton conduction. PVP-DACM protein purified from DACM-treated USMP (for details see Zanotti *et al.*, 1988.)

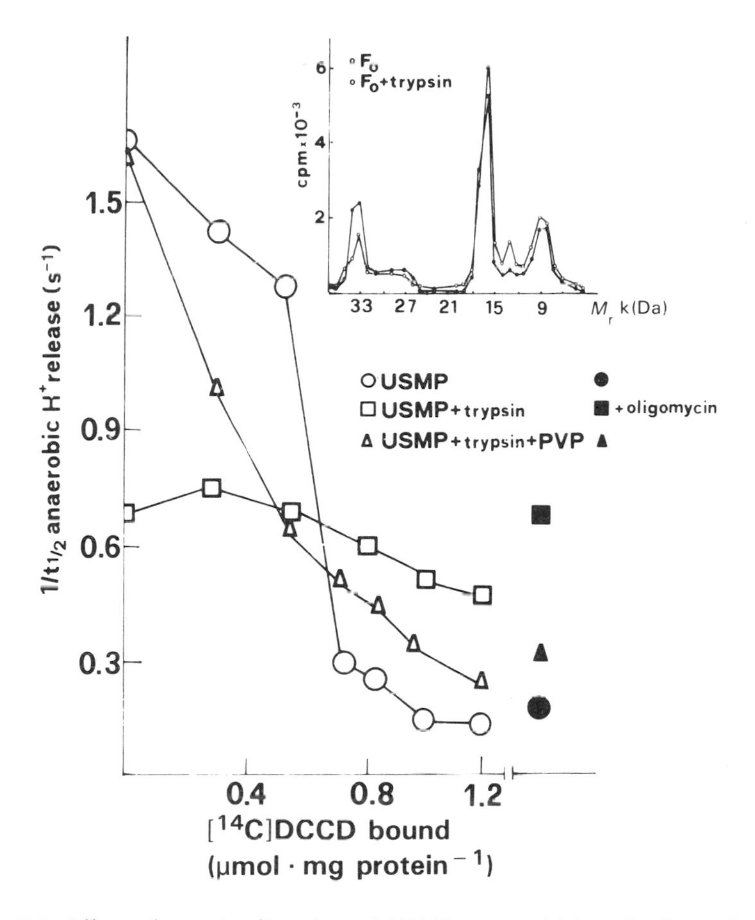

Fig. 2.9 Effect of trypsin digestion of USMP on sensitivity of H^+ conduction to DCCD (open symbols) and oligomycin (closed symbols); effect of reconstitution with isolated PVP protein (triangles). Insert refers to [^{14}C]DCCD binding to F_0 polypeptides extracted from control (•) and trypsin-treated USMP (o). For USMP preparation, isolation of F_0, trypsin treatment of USMP, isolation of PVP protein and analysis of [^{14}C]DCCD binding see Zarotti *et al.* (1988). Symbols: o—o, control (•, + 2 μg oligomycin/mg particle protein); □–□, trypsin-(50 μg/mg protein) treated USMP (■, + 2 μg oligomycin/mg particle protein); Δ—Δ, trypsin-treated USMP + 0.2 μg PVP protein/mg particle protein (▲, + oligomycin, 2 μg/mg particle protein).

Trypsin digestion of F_1-depleted particles did not affect the total binding of [^{14}C]DCCD to F_0 (Guerrieri *et al.*, 1989) or its specific binding to the M_r 8000 protein (Guerrieri *et al.*, 1989; Fig. 2.9). Trypsin digestion resulted, however, at equal levels of [^{14}C]DCCD binding to F_0, in a loss of its inhibitory action on H^+ conduction. Inhibition could then be restored together with oligomycin sensitivity, when the intact purified PVP protein was added back to the digested particles.

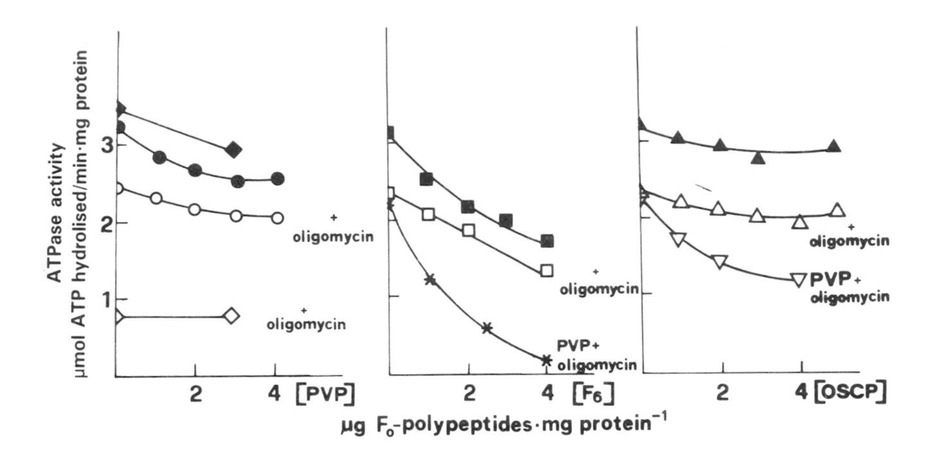

Fig. 2.10 ATPase activity in trypsinized USMP supplemented with purified F_1. For trypsin (50 μg/mg USMP protein) digestion and other conditions see Guerrieri *et al.* (1989), Houstek *et al* (1988) and Zanotti *et al.* (1988). After trypsin digestion isolated PVP protein (•) or F_6(■), or OSCP (▲) were added before F_1 (Guerrieri *et al.*, 1988). ◇—◇, control USMP supplemented with PVP protein. Where indicated, oligomycin (2 μg/mg particle protein) was added (symbols: ◇—◇, control USMP reconstituted with F_1 in the presence of PVP; o—o, trypsinized USMP reconstituted with F_1 in the presence of PVP; □—□, trypsinized USMP reconstituted with F_1 in the presence of F_6; *—*, trypsinized USMP reconstituted with F_1 in the presence of PVP (3 μg/mg particle protein) and F_6 at the concentrations reported in the figure; Δ—Δ, trypsinized USMP reconstituted with F_1 in the presence of OSCP; Δ—Δ, trypsinized USMP reconstituted with F_1 in the presence of PVP (3 μg/mg particle protein) and OSCP (at the concentrations reported in the figure).

Fig. 2.10 illustrates the effect of addition of isolated PVP protein, F_6 and OSCP on the functional binding of soluble F_1 to depleted particles (USMP) digested by trypsin. Functionally correct binding of F_1 to F_0 in the particles is evaluated by reconstitution of oligomycin sensitivity of the ATPase activity of F_1. Trypsin digestion of USMP almost completely abolished the oligomycin sensitivity of the ATPase activity of F_1 added to the particles. In the control, F_1 reconstituted with USMP was 80% inhibited by oligomycin. The addition of PVP protein, OSCP or F_6 which per se caused some inhibition of the ATPase activity of reconstituted F_1 (F_6 was the most effective in this respect), added individually did not promote oligomycin sensitivity. However, combined addition of PVP protein and OSCP and even better of PVP protein and F_6 were effective in restoring oligomycin sensitivity.

Thus OSCP, F_6 and the C-terminal region of the PVP protein are all required for the functionally correct binding of F_1 to F_0 and coupling of the hydrolytic process to transmembrane H^+ conduction. In addition, the C-terminal segment of the PVP protein, which extends at least in part out of the M surface of the membrane and enters into contact with F_1, is essential for a correct functional organization of transmembrane proton channel in F_0. In fact, DCCD binding to the 8 and 16 kDa components of F_0 results in inhibition of H^+ conduction only in the presence of the PVP protein. The PVP protein may thus play a central role in organizing the transmembrane H^+ channel in F_0 and in controlled H^+ conduction from the H^+ channel to catalytic and/or allosteric sites in F_1.

In conclusion, it appears that the ATP synthase of mammalian tissues has acquired during evolution a number of additional components with respect to lower organisms, which enable this enzyme to adapt its activity to the specific energy demands exhibited by the eukaryotic cells in their various physiopathological states.

REFERENCES

Chernyak, B V., Dukhovich, V.F. and Khodiaw, Yu. (1987) *FEBS Lett.* **215**, 300–4.

Erecinska, M. and Silver, I.A. (1989) *J. Blood Flow and Metab.*, **9**, 2–19.

Eya, S., Noumi, T., Maeda, M. and Futai, M. (1988) *J. Biol. Chem.*, **263**, 10056–2.

Fearnley, I.M. and Walker, J.E. (1986) *EMBO J.*, **5**, 2003–8.

Fernandez-Moran, H., Oda, T., Blair, P.V. and Green, D.E. (1964) *J. Cell Biol.*, **22**, 63–9.

Gogol, E.P., Lücken, U. and Capaldi, R.A. (1987) *FEBS Lett.*, **219**, 274–7.

Guerrieri, F. and Papa, S. (1982) *Eur. J. Biochem.*, **128**, 9–13.

Guerrieri, F., Scarfò, R., Zanotti, F. *et al.* (1987a) *FEBS Lett.*, **213**, 67–72.

Guerrieri, F., Zanotti, F., Che, Y.W. *et al.* (1987b) *Biochim. Biophys. Acta*, **892**, 284–93.

Guerrieri, F., Capozza, G., Houstek, J. *et al.* (1989) *FEBS Lett.*, **250**, 60–6.

Hashimoto, T., Yoshida, Y. and Tagawa, K. (1983) *J. Biochem.*, **94**, 714–20.

Hashimoto, T., Yoshida, Y. and Tagawa, K. (1984) *J. Biochem.*, **95**, 131–6.

Hoppe, J. and Sebald, W. (1984) *Biochim. Biophys. Acta*, **7 68**, 1–22.

Houstek, J., Kopecky, J., Zanotti, F. *et al.* (1988) *Eur. J. Biochem.*, **173**, 1–8.

Klein, G., Satre, M., Dianoux, A.C. and Vignais, P.V. (1980) *Biochemistry*, **19**, 2919–25.

Kopecky, J., Guerrieri, F. and Papa, S. (1983) *Eur. J. Biochem.*, **131**, 17–24.

Lightowlers, R.N., Howitt, S.M., Hetch, L. and Cox, G.B. (1988) *Biochim. Biophys. Acta*, **933**, 241–8.

Papa, S. (1990) in *Organelles of Eukaryotic Cells: Molecular Structure and Interactions* (eds J.M. Tager, A. Azzi, S. Papa and F. Guerrieri), pp. 9–46, Plenum Press, New York and London.
Papa, S., Guerrieri, F., Zanotti, F. *et al.* (1988) in *Molecular Basis of Biomembrane Transport* (eds F. Palmieri and E. Quagliariello), Elsevier, Amsterdam, pp. 249–59.
Papa, S., Guerrieri, F., Zanotti, F. *et al.* (1989) *FEBS Lett.*, **249**, 62–66.
Paule, C.R. and Fillingame, R.H. (1989) *Arch. Biochem. Biophys.*, **274**, 270–84.
Pedersen, P.L. and Carafoli, E. (1987) *TIBS*, **12**, 146–8.
Pullman, M.E. and Monroy, G.C. (1963) *J. Biol. Chem.* **238**, 3762–9.
Rouslin, W. and Pullman, M.E. (1987) *J. Med. Cell Biol.* **19**, 661–8.
Sanadi, D.R. (1982) *Biochim. Biophys. Acta*, **683**, 39–56.
Senior, A.E. (1988) *Physiol. Rev.*, **68**, 177–232.
Von Meyenburg, K., Jørgensen, B.B., Michelsen, O. *et al.* (1986) *EMBO J.*, **4**, 2357–62.
Walker, J.E., Runswick, M.J. and Poulter, L. (1987a) *J. Mol. Biol.*, **197**, 89–100.
Walker, J.E., Gey, N.J., Powell, S.J. *et al.* (1987b) *Biochemistry*, **26**, 8613–19.
Yoshida, Y. Wakabayashi, S., Matsubara, H. *et al.* (1984) *FEBS Lett.*, **170**, 135–8.
Zanotti, F., Guerrieri, F., Scarfò, R. *et al.* (1985) *Biochem. Biophys. Res. Commun.*, **132**, 985–90.
Zanotti, F., Guerrieri, F., Che, Y.W. *et al.* (1987) *Eur. J. Biochem.*, **164**, 517–23.
Zanotti, F., Guerrieri, F., Capozza, G. *et al.* (1988) *FEBS Lett.*, **237**, 9–14.

3 Comparative analysis of the amount of subunit I and subunit IV of cytochrome c oxidase mRNAs in the cerebral hemispheres of senescent rat and effect of acetyl-L-carnitine

M.N. GADALETA, V. PETRUZZELLA, M. RENIS, F. FRACASSO AND P. CANTATORE
Department of Biochemistry and Molecular Biology, University of Bari and CSMME, Bari, Italy

Mammalian mitochondrial DNA is a small, double-stranded, circular molecule of about 16 500 bp which carries the information for part of the inner mitochondrial membrane respiratory complexes and for its own protein synthesis machinery (Cantatore and Saccone, 1987). It codes for 13 out of about 60 polypeptides of the mitochondrial respiratory chain, for two rRNAs and for 22 tRNAs. The coordinated expression of mitochondrial (mt)DNA and nuclear (n)DNA is thus essential for the cell energy metabolism. A key role in the degenerative processes of senescence has been assigned to mtDNA by Miquel and Fleming's 'oxygen radical–mitochondrial injury' hypothesis of ageing (Miquel and Fleming, 1986). According to these authors, the fundamental cause of senescence in highly differentiated post-mitotic cells, such as neurones and muscle cells, is the accumulation of damaged mitochondria, due to the release in the organelles of the toxic by-products of respiration. These by-products, besides altering mitochondrial membrane permeability, are mutagenic towards mtDNA. This molecule could represent an effective target for oxygen radicals since it is in close spatial relationship to the mitochondrial inner membrane and is unprotected by histone-like proteins. The lack of mtDNA repair mechanisms should lead to an accumulation of mutated mtDNA molecules with consequent impaired synthesis of mtDNA-coded polypeptides, decline in ATP synthesis, and in physiological performance.

Other authors (Richter, 1988; Linnane *et al.*, 1989; Grivell, 1989) have more recently stressed the relevance of mtDNA as a target for somatic gene mutation for several reasons: in particular the mtDNA high mutation rate (10–12 times higher than nuclear DNA) and the compactness of its gene organization lead any mutational event to potentially affect a functionally important part of the molecule. Although mtDNA mutation can occur continuously, in somatic cells the effects of potentially harmful mutations may not be immediately obvious. Cells normally contain thousands of mtDNA molecules and phenotypic expression will depend both on the extent of the segregation of the mixed population of normal and mutant mtDNAs as well as on the cell's need for mitochondrial function. As a consequence a range of cells of varying bioenergetic capacity will appear. The effect of the accumulation of bioenergetically deficient cells will have a different outcome in various tissues according to their dependence on aerobic metabolism. Studies with respiratory inhibitors have shown that the tissues most likely affected are brain and, to a lesser extent, skeletal muscle, heart, kidney and liver (Wallace, 1989). This would explain why declining mental capacity, muscle weakness and non-atherosclerotic heart dysfunction (presbycardia) are the phenotypic events associated with old age.

A search for age-dependent modifications in mtDNA has not yet given consistent results. The most recent report on this subject is an electron microscope analysis of denatured–reannealed mtDNA duplexes (Pikò *et al.*, 1988). The authors found an increased frequency of small loops and knobs in senescent mouse liver, suggesting deletions/additions of about 400 nucleotides. Other indirect data supporting the hypothesis of extensive damage to mtDNA in ageing are the degeneration and loss of mitochondria observed during senescence in insects (Miquel and Fleming, 1986) and a prevalent oxidative damage to mtDNA compared with nDNA (Richter *et al.*, 1988). Recently (Gadaleta *et al.*, 1990), we reported data on the mitochondrial steady-state level of the 12 S rRNA and of the mRNA for the subunit 1 of cytochrome c oxidase (CoI mRNA) in various tissues of 28-month-old Fisher rats. We found a lower number of both RNAs in senescent rat. The reduction was most marked in cerebral hemispheres and in the cerebellum mitochondria, lower in heart, but absent in the liver. The decrease in the mitochondrial content of the two RNAs was, however, erased by 1 h pre-treatment of senescent rat with acetyl-L-carnitine (300 mg/kg body weight), thus excluding an irreversible damage to the transciptional apparatus in senescent rat.

It is well known that mtDNA codifies only for three of the cytochrome c oxidase subunits, whereas the other eight, some of which are tissue specific (Kuhn-Nentwig and Kadenbach, 1985), are nDNA coded. A comparison of the age-dependent variation in the content of the mRNAs for nDNA- and mtDNA-coded subunits should be useful to show whether mtDNA expression is preferentially impaired during ageing. Moreover, it will add new data about the relationship between mitochondrial and nuclear gene expression. Here, we have measured the cellular level of CoI and CoIV mRNAs in the brain of 26-month-old Fisher rats. The effect of the acetyl-L-carnitine pre-treatment was also investigated.

3.2 EXPERIMENTAL PROCEDURES

3.2.1 Isolation of mitochondria and nucleic acid extraction

Fisher rats, 9 and 26 months old, were used. They were kept in an animal room under controlled conditions up to the moment of use. The brain areas from cerebral hemispheres were minced, washed repeatedly and hand homogenized in about 20 volumes of 0.32 M sucrose, 1 mM K^+-ethylenediaminetetraacetic acid (EDTA), 10 mM Tris-HCl pH 7.4 (medium A), with 10–12 up and down strokes in a Dounce glass homogenizer. The homogenate was centrifuged at 1200 *g* for 5 min. The nuclear pellet was washed once at 1200 *g* for 3 min and the combined supernatants were centrifuged at 18 000 *g* for 10 min to yield a crude mitochondrial pellet. After careful removal of the fluffy layer, the pellet was suspended with medium A (3 ml/g of fresh tissue), layered onto 25 ml of a discontinuous Ficoll gradient (7.5–13% in medium A) and centrifuged at 99 000 *g* for 30 min. At the end of the run three fractions were visible: a myelin-containing fraction, placed at the top of the gradient; a synaptosomal fraction located at the interphase between the two Ficoll solutions; and a pellet, made mainly of free mitochondria. The purity of the synaptosomal fraction was checked both by electron microscopy and by determination of marker enzymes activities. The enzymes assayed were reduced nicotinamide adenine dinucleotide (NADH)–cytochrome c reductase (rotenone insensitive), as marker of the external membrane and therefore of free mitochondrial contamination, and lactate dehydrogenase as indicator of synaptosomal membrane integrity (Lai and Clark, 1976; Booth and Clark, 1978). By using both structural and functional criteria, the purity of the synaptosomes was never below 70–80%. To isolate the mitochondria, the synaptosomal fraction was washed in medium A, centrifuged at 18 000 *g* for 15 min, carefully resuspended in 6 mM Tris-HCl pH 8.1 (about 10 ml/area) and centrifuged at 12 000 *g* for 10 min. The pellet was finally resuspended in medium A. Nucleic acid extraction from synaptic mitochondria and from aliquots of post-nuclear supernatant was carried out as previously reported (Cantatore *et al.*, 1986). DNA and RNA concentrations were determined by diphenylamine (Burton, 1956) and orcinol respectively. The estimation of mtDNA contained in the homogenate and in the mitochondrial nucleic acids was carried out by a quantitative hybridization procedure reported elsewhere (Gadaleta *et al.*, 1990). Protein content was determined by the Waddel method (Waddel, 1956).

3.2.2 Probe labelling and hybridization

To prepare single-stranded hybridization probes, mtDNA fragment coding for rat CoI mRNA and for human CoIV mRNA were cloned in a plasmid vector (Bluescribe–Stratagene) which has the T_3 and T_7 RNA polymerase promoters adjacent to the polylinker cloning sites. The probes used were a fragment of 855 bp (CoI probe) derived from the digestion of *Eco* RI C with *Taq*I (Cantatore *et al.*, 1984) and a fragment of 700 bp derived from the digestion of the plasmid pCOX4.111, containing 700 bp of the CoIV coding sequence (Zeviani *et al.*, 1987), with *Eco* RI. Cloning sites were *Acc*I and *Sma*I for CoI and CoIV probes respectively. For the transcription, 0.4 μg of each recombinant plasmid was first linearized with *Hind* III and *Sac* I respectively. To eliminate the 3′ protruding ends generated by *Sac* I digestion, the CoIV probe labelling was preceded by a treatment with Klenow–DNA polymerase (2.5 units/μg DNA) in the presence of 0.2 mM dTTP, 50 mM Tris-HCl pH 7.2, 10 mM $MgSO_4$, 1 mM dithiothreitol and 50 μg/ml bovine serum albumin for 1 h at 22 °C. Then the mixture was incubated in the presence of 0.4 mM ATP, 0.4 mM CTP, 0.4 mM GTP, 0.02 mM UTP, 200 μCi of [α-^{32}P] UTP (800 Ci/mmol), 40 mM Tris-HCl pH 8.0, 8 mM $MgCl_2$, 2 mM spermidine, 50 mM NaCl, 10 mM dithiothreitol, 30 units of RNasin (Boheringer) and 10 units of T_3 RNA polymerase (Stratagene) for 1 h at 37 °C. CoI probe labelling was carried as reported by Gadaleta *et al.* (1990). At the end of the RNA polymerase reaction, 30 units of RNasin and 20 μg/ml of DNase (RNase-free Promega) were added. After a further incubation for 10 min at 37 °C, the mixture was phenol extracted and ethanol precipitated several times. The RNA pellet was resuspended in a small volume of 94% formamide, 60 mM PIPES pH 7.0 and used for solution hybridization experiments. These were carried out by combining the RNA with increasing amounts of labelled riboprobes in a final volume of 20–40 μl and incubating for 20 h at 50 °C in the presence of 80% formamide, 40 mM PIPES pH 7.0, 1 mM EDTA, 0.4 M NaCl and 10 μg of *Escherichia coli* tRNA. At the end of the reaction, the samples were diluted ten times with 0.3 M NaCl, 10 mM Tris-HCl pH 7.8, 10 mM EDTA and digested with 0.075 units/ml of preboiled RNase A (Boheringer) and 1500 units/ml of RNase T_1 (BRL) for 30 min at 30 °C. Preliminary tests (not shown) demonstrated that under these conditions hybrids of the expected length were produced. The amount of the hybrid was determined by the difference between the amount of trichloracetic acid-insoluble radioactivity in the mitochondrial nucleic acids or RNA-containing samples and the controls. These contained the same components of the hybridization mixture except the cold RNA. In all the experiments at least two different amounts of mitochondrial and total RNA were used. Each experimental point was obtained in triplicate. At saturation, the RNase-resistant radioactivity was about 5% of the input in the mitochondrial nucleic acids or cytoplasmic RNA-containing samples, and 1% of the input in the controls. Statistical significance of differences in RNA concentration between adult and senescent tissues was determined by Student's *t*-test.

3.3 RESULTS AND DISCUSSION

Although it occupies only 2% of the body weight, the brain consumes at least 20% of the total oxygen intake of the body at rest. In fact, brain uses ATP for active ion pumping, to maintain the resting membrane potential and the fast axoplasmic transport and for the synthesis of macromolecules and neurotransmitters. Cytochrome c oxidase is the terminal enzyme of the electron transport chain. Its activity is linked to the generation of ATP through the coupled process of oxidative phosphorylation. Therefore its activity is essential for those organs which critically depend on oxidative metabolism. Indeed, cytochrome c oxidase activity is considered a valuable endogenous metabolic marker for neurones, since maintenance of ion balance is their major energy-consuming function (Wong-Riley, 1989).

An age-dependent decrease of cytochrome c oxidase activity has been reported both in mitochondria isolated from liver (Weindruch *et al.*, 1980), heart (Abu-Erreish and Sanadi, 1978), and brain (Benzi *et al.*, 1980), as well as in the homogenates of striated muscle of rat (Hansford and Castro, 1982) and human (Trounce *et al.*, 1989). Furthermore, a 50% decrease of a-type cytochromes in synaptic mitochondria, but not in free mitochondria, of 28–30-month-old Fisher rats has been reported (Harmon *et al.*, 1987). A lower content of respiratory units in the mitochondria of old animals has been suggested by Abu-Erreish and Sanadi (1978).

Recently, we performed a detailed study on the level of CoI mRNA in various tissues of young and senescent rats (Gadaleta *et al.*, 1990). A consistent reduction of CoI mRNA in the synaptic mitochondria of 28-month-old rat cerebral hemispheres was observed. As a further step in this study, we decided to measure the cellular level of the mRNA for a nuclear coded cytochrome c oxidase subunit, in adult and senescent rat. The quantification was performed by hybridizing in solution RNA extracted from the cytoplasm of the cerebral hemispheres of adult and senescent rat with a human probe for the nuclear coded subunit IV. Hybridization conditions were as for the homologous hybridization since human and rat CoIV coding sequences have a high level of similarity (about 84%). Fig. 3.1 shows the comparison of the cellular level of CoI and CoIV mRNAs normalized with respect to the total RNA content. It appears that CoI mRNA is about one order of magnitude more abundant

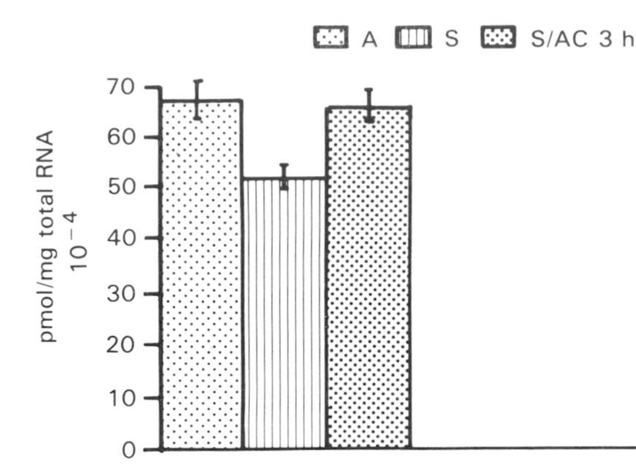

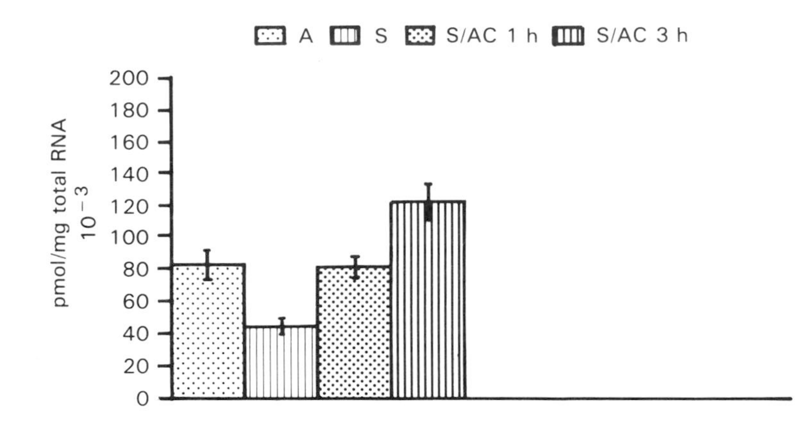

Fig 3.1 CoIV mRNA (top) and CoI mRNA (bottom) content in cerebral hemispheres of adult, senescent and acetyl-L-carnitine-treated senescent rat. CoIV mRNA determination was carried out on cytoplasmic RNA and then refined to total cellular RNA, CoI mRNA determination was carried out on synaptic mitochondria and then refined to total cellular RNA. The values are the mean of four determinations, run in triplicate. Acetyl-L-carnitine at a dose of 300 mg/kg body weight was injected intraperitoneally to senescent rats. A, adult rat; S, senescent rat; S/A.C. (1 h), senescent rat 1 h after injection of acetyl-L-carnitine; S/A.C. (3 h), senescent rat 3 h after injection of acetyl-L-carnitine.

than CoIV mRNA. This is a minimum estimate since it reflects only the CoI mRNA content of the synaptosomal population of the cerebral hemispheres. The amount of both RNAs is lower in senescent rat than in adult rat. However, the reduction is greater for the mtDNA-coded transcript: CoI mRNA content of senescent tissue is about 45% of that found in the adult tissue, whereas CoIV mRNA is only about 25% less. The difference in the decrease of the two mRNAs could be due either to a higher instability of mitochondrial transcripts or to the fact that the transcription process is impaired more in mitochondria than in nuclei. If the second hypothesis is correct, a possible explanation – in agreement with Miquel and Fleming's hypothesis (1986) – is that the primary structure of mtDNA is affected by ageing much more than that of nDNA. The results obtained after the administration to senescent rat of acetyl-L-carnitine do not support such a conclusion (Fig. 3.1). As reported in Figure 3.1, acetyl-L-carnitine greatly increases the content of CoI mRNA in senescent animals 3 h after injection but has less effect on the content of CoIV mRNA. Therefore, the age-related decrease of CoI mRNA is clearly reversible. This excludes severe mtDNA damage in ageing rat and suggests that the alteration of the mitochondrial transcriptional apparatus could be due to some metabolic impairment, such as an alteration in the ATP and/or in the cation concentration. Both ATP (Gaines *et al.*, 1987) and Ca^{2+} ions (Attardi, personal communication) have been recently reported to play a role in mtDNA transcription. Since acetyl-L-carnitine pre-treatment of adult rat does not produce any variation in the level of mtRNAs (Gadaleta *et al.*, 1990), the effect of this substance on mitochondrial transcription is probably related to changes produced by ageing. The acetyl-L-carnitine can increase the efficiency of oxidative phosphorylation, or can reduce the alteration of membrane permeability induced by ageing (Vorbeck *et al.*, 1982; Paradies and Ruggiero 1990). An effect of acetyl-L-carnitine on the cholesterol content in cerebral hemispheres of senescent rat has been recently found (F. Ruggiero, personal communication). An action at membrane level could explain why the acetyl-L-carnitine treatment is more effective on mtDNA than on nDNA transcription.

As far as the coordinated expression of mitochondrial and nuclear genomes is concerned, the data here reported show that during ageing the level of CoI mRNA is much more prone to be altered by metabolic perturbations than that of the DNA-coded CoIV mRNA. Similar data have been recently reported in the literature. A chronic stimulation of rat skeletal muscle induces coordinate increases of CoIII and CoVIc mRNAs (Hood *et al.*, 1989). Moreover, thyroid hormones are able to induce CoI, CoII and CoVI polypeptide accumulation in hypothyroid rats (Mutvei and Nelson, 1989). In both cases, the effects at the level of mtDNA-coded products were more pronounced. Finally, in a patient affected by Kearn-Sayre mitochondrial myopathy (Mita *et al.*, 1989), the level of cytochrome c oxidase subunit II is deficient in muscle fibres lacking the cytochrome c oxidase activity, whereas the cytochrome c oxidase subunit IV level remains unchanged.

Recently several groups have found in brain and muscle mtDNA isolated from normal aged individuals, a deletion identical as for dimension and breakpoints to that observed in the majority of patients with KSS and PEO, although in a very low concentration [Ozawa, T. *et al.* (1990) and Cortopassi *et al.* (1990)].

ACKNOWLEDGEMENTS

This work has been accomplished partly with funds from MPI (40% and 60%) and partly from the Sigma-Tau, Italy. The authors would like to thank Professor S. Di Mauro for the gift of the human CoIV probe, Mr D. Munno for excellent technical assistance and Ms F. de Palma for typing and word-processing.

REFERENCES

Abu-Erreish, C.M. and Sanadi, D.R. (1978) *Mech. Ageing Dev.*, **7**, 425–32.
Benzi, G., Arrigoni, E., Dagani, F. *et al.* (1980) *Aging. ISS Aging Brain Dementia* (NY), **13**, 113–17.
Booth, R.F.G. and Clark, J.B. (1978) *Biochem. J.*, **176**, 365–70.
Burton, K. (1956) *Biochem. J.*, **62**, 315–23.
Cantatore, P. and Saccone, C. (1987) *Int. Rev. Cytol.*, **108**, 149–208.
Cantatore, P., Gadaleta, M.N. and Saccone, C. (1984) *Biochem. Biophys. Res. Commun.*, **118**, 284–91.
Cantatore, P., Loguercio Poloa, P., Fracasso, F., Flagella, Z. and Gadaleta. M.N. (1986) *Cell Differ.*, **19**, 125–32.
Cortopassi, G.A. and Arnheim, N. (1990) *NAR*, **18**, 6927–33.
Gadaleta, M.N., Petruzzella, V., Renis, M. *et al.* (1990) *Eur. J. Biochem.*, **187**, 501–6.
Gaines, G., Rossi, C. and Attardi, G. (1987) *J. Biol. Chem.*, **262**, 1907–15.
Grivell, L.A. (1989) *Nature*, **341**, 569–71.
Hansford, R.G. and Castro, F. (1982) *Mech. Ageing Dev.*, **19**, 5–13.
Harmon, H.J., Nank, S. and Floyd, R.A. (1987) *Mech. Ageing Dev.*, **38**, 167–77.
Hood, D., Zak, R. and Pette, D. (1989) *Eur. J. Biochem.*, **179**, 275–80.
Kuhn-Nentwig, L. and Kadenbach, B. (1985) *Eur. J. Biochem.*, **149**, 147–58.
Lai, J.C.K. and Clark, J.B. (1976) *Biochem. J.*, **154**, 423–32.
Linnane, A.W., Marzuki, S., Ozawa, T. and Tanaka, M., (1989) *Lancet*, **i**, 642–45.
Miquel, J. and Fleming, J. (1986) in *Free Radicals, Ageing and Degenerative Diseases* (eds J.E. Johnson Jr *et al.* (*Modern Ageing Research*, Vol. 8), Liss, New York, pp. 51–74.

Mita, S., Schmidt, B., Schon, E. *et al.* (1989) *Proc. Natl. Acad. Sci. USA*, **86**, 9509–13.
Mutvei, A. and Nelson, B.B. (1989) *Arch. Biochem. Biophys.*, **268**, 215–20.
Ozawa *et al.* (1990) *Bioch. Biophys. Res. Commun.*, **172**, 483–9.
Paradies, G. and Ruggiero, F.M. (1990) *Biochim. Biophys. Acta*, **1016**, 207–12.
Pikó, L., Hougham, A. and Bullpitt, K. (1988) *Mech. Ageing Dev.*, **43**, 279–93.
Richter, C. (1988) *FEBS Lett.*, **241**, 1–5.
Richter, C.H., Perk, J.W. and Ames, B. (1988) *Proc. Natl. Acad. Sci. USA*, **85**, 6465–67.
Trounce, I., Byrne, E. and Marzuki, S. (1989) *Lancet*, **i**, 637–9.
Vorbeck, M., Martin, A.P., Lang, J.W.J. *et al.* (1982) *Arch. Biochim. Biophys.*, **277**, 351–61.
Waddel, W.J. (1956) *J. Lab. Clin. Med.*, **48**, 311–14.
Wallace, D.C. (1989) *Trends Genet.*, **5**, 9–13.
Weindruch, R.H., Cheng, M.K., Verity, M.A. and Walford, R. (1980) *Mech. Ageing Dev.*, **12**, 375–93.
Wong-Riley, M.T.T. (1989) *Trends Neurosci.*, **12**, 94–100.
Zeviani, M., Nakagawa, M., Herbert, J. *et al.* (1987) *Gene*, **55**, 205–17.

4 Cytochemical and immuno-cytochemical investigation of respiratory complexes in individual fibres of human skeletal muscle

M.A. JOHNSON
School of Neurosciences, University of Newcastle upon Tyne, UK

4.1 INTRODUCTION

The initial detection of mitochondrial abnormalities in human muscle biopsies has always relied heavily on the use of cytochemical techniques. Early descriptions of mitochondrial myopathies were largely based on morphological data, with cytochemical methods being used primarily to detect characteristic abnormalities of mitochondrial distribution (Olson *et al.*, 1972). It has since been demonstrated that many mitochondrial myopathies are due to specific defects within individual complexes of the respiratory chain; in some disorders single respiratory complexes are involved, whereas in others more than one complex is involved (for review see DiMauro *et al.*, 1985). It has also been shown that while some mitochondrial defects appear to be expressed in skeletal muscle in a homogenous manner, with all muscle fibres similarly affected, other disorders are characterized by a mosaic distribution of the defect, with normal and abnormal fibres coexisting within the same muscle fibre population (Johnson *et al.*, 1983).

In modern diagnostic practice the traditional role of cytochemistry in preliminary screening protocols has been extended to include techniques which have as their specific aim the investigation of individual respiratory complexes in situ. Assays of the catalytic activity of many mitochondrial oxidative enzymes in single muscle fibres have been made possible by means of the development of kinetic microphotometric techniques (Pette 1981; Reichmann and Pette 1982). Microphotometric enzyme assay (MEA) is based on the measurement of the final reaction product of a cytochemical reaction in tissue sections or cell cultures and thus provides a method of estimating enzyme activities in single cells or muscle fibres. The detailed protein subunit composition of individual respiratory complexes can also be studied in situ by means of immunocytochemical techniques (ICC) which provide a useful adjunct to investigations by immunoblotting.

A particular advantage of a cytochemical approach to the investigation of mitochondrial disorders is that it allows the mosaic distribution of certain of these defects to be detected, whereas the tissue homogenization involved in conventional enzyme assays or immunoblotting precludes this. A further advantage of MEA or ICC is that only small amounts of tissue are needed, which is important since many of the affected patients are infants or small children. The main aim of this communication is to outline ways in which these techniques can be used in the diagnosis and further investigation of mitochondrial disorders. Reference will be made not only to those situations in which MEA and ICC offer advantages over standard enzyme assays and immunoblotting but also to contexts in which the reverse applies.

4.2 MATERIALS AND METHODS

Muscle biopsies for cytochemical investigation were snap-frozen using isopentane cooled to −150 °C in liquid nitrogen. Samples were stored in heat-sealed polythene packets in the vapour phase of liquid nitrogen containers.

4.2.1 Microphotometric enzyme assays

Frozen sections 8 μm thick were cut using a Reichert–Jung Frigocut cryostat microtome equipped with motor-driven cutting action to maintain maximal reproducibility of section thickness. Sections were picked up on microscope slides and air-dried for 15 min at room temperature. A coverglass was placed over each section, using parallel strips of electrical insulating tape to maintain the required distance between slide and coverslip of assay media. Sections were stored at + 4 °C for up to 2 h until required.

The optimal composition of assay media for the estimation of succinate dehydrogenase (SDH) and cytochrome c oxidase (CCO) activities was determined experimentally (Old and Johnson, 1989). SDH activity was assayed using 130 mM succinate, 1.5 mM Nitro Blue tetrazolium (NBT), 0.2 mM phenazine methosulphate (PMS) and 1.0 mM azide in 100 mM phosphate, pH 7.0. In this medium, NBT functions as final electron acceptor and on reduction its corresponding formazan is measured at 574 nm. CCO activity was assayed in a medium containing 100 μM cytochrome c (cyt. c) and 4 mM 3,3′-diaminobenzidine tetrahydrochloride (DAB) in 100 mM phosphate, pH 7.0. DAB acts as electron donor in the respiratory chain at the level of cyt. c and undergoes oxidative polymerization and cyclization to form an indamine polymer (Seligman *et al.*, 1968), which is measured at its absorption maximum of around 450 nm.

A Zeiss UMSP30 microphotometer equipped with a computer-controlled scanning stage was used to make absorbance measurements in up to 30 muscle fibres in any one measuring cycle. Absorbance measurements were made immediately after the addition of assay medium to the tissue section and at timed intervals during the linear phase of the cytochemical reaction. In the case of SDH (Fig. 4.1) linearity was maintained for more than 10 min in human muscle sections; and with CCO for more than 15 min. Enzyme reaction rates were computed as increase in absorbance per unit time ($\Delta A \times s^{-1}$) and regression analysis was used to monitor linearity of reactions.

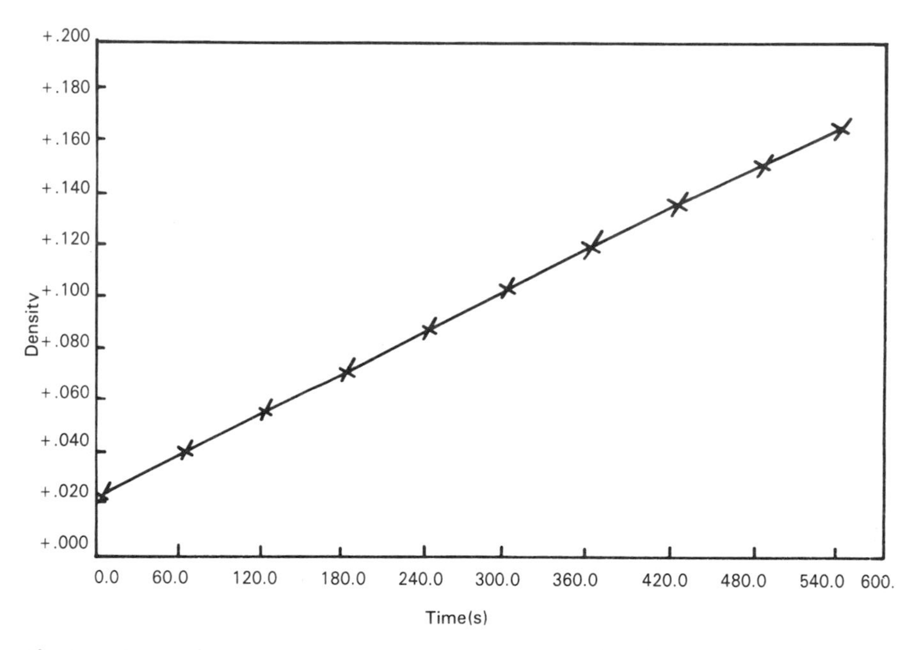

Fig. 4.1 Original computer graph of increase in optical density (ΔA) due to formazan production in the microphotometric assay of SDH activity in a single muscle fibre.

Routinely >100 muscle fibres were assayed per patient, using four or five serial tissue sections. Since mitochondrial oxidative enzyme activity varies according to fibre type, individual fibres were retrospectively identified as type 1 (slow oxidative), type 2A (fast oxidative/glycolytic) or type 2B (fast glycolytic) by reference to serial sections in which the activity of myofibrillar ATPase (mATPase) was demonstrated (Brooke and Kaiser, 1970). Assay results from individual patients were most appropriately presented in the form of histograms (Fig. 4.2). this allows the assessment of any departure from unimodality of enzyme activities in any fibre population. In order to compare MEA results directly with those of standard enzyme assays, it is necessary to calculate the relative areas of tissue sections occupied by types 1, 2A, 2B fibres and to use these data to calculate mean activities for the tissue as a whole.

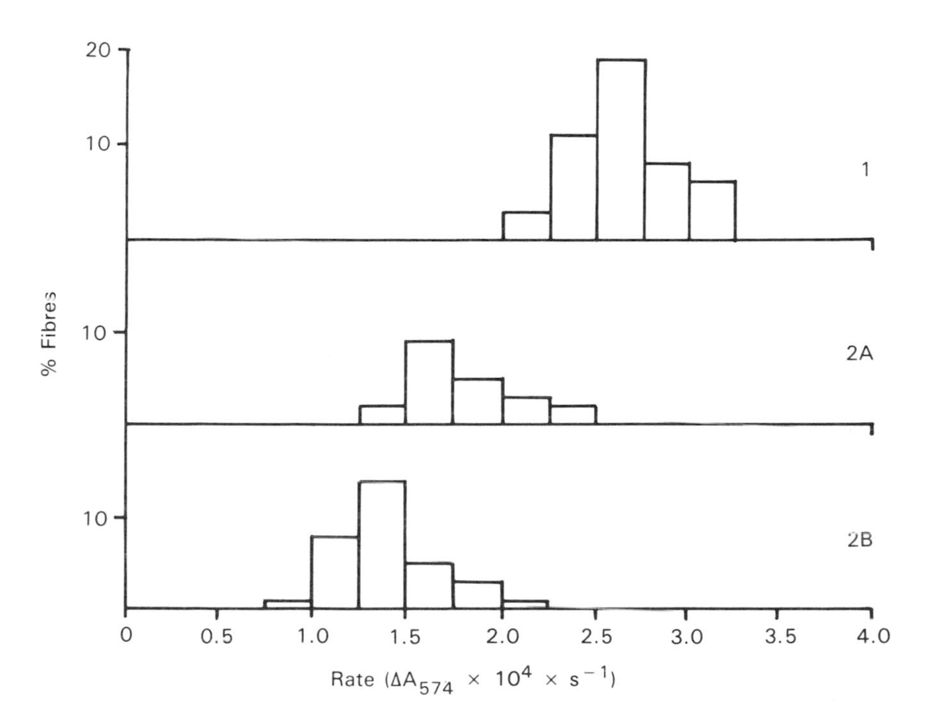

Fig. 4.2 Histogram showing SDH activities in single muscle fibres of normal human m. quadriceps. Note the gradient of activities: type 1 fibres > type 2A fibres > type 2B fibres

4.2.2. Immunocytochemical techniques

Much of the methodology involved in the immunolabelling of tissue proteins is well documented (Sternberger, 1979) but the demonstration of intramitochondrial antigens poses some specific problems. One of these concerns the resolution of individual mitochondria, which is more satisfactory using immunoenzyme labelling than by means of immunofluorescence, in which details of individual mitochondria tend to be lost.

The problem of permeability of mitochondrial membranes to immunoglobulins must also be overcome. Sometimes prolonged incubation with primary antibody will result in satisfactory immunolabelling but it is often preferable to use some form of pre-treatment designed to permeabilize tissue sections or cultured cells (Miranda *et al.*, 1989). The following procedure has been used successfully in our laboratory for the immunolocalization of many respiratory chain antigens.

Tissue sections 6 μm thick were air-dried for 2 h and then fixed in formol–calcium (4% aqueous formaldehyde containing 100 mM calcium chloride, pH 7.0) at room temperature for 1 h. After washing they were passed through a graded ethanol series to 100% ethanol for 20 min. Sections were then rehydrated and rinsed in the buffer used for immunolabelling (12.5 mM Tris-buffered saline, pH 7.6; TBS). A standard peroxidase–antiperoxidase (PAP) immunolabelling method was used (Sternberger *et al.*, 1970). Primary rabbit polyclonal antibody, optimally diluted, was applied to tissue sections in 50 μl aliquots for 1 h. Sections were washed for 10 min × 3 in TBS prior to incubation for 1 h with swine anti-rabbit immunoglobulins (1/100, Dako Z196). Further thorough washing was followed by treatment with rabbit PAP complex (1/100, Dako Z113). Visualization of the peroxidase label was done using a medium containing 0.05% DAB and 0.01% H_2O_2 in 100 mM Tris, pH 7.6 for 10 min. Sections were fianlly dehydrated in a graded ethanol series, cleared and mounted in DPX.

Normal levels of particular respiratory chain antigens were defined using muscle sections from at least six control subjects. The highest dilution of primary antibody which gave consistent particulate immunolabelling of all muscle fibres was used in preliminary screening procedures. At least one normal muscle sample was included with each batch of sections from patients to be immunolabelled. In patients found to show absent or decreased immunoreactivity at the screening dilution, a further series of primary antibody dilutions was used to determine the extent of the decrease.

4.3 APPLICATIONS OF CYTOCHEMICAL TECHNIQUES IN THE INVESTIGATION OF INDIVIDUAL RESPIRATORY COMPLEXES

A primary objective of current cytochemical and immunocytochemical research has been to develop techniques which will enable each of the respiratory complexes to be studied at the cellular level, both from the standpoint of catalytic activity and of antigenic composition. This objective is only partially achieved at present. However, the applications outlined in this communication illustrate ways in which this approach can yield information not obtainable by other means. In particular, cytochemical techniques allow the observer to determine whether a respiratory chain defect shows a generalized or mosaic distribution; this has important genetic implications which will be discussed later.

4.3.1 Complex I (NADH-ubiquinone reductase)

Deficiencies of complex I have been reported by several groups of workers, notably by Morgan-Hughes and colleagues (Morgan-Hughes *et al.*, 1979; Land *et al.*, 1981) and more recently by Morcadith *et al.* (1987) and Watmough *et al.* (1989). Some patients have abnormalities predominantly affecting the central nervous system but a significant number have presented with exercise intolerance. Studies of skeletal muscle from these patients using immunoblotting may show a generalized decrease in all detectable subunits but specific deficiencies of certain iron–sulphur proteins have also been recorded (Schapira *et al.*, 1988).

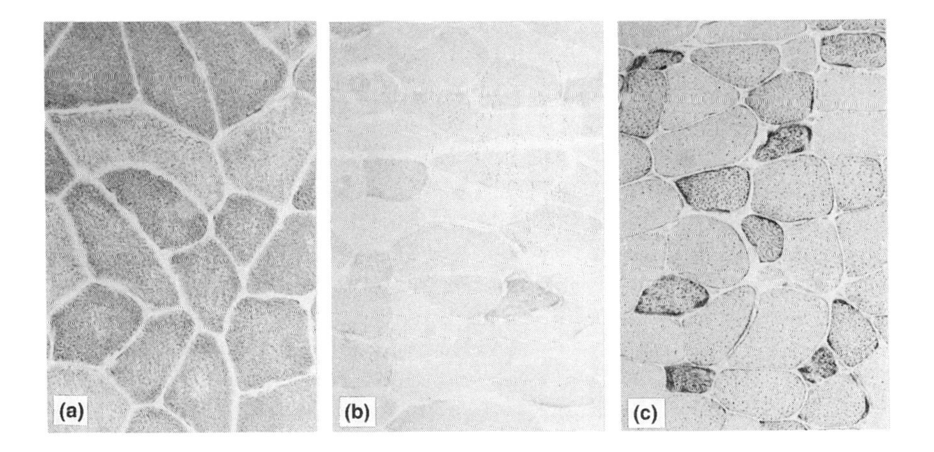

Fig 4.3 (a) Holocomplex I immunolabelling in normal skeletal muscle. (b) Severe decrease in labelling in all muscle fibres of patient aged 14 years with muscle-specific complex I deficiency. (c) Only some muscle fibres in this patient show 'ragged-red' changes. Immunoperoxidase (a, b); SDH (c).

Fig. 4.3 shows the results of immunocytochemical labelling, using a primary antibody raised to holocomplex I, in a normal control subject and in a patient with severe complex I deficiency (Watmough *et al.*, 1989). In this patient the defect appeared to be confined to skeletal muscle and was not expressed in liver. Immunoblotting showed a marked increase in all detectable subunits and ICC demonstrated a decrease of immunoreactivity in situ which was virtually homogeneous throughout the whole muscle fibre population. In spite of this apparently uniform distribution of the defect within the skeletal muscle, only scattered fibres showed morphological mitochondrial abnormalities in the form of 'ragged red' fibres (RRFs). Since the defect in this patient was confined to complex I, and complex IV activity was normal, all 'ragged-red' fibres were CCO positive on cytochemical testing, as were the rest of the fibre population.

In this patient, ICC labelling using holocomplex I antisera clearly demonstrated a severe decrease in immunoreactive protein, which correlated with the generalized loss of labelling of subunits detectable by immunoblotting. If one or more immunoreactive subunits of the complex had been retained in near-normal amounts it is likely that this would have masked decreases in other subunits in this immunocytochemical system. It is important to recognize this limitation of ICC when only holocomplex antisera are available. In immunoblotting, where subunit proteins have been separated previously by sodium dodecyl sulphate – polyacrylamide gel electrophoresis (SDS–PAGE), deficiencies of specific subunits may be detected by means of holocomplex antisera. In any immunocytochemical system, detection of subunit-specific defects requires subunit-specific antisera. In the case of complex I with over 30 subunits this would be so time-consuming as to be impractical even if all the requisite antisera were available.

Further investigations of complex I in single muscle fibres require the development of a valid cytochemical assay of its catalytic activity. Oxidation of reduced nicotinamide adenine dinucleotide (NADH) linked to the reduction of a tetrazolium salt can be demonstrated in tissue sections by means of the NADH–tetrazolium reductase (NADH–TR) reaction (Scarpelli *et al.*, 1958). However, this reaction demonstrates not only NADH–ubiquinone reductase (complex I) activity but also NADH–cyt. b_5 reductase activity from outer mitochondrial membranes and sarcoplasmic reticulum and possibly the activities of other NADH-oxidizing enzymes as well. A valid assay procedure for complex I involves demonstration of the sensitivity of this component of the NADH–TR reaction to rotenone or to some other specific inhibitor. However, rotenone has no significant effect on the NADH–TR activity demonstrable in skeletal muscle. This is probably because all the commonly used tetrazolium salts accept electrons from a site in complex I upstream of the site of rotenone inhibition. Another complex I inhibitor, rhein (4,5-dihydroxyanthraquinone-2-carboxylic acid), (Kean, 1970) exerts its effect close to the substrate binding site of the complex and may offer a feasible alternative approach. However, the inhibitory effect to rhein decreases substantially in media with high protein content, which would severely limit its efficacy in any cytochemical system.

4.3.2 Complex II (succinate-ubiquinone reductase)

Defects involving complex II are rare and are frequently not limited to this complex only (Rivner *et al.*, 1989; Desnuelle *et al.*, 1989). The two reports cited here describe two patients in whom complex II was unequivocally the most severely affected complex. Both patients were young adults, one presenting with an atypical Kearns–Sayre syndrome and the other with progressive muscle weakness and encephalopathy.

The microphotometric assay for SDH activity is of potential value in rare cases such as these, but its applications are not limited to this context. Precisely because complex II deficiencies are so rarely encountered, SDH has been regularly used as a 'reference' enzyme, with activities of an affected enzyme being expressed relative to those of SDH rather than as absolute values. This approach may also be adopted in a cytochemical system. Parallel assays using identical samples of muscle fibres may be made in serial tissue sections. Ratios of the activity of a particular enzyme to that of SDH may then be calculated in each individual muscle fibre.

4.3.3 Complex III (ubiquinol-cytochrome c reductase)

Deficiencies of this complex have been recorded in patients with varying patterns of clinical presentation. One group includes those adult patients with multisystem involvement described by Morgan-Hughes *et al.*, (1977, 1982) in whom the primary defect appeared to be lack of reducible cyt. b. These patients often showed relatively minor abnormalities on routine cytochemical screening, using SDH as a 'marker' enzyme to monitor mitochondrial distribution. Accentuation of the normal mitochondrial clusters found adjacent to muscle capillaries was a more common feature than classical RRFs.

In a patient presenting in childhood with exercise intolerance and muscle weakness (Darley-Usmar *et al.*, 1983) decreased amounts of core proteins, non-haem iron–sulphur (nFeS) protein and subunit VI were recorded. Recently a patient presenting in the neonatal period with fatal lactic acidosis, a severe defect of complex III and a partial defect in complex IV, was described (Birch-Machin *et al.*, 1989). In this case no morphological evidence of an underlying respiratory chain defect was found on routine cytochemical screening. Immunocytochemical investigation using subunit-specific and holocomplex antisera demonstrated a significant decrease in immunoreactivity in tissue sections which was particularly marked in the case of the nFeS protein (Fig. 4.4). These findings were essentially the same as those obtained by immunoblotting but yielded extra information that the

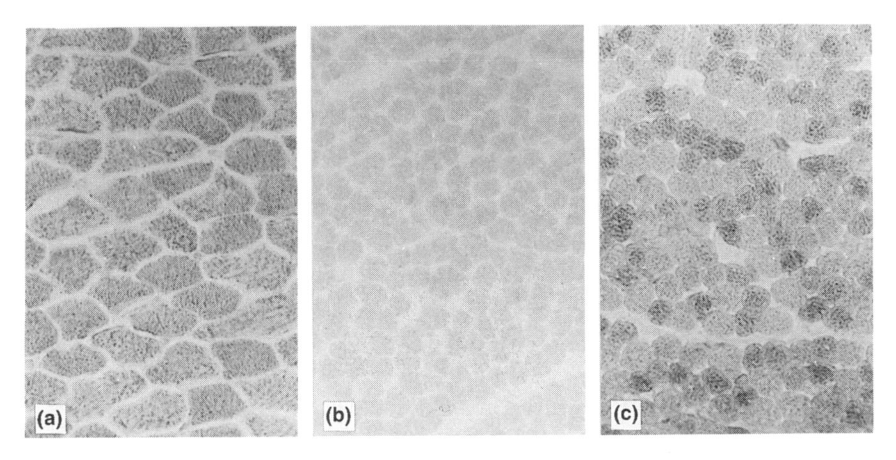

Fig. 4.4 (a) Complex III immunolabelling (nFeS protein) in skeletal muscle from a normal child aged 20 months. (b) Severe decrease in immunolabelling in muscle of infant with complex III deficiency. (c) No morphological mitochondrial abnormalities were apparent in this patient. Immunoperoxidase (a, b); SDH (c).

defect appeared to be uniform across the whole muscle fibre population.

4.3.4 Complex IV (cytochrome c oxidase)

Defects in complex IV are found in patients with a very wide range of clinical presentations of which only the major types will be outlined here.

(a) Fatal infantile cytochrome oxidase deficiency

This is characterized by total absence of complex IV activity in skeletal muscle (Van Biervliet *et al.*, 1977), frequently accompanied by a partial deficiency of cyt. b. Cytochemical assay of cytochrome oxidase activity is a convenient way of screening these infants, who present with severe lactic acidosis and hypotonia.

(b) Benign infantile cytochrome oxidase deficiency

This must be carefully differentiated from the fatal form of the disorder. Presentation with profound hypotonia in the neonatal period is followed by gradual resolution of the motor deficit, accompanied by acquisition of increasing cytochrome oxidase activity in skeletal muscle. Serial biopsy is necessary for monitoring the course of this disorder (DiMauro *et al.*,

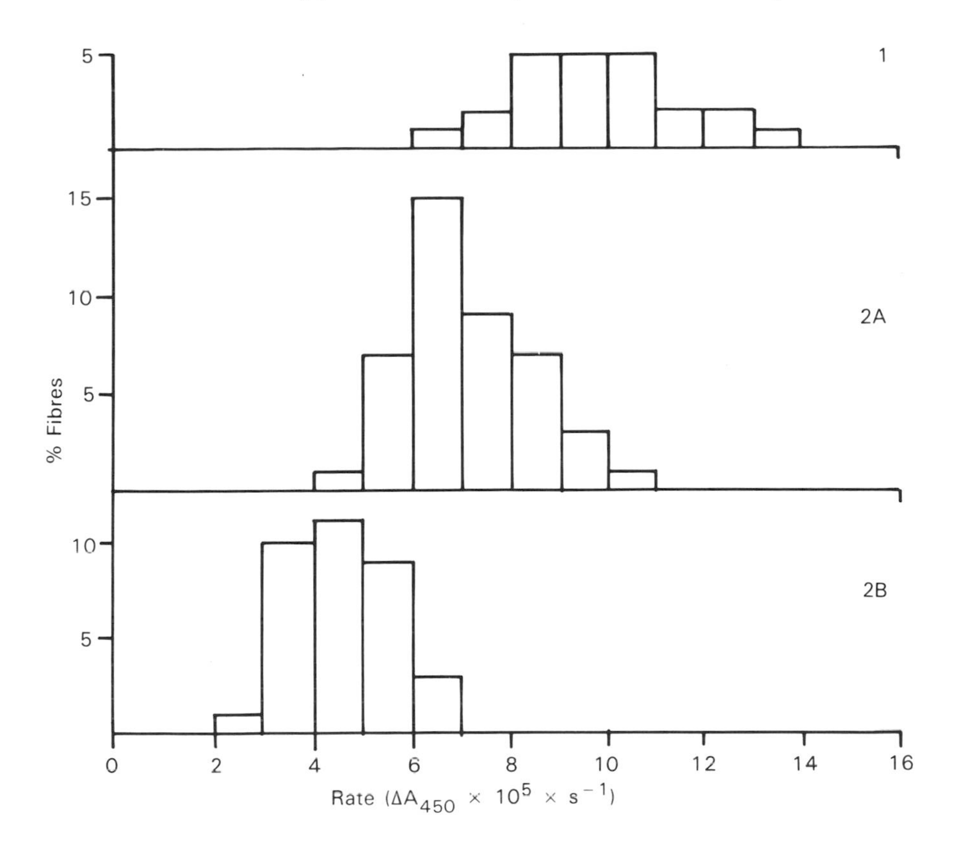

Fig. 4.5 Histogram showing the activities of cytochrome oxidase in single muscle fibres of normal human m. quadriceps.

1983) but should be minimally invasive if small needle biopsies are used in conjunction with MEA.

(c) Leigh's disease (subacute necrotizing encephalomyelopathy)

This may be associated with severe cytochrome oxidase deficiency or with a defect of the pyruvate dehydrogenase complex (PDHC). Many of the patients are small children and MEA represents an appropriate screening method for the identification of those cases with defects in complex IV. Figs 4.5 and 4.6 illustrate the results of MEA of cytochrome oxidase in a normal control subject, and the severe decrease in cytochrome oxidase activities in all muscle fibres of a child with Leigh's disease (Shepherd *et al.*, 1988).

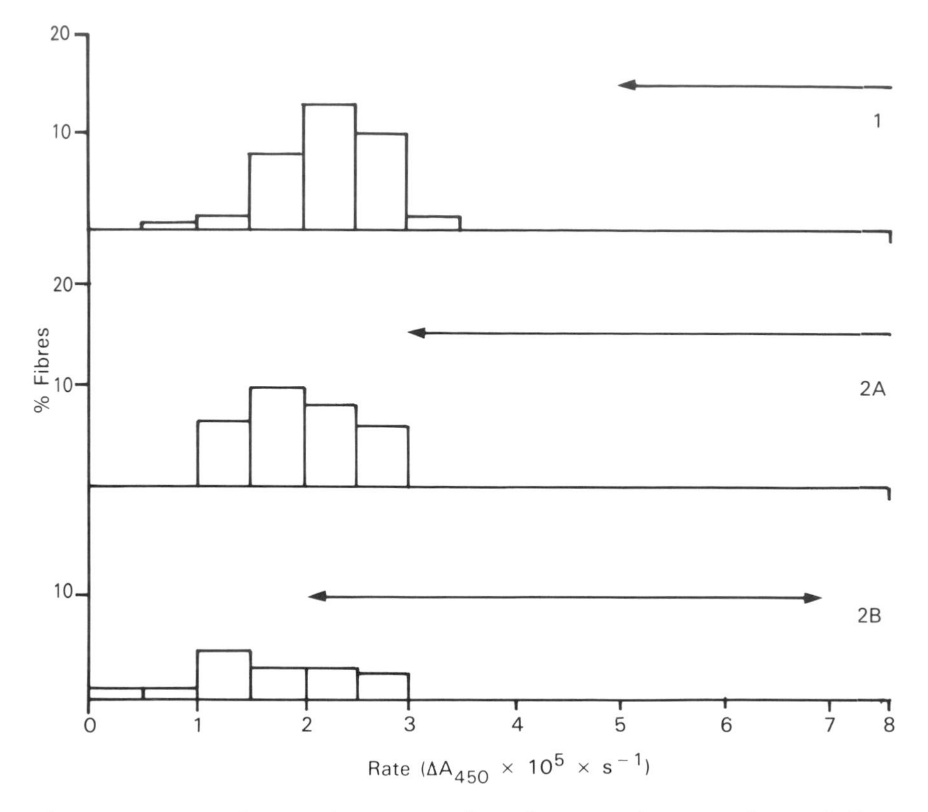

Fig. 4.6 Histogram showing the activities of cytochrome oxidase in single muscle fibres from a child with Leigh's disease. (Arrows indicate normal range.) Virtually all fibres show severely decreased activity.

4.3.5 Multiple complex defects

Patients with Kearns–Sayre syndrome (KSS) and the more restricted syndrome of chronic progressive external ophthalmoplegia (CPEO) comprise a major proportion of all those presenting with some form of mitochondrial myopathy. The identification of the biochemical defects in these patients has been complicated by the fact that only subpopulations of muscle fibres are involved (Johnson *et al.*, 1983). In severely affected KSS and CPEO patients significant decreases in complex IV have been recorded and in some patients complexes I and III can also be shown to be deficient (Sherratt *et al.*, 1986; D.M. Turnbull, unpublished observations). Since complexes I, III and IV all have one or more subunits coded on mitochondrial DNA (mtDNA) (Anderson *et al.*, 1981), there were strong grounds for suspecting a primary defect of the mitochondrial genome, evidence of which has been provided subsequently by the work of Holt *et al.* (1988), Lestienne and Ponsot (1988), Zeviani *et al.* (1988) and Poulton *et al.* (1989).

With the exception of Southern blotting, which identifies dual populations of mtDNA, investigation of skeletal muscle in these patients by any method involving tissue homogenization, e.g. standard biochemical assays or immunoblotting, may yield inconclusive results, especially if the proportion of affected muscle fibres is low. Cytochemical methods are, however, ideally suited for the investigation of phenotypic expression in these disorders. Fig. 4.7 shows the results of estimating cytochrome oxidase activities in the muscle of a CPEO patient using MEA. It can be seen that the patient has muscle fibres in which cytochrome oxidase activity is very low indeed, whereas in other fibres it is well within the

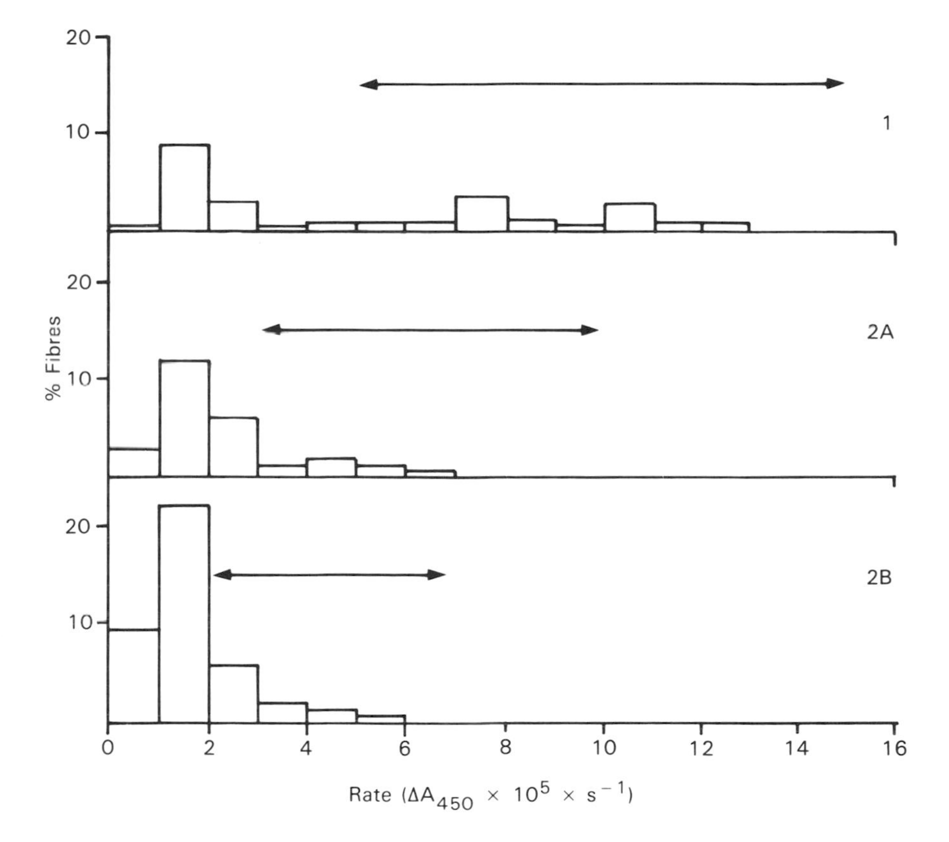

Fig. 4.7 Histogram showing the activities of cytochrome oxidase in single muscle fibres from a patient with CPEO. (Arrows indicate normal range.) Note the presence of fibres with normal activity and fibres with severely decreased activity within the same fibre population.

normal range. Follow-up biopsies have been obtained in a limited number of patients and provide some direct evidence of the progressive nature of KSS and CPEO, which appears to be associated with increasing proportions of abnormal muscle fibres (Bresolin *et al.*, 1987; M. A. Johnson, unpublished observations).

In KSS and CPEO patients, immunocytochemical studies provide a more logical approach to the investigation of possible specific subunit involvement in affected respiratory complexes than does standard immunoblotting. In a study of in situ immunolabelling in 17 patients (Johnson *et al.*, 1988) a panel of subunit-specific antisera to both mitochondrially coded and nuclear-coded subunits of complex IV were used (Kuhn-Nentwig and Kadenbach, 1986). In almost all patients there was a high degree of correlation between loss of catalytic activity and decreased immunolabelling of the mitochondrially coded subunits II and III (Fig. 4.8). In serial sections it was demonstrated that immunoreactivity of the nuclear-coded subunit IV was generally retained, with varying degrees of loss of immunoreactivity being shown by other nuclear-coded subunits.

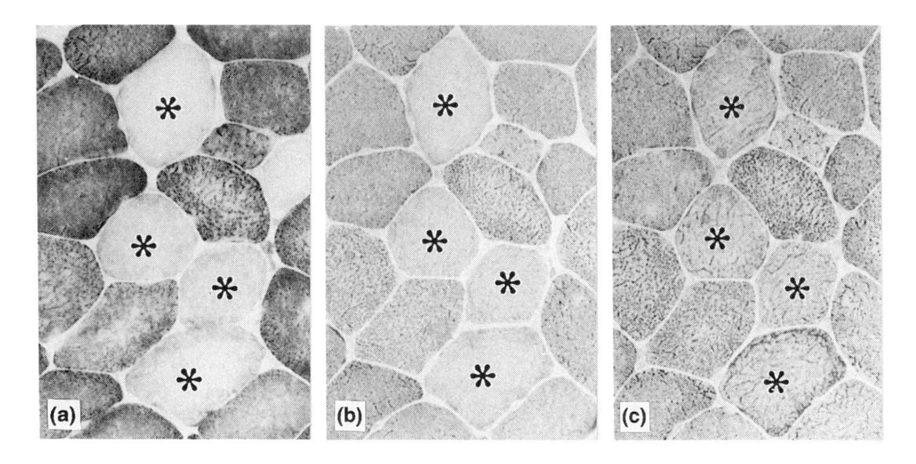

Fig. 4.8 (a) Severely decreased catalytic activity of cytochrome oxidase is seen in some muscle fibres (asterisks) of patient with CPEO. (b) Immunolabelling of subunits II and III (mitochondrially coded) shows a parallel decrease. (c) Immunolabelling of subunit IV (nuclear-coded) is preserved.

Due to the intricate way in which individual respiratory complexes are arranged within the inner mitochondrial membrane, and the structural interdependence of many of their component subunits, it is likely that primary disorders of the mitochondrial genome may exert secondary effects on nuclear-coded respiratory chain components. Immunocytochemical labelling provides a feasible method for the further investigation of these interactions in complexes I, III and IV.

4.4 DISCUSSION

Demonstration of the activity of respiratory chain enzymes by MEA or of tissue immunoreactivity by ICC gives a reliable indication of whether expression of particular defects is homogeneous or mosaic. However, the distribution of RRFs and other morphological abnormalities gives no such indication and may be misleading. For example, only sporadic fibres, normally type 1, showed RRF characteristics in a complex I deficiency shown to be expressed in all muscle fibres (Watmough *et al.*, 1989). In the fatal form of infantile cytochrome oxidase deficiency, morphological changes detectable at the light microscope level are confined to scattered fibres, while the biochemical defect is homogeneously expressed in skeletal muscle (DiMauro *et al.*, 1980; Minchom *et al.*, 1983). It would seem that RRF and other morphological abnormalities are simply non-specific indications of an underlying mitochondrial disturbance, their distribution merely indicating which muscle fibres have responded more adversely than others to a particular biochemical defect.

In contrast, MEA can provide precise information on the catalytic activity in individual muscle fibres and ICC techniques have proved to be of value in the analysis of specific subunit involvement. The rapidly growing field of in situ hybridization (Chapter 20) represents a natural extension of cytochemical technology and allows investigations of abnormalities of the mitochondrial genome and its transcription to be made in single muscle fibres.

Studies of interrelationships between nuclear and mitochondrial genomes indicate that abnormalities of mtDNA may, in some instances, be due to mutant nuclear-coded *trans*-acting factors (Zevani *et al.*, 1989). In most patients with deletions of mtDNA, however, the cause of the molecular lesions is unclear but appears to be related to the existence of 'hot-spots' (Schon *et al.*, 1989).

The cytochemical demonstration of mosaic abnormalities appears to correlate well with the presence of abnormalities in the mitochondrial genome, whether primary or secondary to mutations in nuclear-coded proteins. Conversely, in those disorders involving nuclear mutations but no mtDNA heteroplasmy, the tissue expression of the gene defect appears homogeneous in muscle. Examples of this would include the complex IV deficiency seen in Leigh's syndrome (Miranda *et al.*, 1989) and, by inference, probably other respiratory chain defects where involvement of tissue-specific isoenzymes probably indicates a nuclear mutation

(Watmough *et al.*, 1989). Until proven otherwise, mosaic expression of a respiratory chain defect may be taken to indicate mtDNA heteroplasmy, whereas nuclear mutations in association with mtDNA homoplasmy are likely to underlie disorders of respiratory complexes which show a uniform tissue expression in skeletal muscle.

ACKNOWLEDGEMENTS

The primary antisera used in this work were the generous gifts of collaborators whose contributions are gratefully acknowledged: Dr I. Ragan, Merck Sharp & Dohme, UK (complex I antisera); Dr V. Darley-Usmar, Wellcome Foundation (complex III antisera); Professor B. Kadenbach, University of Marburg, Germany (complex IV antisera). The author would also like to thank her colleagues in the University of Newcastle upon Tyne, Dr D.M. Turnbull and Dr H.S.A. Sherratt for much helpful advice and discussion. This work has received financial support from the Medical Research Council.

REFERENCES

Anderson, S., Bankier, A.T., Barrell, B.G. *et al.* (1981) Sequence and organisation of the human mitochondrial genome. *Nature*, **290**, 457–64.

Birch-Machin, M.A., Shepherd, I.M., Watmough, N.J. *et al.* (1989) Fatal lactic acidosis in infancy with a defect of complex III of the respiratory chain. *Pediatr. Res.*, **25**, 553–9.

Bresolin, N., Moggio, M., Bet. L. *et al.* (1987) Progressive cytochrome *c* oxidase deficiency in a case of Kearns–Sayre syndrome: morphological immunological and biochemical studies in muscle biopsies and autopsy tissues. *Ann. Neurol.*, **21**, 564–72.

Brooke, M.H. and Kaiser, K.K. (1970) Muscle fiber types: how many and what kind? *Arch. Neurol.*, **23**, 369–79.

Darley-Usmar, V.M., Kennaway, N.G., Buist, N.R.M. and Capaldi, R.A. (1983) Deficiency in ubiquinone–cytochrome *c* reductase in a patient with mitochondrial myopathy and lactic acidosis. *Proc. Natl. Acad. Sci. USA*, **80**, 5103–6.

Desnuelle, C., Birch-Machin, M., Pellisier, J.F. *et al.* (1989) Multiple defects of the respiratory chain including complex II in a family with myopathy and encephalopathy. *Biochem. Biophys. Res. Commun.*, **163**, 695–700.

DiMauro, S., Mendell, J.R. Sahenk, Z. *et al.* (1980) Fatal infantile mitochondrial myopathy and renal dysfunction due to cytochrome-c-oxidase deficiency. *Neurology*, **30**, 795–804.

DiMauro, S., Nicholson, J.F., Hays, A.P. *et al.* (1983) Benign infantile mitochondrial myopathy due to reversible cytochrome *c* oxidase deficiency. *Ann. Neurol.*, **14**, 226–34.

DiMauro, S., Bonilla, E., Zeviani, M. *et al.* (1985) Mitochondrial myopathies. *Ann. Neurol.*, **17**, 521–38.

Holt, I.J., Harding, A.E. and Morgan-Hughes, J.A. (1988) Deletions of muscle mitochondrial DNA in patients with mitochondrial myopathies. *Nature*, **331**, 717–19.

Johnson, M.A., Turnbull, D.M., Dick, D.J. and Sheratt, H.S.A. (1983) A partial deficiency of cytochrome *c* oxidase in chronic progressive external ophthalmoplegia. *J. Neurol. Sci.*, **60**, 31–53.

Johnson, M.A., Kadenbach, B., Droste, M. *et al.* (1988) Immunocytochemical studies of cytochrome oxidase subunits in skeletal muscle of patients with partial cytochrome oxidase deficiencies. *J. Neurol. Sci.*, **87**, 75–90.

Kean, E.A. (1970) Inhibitory action of rhein on the reducted nicotinamide adenine dinucleotide–dehydrogenase complex of mitochondrial particles and on other dehydrogenases. *Biochem. Pharmacol.*, **19**, 2201–10.

Kuhn-Nentwig, L. and Kadenbach, B. (1986) Isolation and characterisation of human heart cytochrome *c* oxidase. *J. Bioenerg. Biomembranes*, **18**, 307–14.

Land, J.M., Morgan-Hughes, J.A. and Clark, J.B. (1981) Mitochondrial myopathy: biochemical studies revealing a deficiency of NADH-cytochrome *b* reductase activity. *J. Neurol. Sci.*, **80**, 1–13.

Lestienne, P. and Ponsot, G. (1988) Kearns-Sayre syndrome with muscle mitochondrial DNA deletion. *Lancet*, **i**, 885.

Minchom, P.E., Dormer, R.L. Hughes, I.A. *et al.* (1983) Fatal infantile mitochondrial myopathy due to cytochrome *c* oxidase deficiency. *J. Neurol. Sci.*, **60**, 453–63.

Miranda, A.F., Ishii, S., DiMauro, S. and Shay, J.W. (1989) Cytochrome *c* oxidase deficiency in Leigh's syndrome: genetic evidence for a nuclear DNA-encoded mutation. *Neurology*, **39**, 697–702.

Moreadith, R.W., Cleeter, M.J.W., Ragan, C.I. *et al.* (1987) Congenital deficiency of two polypeptide subunits of the iron-protein fragment of mitochondrial complex I. *J. Clin. Invest.*, **79**, 463–7.

Morgan-Hughes, J.A., Darveniza, P., Kahn, S.N. *et al.* (1977) A mitochondrial myopathy characterised by a deficiency in reducible cytochrome *b*. *Brain*, **100**, 617–40.

Morgan-Hughes, J.A., Darveniza, P., Landon, D.N. *et al.* (1979) A mitochondrial myopathy with a deficiency of respiratory chain NADH–CoQ reductase activity. *J. Neurol. Sci.*, **43**, 27–46.

Morgan-Hughes, J.A., Hayes, D.J., Clark, J.B. *et al.* (1982) Mitochondrial encephalomyopathies: biochemical studies in two cases revealing defects in the respiratory chain. *Brain*, **105**, 553–82.

Old, S.L. and Johnson, M.A. (1989) Methods of microphotometric assay of succinate dehydrogenase and cytochrome *c* oxidase activities for use on human skeletal muscle. *Histochem. J.*, **21**, 545–6.

Olson, W., Engel, W.K., Walsh, G.O. and Einaugler, R. (1972) Oculocraniosomatic neuromuscular disease with 'ragged-red' fibers: histochemical and ultrastructural changes in limb muscles of a group of patients with idiopathic progressive external ophthalmoplegia. *Arch. Neurol.*, **26**, 193–211.

Pette, D. (1980) Microphotometric measurement of initial reaction rates in quantitative enzyme histochemistry *in situ*. *Histochem. J.*, **13**, 319–27.

Poulton, J., Deadman, M.E. and Gardiner, R.M. (1989) Duplications of mitochondrial DNA in mitochondrial myopathy. *Lancet*, **i**, 236–40.

Reichmann, H. and Pette, D. (1982) A comparative microphotometric study of succinate dehydrogenase activity levels in Type I, IIA and IIB fibres in mammalian and human muscles. *Histochemistry*, **74**, 27–41.

Rivner, M.H., Shamsnia, M., Swift, T.R. *et al.* (1989) Kearns-Sayre syndrome and complex II deficiency. *Neurology*, **39**, 693–6.

Scarpelli, D.G., Hess, R. and Pearse, A.G.E. (1958) The cytochemical localisation of oxidative enzymes. I. Diphosphopyridine nucleotide diaphorase and triphosphopyridine diaphorase. *J. Biophys. Biochem. Cytol.*, **4**, 747–59.

Schapira, A.H.V., Cooper, J.M., Morgan-Hughes, J.A. *et al.* (1988) Molecular basis of mitochondrial myopathies: polypeptide analysis in complex I deficiency. *Lancet*, **i**, 500–3.

Schon, E.A., Rizzuto, R., Moraes, C.T. *et al.* (1989) A direct repeat is a hotspot for large-scale deletion of human mitochondrial DNA. *Science*, **244**, 346–9.

Seligman, A.M., Karnovsky, M.J., Wasserkrug, H.L. and Hanker, J.S. (1968) Non-droplet ultrastructural demonstration of cytochrome oxidase activity with a polymerising osmiophilic reagent diaminobenzidine (DAB). *J. Cell Biol.*, **38**, 1–14.

Shepherd, I.M., Birch-Machin, M.A., Johnson, M.A. *et al.* (1988) Cytochrome oxidase deficiency: immunological studies of skeletal muscle mitochondrial fractions. *J. Neurol. Sci.*, **87**, 265–74.

Sherratt, H.S.A., Johnson, M.A. and Turnbull, D.M. (1986) Defects of complex I and complex IV in skeletal muscle from patients with chronic progressive external ophthalmoplegia. *Ann. NY Acad. Sci.*, **488**, 508–10.

Sternberger, L.A. (1979) *Immunocytochemistry*, 2nd edn, Wiley, New York.

Sternberger, L.A., Hardy, P.H., Cuculis, J.J. and Meyer, H.G. (1970) The unlabelled antibody–enzyme method of immunohistochemistry: preparation and properties of soluble antigen–antibody complex (horseradish peroxidase–antihorseradish peroxidase) and its use in identification of spirochaetes. *J. Histochem. Cytochem.*, **18**, 315–33.

Van Biervliet, J.P.G.M., Bruinvis, L., Ketting, D. *et al.* (1977) Hereditary mitochondrial myopathy withh lactic acidaemia, a de Toni–Fancolni–Debré syndrome and a defective respiratory chain in voluntary striated muscles. *Pediatr. Res.*, **11**, 1088–92.

Watmough, N.J., Birch-Machin, M.A., Bindoff, L.A. *et al.* (1989) Tissue specific defect of complex I of the mitochondrial respiratory chain. *Biochem. Biophys. Res. Commun.*, **160**, 623–7.

Zeviani, M., Moraes, C.T., DiMauro, S. *et al.* (1988) Deletions of mitochondrial DNA in Kearns–Sayre syndrome. *Neurology*, **38**, 1339–46.

Zeviani, M., Servidei, S., Gellera, C. *et al.* (1989) An autosomal dominant disorder with multiple deletions of mitochondrial DNA starting at the D-loop region. *Nature*, **339**, 309–11.

5 Native myelin proteins and myelin assembly in the central nervous system

P. RICCIO, A. BOBBA,[1] G.M. LIUZZI,
T. ZACHEO[2] and E. QUAGLIARIELLO
Dipartimento di Biochimica e Biologia Molecolare, Università di Bari, Italy; [1]Centro Studio sui Mitocondri e Metabolismo Energetico, CNR, Bari, Italy; [2]Istituto di Nematologia Agraria, CNR, Bari, Italy

Myelin is a membrane required for impulse conduction in the nervous system. In line with its unique function, myelin possesses properties which cannot be compared with other biological structures (Raine, 1984).

First of all, myelin is the result of the close relationship between two different types of cells: in the central nervous system these are the oligodendrocyte and the neurone. Each oligodendrocyte forms several myelin processes from its membrane, each of which covers a segment of the axon in a spiral fashion (Gregson, 1983; Kirshner *et al.*, 1984). The resulting compact sheath consists of many membrane pairs bound to each other. One example in Fig. 5.1 shows an electron micrograph of purified bovine brain myelin.

Although the way in which this unique assembly of membranes is formed and maintained is not yet clear, myelin is known to be very rich in lipids (70–75% of dry weight), with relatively limited protein composition: only two proteins account for about 80% of the total protein fraction (proteolipid protein, 30 kDa, with a share of 50%, and basic protein (MBP), 18 kDa, with a share of 30% (Carnegie and Moore, 1980; Lees and Brostoff, 1984). One of these major components of myelin, probably one of the two proteins, may account for its multilamellar structure.

Various sources indicate that MBP may have a major role in the stabilization of myelin structure (Carnegie and Dunckley, 1975; Smith, 1977), acting as a sort of biological glue at the cytoplasmic surface of the closely apposed membranes, i.e. at the major dense line regions of Fig. 5.1 (Golds and Braun, 1976; Omlin *et al.*, 1982).

MBP is the most well-known and studied myelin protein because it is highly encephalitogenic (Kies, 1985) and commonly used in the treatment, study and detection of demyelinating diseases. MBP is widely known as a lipid-free, water-soluble and unfolded protein, i.e. as it was

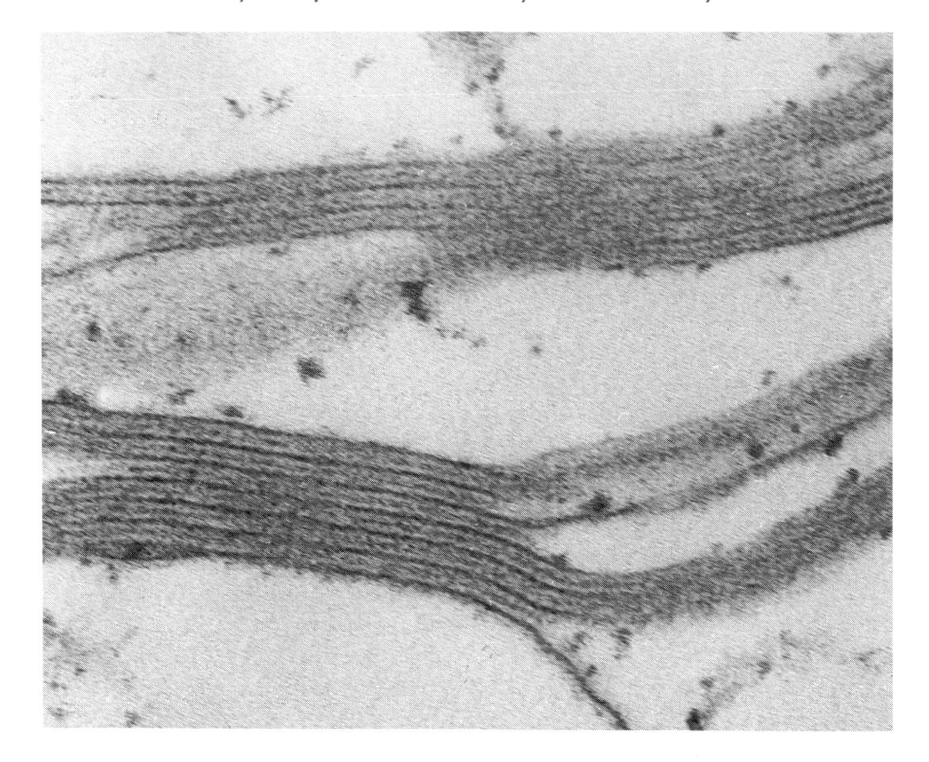

Fig. 5.1 Electron micrograph of bovine brain myelin prepared according to Norton (1974) but in the presence of 0.25 mM phenylmethylsulphonyl fluoride (PMSF). Two additional sucrose gradient centrifugations were also performed (Cruz and Moscarello, 1985). Electron microscopy: myelin fragments, washed three times with water, were processed for electron microscopy as described (Carnegie and Moore, 1980). In particular, the fragments were fixed in 5% glutaraldehyde at 4 °C in 0.1 M cacodylate buffer at pH 7.4 for 2 h. Fixed material was washed with buffer, post-fixed in 2% osmium tetroxide and dehydrated in an ascending series to absolute ethanol and embedded in Spurr's medium. Ultrathin sections were stained with uranyl acetate and lead citrate and examined under a Philips 400T transmission electron microscope. × 250 000.

obtained after isolation based on the use of organic solvents and HCl at pH below 3 (Oshiro and Eylar, 1970; Deibler *et al.*, 1972). In this form MBP is probably denatured and rather unsuitable for the study of myelin assembly.

As we have not been closely involved in the study of myelin assembly what can be stated in this field is either based on our studies on myelin protein-lipid interactions or derives from our experience with those proteins of myelin, in particular MBP, which have been purified in a native-like form. To this end, we have developed a mild procedure based on the use of detergents, mainly non-ionic *n*-octylpentaoxyethylene (octyl-POE), and adsorbing resin hydroxyapatite (Riccio *et al.*, 1984; Riccio and Quagliariello, 1985; Riccio *et al.*, 1985; Riccio *et al.*, 1986; Riccio *et al.*, 1991).

The most interesting feature of the proteins we have purified from bovine brain or guinea-pig spinal cord myelin is their binding to myelin lipids. MBP in particular was found to retain the binding to all the various myelin lipids. When using thin-layer chromatography no difference was observed between myelin and MBP preparations. In fact MBP was found to be associated with cholesterol, cerebrosides, sulphatides, phosphatidylethanolamine, phosphatidylserine, phosphatidylcholine and other lipids as schematically represented in Fig. 5.2(a). This hypothetical representation derives from our observation using electron microscopy of the lipid–MBP–detergent complex purified by our procedure, after removal of the detergent by dialysis (Fig. 2(b) and 2(c)). This step is necessary in order to study MBP interaction with cells, or, in general, with both simple and supramolecular lipophilic structures. MBP was found to be in the lipid-bound form even after gel filtration, dialysis, analytical centrifugation or precipitation by ammonium sulphate. On the contrary, lipids could be released when the protein is frozen and thawed.

Lipid binding to MBP appears to be at least partly of an ionic nature. This assertion is based on the observation of the differing chromatagraphic behaviour of MBP when applied to a hydroxyapatite column in the presence or absence of lipids. In fact, MBP binds to the resin only in the lipid-free form. It is worth mentioning that protein adsorption on hydroxyapatite is a process of an ionic nature which occurs between charge clusters on the two counterpart surfaces (Bernardi, 1971). Intrinsic membrane proteins such as the mitochondrial anion carriers are not adsorbed on hydroxyapatite as is the case for lipid-bound MBP (Riccio, 1989). On the other hand, MBP is widely believed to be an extrinsic membrane protein. Such proteins should bind to lipids ionically but should not require detergents for extraction. Thus, the case of MBP merits further investigation, also in the general light of membrane assemblies.

On the other hand, we have studied the structure of lipid-free MBP in buffer and lysolecithin complexes by using the emission properties of the single tryptophan residue of the protein (Trp 115). In relation to the free protein, MBP bound to lysolecithin showed an increase in fluorescence intensity and a blue-shift in the maximum emission wavelength (Cavatorta *et al.*, 1988), indicating that the Trp-containing region of MBP can be embedded in a lipid matrix. The binding of the protein to lysolecithin appears to be similar to that predicted for the interaction of amphipathic helices with non-polar lipids.

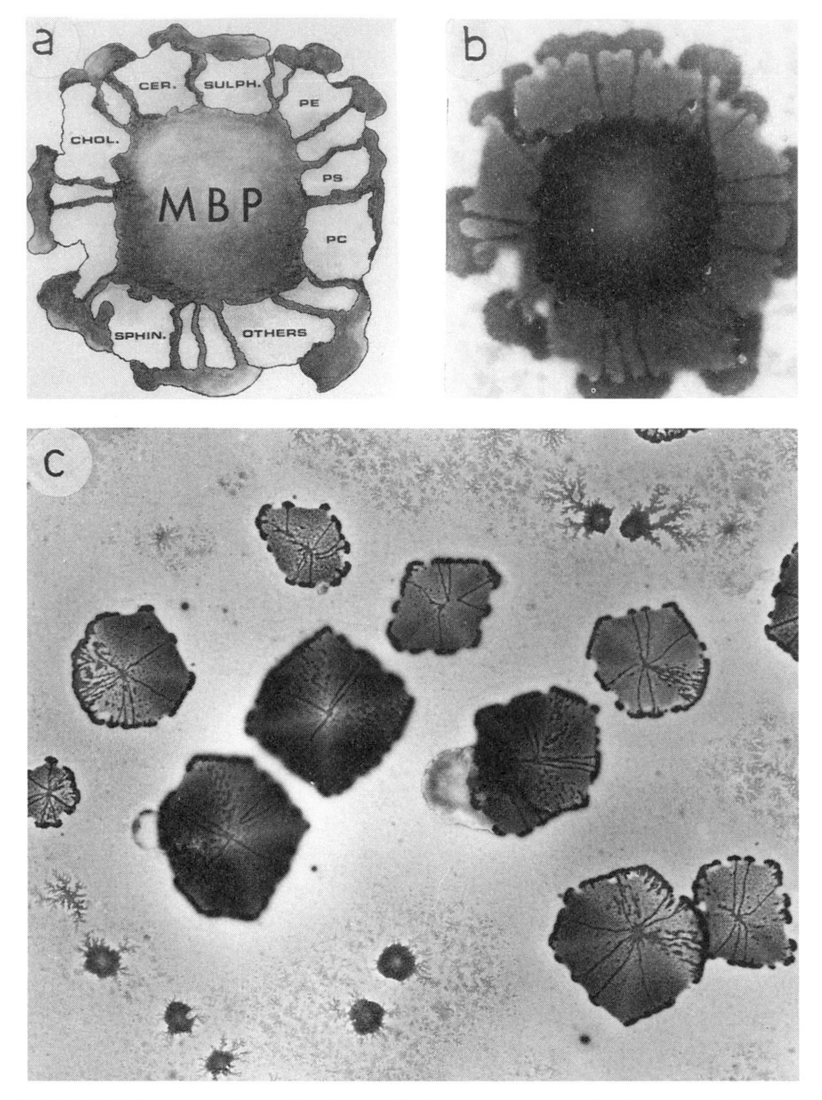

Fig. 5.2 (a) Artist's conception of brain myelin basic protein in the form retaining binding to all myelin lipids. (b), (c) Electron micrographs of purified lipid-bound MBP obtained from bovine brain myelin. Preparation of the sample for electron microscopy: after the gel filtration step (Riccio *et al.*, 1984) the mean fraction was concentrated by ultrafiltration on the Amicon YM5 membrane and applied to dialysis at pH 7.0 and room temperature against about 3000 volumes of the medium containing 0.5% octyl-pentranoxyethylene (octyl-POE), 20 mM morpholinopropane sulphonic acid (MOPS, 0.2 mM EDTA, 0.2 mM butylated hydroxytoluene (BHT), 3 mM NaN_3. The dialysis bag was Spectrapor 8000 and the protein concentration was 6.2 mg/ml. After 16 h, dialysis was continued against a medium as indicated above but without the detergent. After 24 h the procedure was concluded with 4 h dialysis against water. The protein sample thus obtained was refrigerated at about 4–8 °C, stained 6 weeks later in neutral sodium phosphotungstate and then observed by electron microscopy. (b) ×42 600; (c) ×13 100.

Interestingly, when MBP was added to lysolecithin micelles in the lipid-bound form, it was found to induce the structural transition of micellar lysolecithin into bilayers with the formation of mono- and multilaminar vesicles, a result which was not obtained with either lipid-free MBP or free myelin lipids (Riccio *et al.*, 1986).

Both its binding to all myelin lipids and its ability to form bilayers could indicate that, in addition to stabilizing myelin structure through binding to lipids, MBP might also have a role in its organization. In previous reports, lipid-free MBP was found to bind only anionic lipids (see Carnegie and Moore, 1980, for a review) and to cross-linked pre-formed bilayers (Smith, 1977; Epand *et al.*, 1984).

Thus, on the basis of these results, lipid-bound MBP may be considered in a more intact and functional state than lipid-free MBP used in myelin studies to date. This observation is confirmed by other studies in which lipid-bound MBP was compared to the corresponding lipid-free preparation, where lipid-bound MBP appears as an oblate ellipsoid about five times more compact in size than detergent-free and lipid-free MBP (Riccio *et al.*, 1985b; Riccio *et al.*, 1991).

In addition, circular dichroism (CD) clearly shows that the two MBP forms possess different conformations (Cavatorta *et al.*, 1989). Although lipid-free MBP appears to be unfolded, it is not easy to assign a particular secondary structure to lipid-bound MBP, since computer programs used for the evaluation of CD spectra may be unreliable (Bobba *et al.*, 1990).

Thus, MBP may at this point be seen in a new light, appearing as it now does to be a protein, which is not water soluble (it requires detergents), binds to all myelin lipids (no other similar examples are on record), it is able to form bilayers and is much less extended than previously believed. These properties, observed in a protein with functions otherwise not yet known, appear more appropriate for a membrane protein.

With a procedure similar to that used in the case of lipid-bound MBP, we purified a previously unknown small protein of 13 kDa which we call MSP, myelin small protein (Riccio *et al.*, 1985a). This new protein has two main characteristics: (1) the electrophoretic pattern in sodium dodecyl sulphate (SDS) is anomalous with many bands in the high-molecular-weight region; (2) MSP is co-purified with phosphatidylserine.

We do not as yet have information about either the role of this protein in myelin assembly or its location in the sheath. However, MSP-specific binding to phosphatidylserine, along with our preliminary observation that at least part of the proteolipid is able to bind some myelin lipids, sheds some light on possible protein–lipid interaction in CNS myelin.

Furthermore, lipid-binding ability may be related to the native form of myelin proteins. In this respect, other studies on the in vivo

(Zehetbauer *et al.*, 1991) and in vitro (Riccio *et al.*, 1987) interaction of MBP with the immune system provide evidence that lipid-bound MBP is in a native-like, more functional form. In fact, only in the lipid-bound form is MBP clearly able to bind to purified subsets of immunocompetent cells (Riccio *et al.*, 1988). In addition, the less specific binding of MSP to most immunocompetent cells (Riccio *et al.*, 1989) was found to depend on the more native state of the protein (Bobba *et al.*, 1987). Taken as a whole, these and the previously cited evidence clearly show that lipid-bound MBP may effectively be considered to be in a native-like form.

If this is the case, the binding of MBP to all myelin lipids as well as its ability to form and to cross bilayers may be considered as properties of the protein as it is in the membrane, which are needed for formation and stability of the myelin sheath. A possible representation could thus be that of Fig. 5.3, with MBP mainly in the monomeric form as in our preparations. On the other hand, the presence of dimeric MBP cannot be excluded (Golds and Braun, 1978). The role of a trimeric (Riccio

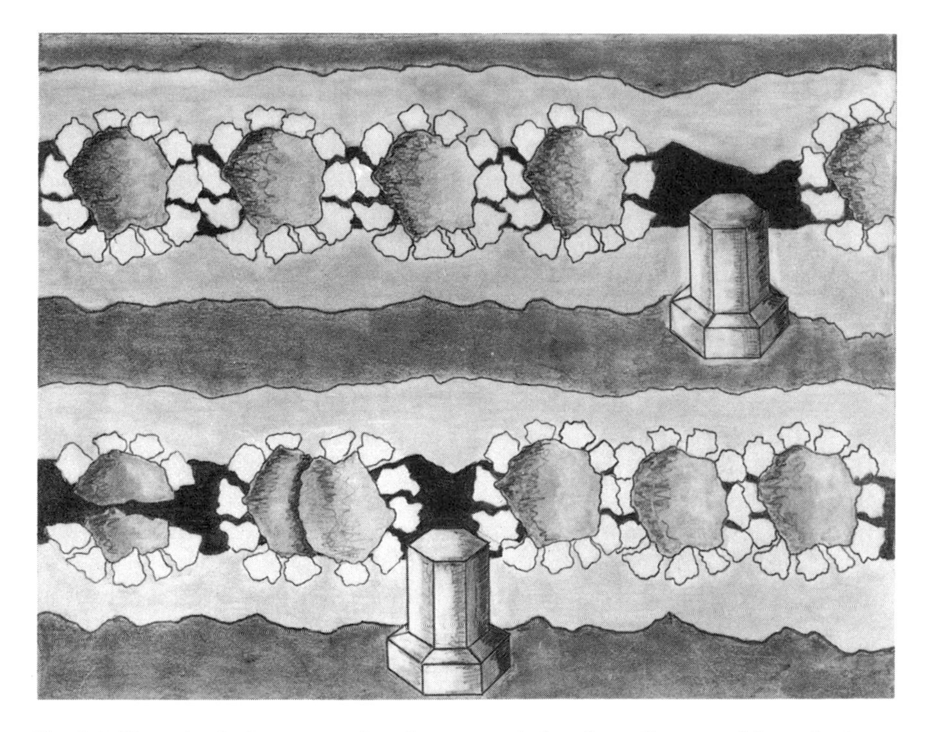

Fig. 5.3 Hypothetical cross-sectional representatin of myelin assembly assigning a major role to MBP in its lipid-bound form. The hexagonal transmembrane protein is the proteolipid.

and Quagliariello, 1985) or hexameric (Smith *et al.*, 1984) transmembrane proteolipid protein might be of minor importance with regard to myelin assembly, and more relevant in such processes as ion channelling.

ACKNOWLEDGEMENTS

The authors are grateful to Francesco Arrivo, set designer of the Academy of Fine Arts, Bari, for his computer drawings; to Roberto Lerario, of the Istituto di Nematologia Agraria, for the photographic work; and to Professor Cecilia Saccone and the Bioinformatic Group of our department for computer facilities.

REFERENCES

Bernardi, G. (1971) *Methods Enzymol.*, **22**, 325–42.
Bobba, A., Caretto, G., Jirillo, E. *et al.* (1987) *Med. Sci. Res.*, **15**, 1191–2.
Bobba, A., Cavatorta, P., Attimonelli, M. *et al.* (1990) *Protein Seq. Data Anal.*, **3**, 7–10.
Carnegie, P.R. and Dunckley, P.R. (1975) in *Advances in Neurochemistry* (eds B.W. Agranoff and M.H. Aprison), Plenum Press, New York, p. 95.
Carnegie, P.R. and Moore, W.J. (1980) in *Proteins of the Nervous System* (eds R.A. Bradshaw and M.D. Schneider), Raven Press, New York, pp. 119–143.
Cavatorta, P., Masotti, L., Szabo, A.G. *et al.* (1988) *Cell Biophys.*, **13**, 201–15.
Cavatorta, P., Masotti, L., Szabo, A.G. *et al.* (1989) in *Fluorescent Biomolecules: Methodologies and Applications* (eds D.M. Jameson and G.D. Reinhart), Academic Press, New York, pp. 383–8.
Cruz, T.F. and Moscarello, M.A. (1985) *J. Neurochem.*, **44**, 1411–18.
Deibler, G.E., Martenson, R.E. and Kies, M.W. (1972) *Prep. Biochem.*, **2**, 139–65.
Epand, R.M., Dell, K., Surewicz, W.K. and Moscarello, M.A. (1984) *J. Neurochem.*, **43**, 1550–5.
Golds, E.E. and Braun, P.E. (1976) *J. Biol. Chem.*, **251**, 4729–35.
Golds, E.E. and Braun, P.E. (1978) *J. Biol. Chem.*, **253**, 8171–7.
Gregson, N.A. (1983) in *Multiple Sclerosis, Pathology, Diagnosis and Management* (eds J.F. Hallpike, C.W.M. Adamy, and W. Tourtellotte), Chapman and Hall, London, pp. 1–27.
Kies, M.W. (1985) in *Handbook of Neurochemistry*, (ed. A. Lajtha), Plenum Press, New York, pp. 533–52.
Kirschner, D.A., Ganser, A.L. and Caspar, D.L.D. (1984) in *Myelin*, 2nd edn (ed. P. Morell), Plenum Press, New York, pp. 51–95.
Lees, M.B. and Brostoff, S.W. (1984) in *Myelin*, 2nd edn (ed. P. Morell), Plenum Press, New York, pp. 197–224.
Norton, W.T. (1974) *Methods Enzymol.* **31**, 435–44.
Omlin, F.X., Webster, H. de F., Palkovits, C.G. and Cohen, S.R. (1982) *J. Cell Biol.*, **95**, 242–8.
Oshiro, Y. and Eylar, E.H. (1970) *Arch. Biochem. Biophys.*, **138**, 392–6.

Raine, C.S. (1984) in *Myelin*, 2nd edn (ed. P. Morell), Plenum Press, New York, pp. 1–50.
Riccio, P. (1989) in *Anion Carriers of Mitochondrial Membranes* (eds A. Azzi, K.A. Nałecz, M.J. Nałęcz and L. Wojtczak), Springer-Verlag, Berlin, pp. 35–44.
Riccio, P. and Quagliariello, E. (1985) *Proceedings of the 16th FEBS Congress*, Part B, VNU Science Press, Moscow, pp. 545–50.
Riccio, P., Rosenbusch, J.P. and Quagliariello, E. (1984) *FEBS Lett.*, **177**, 236–40.
Riccio, P., Tsugita, A., Bobba, A. *et al.* (1985a) *Biochem. Biophys. Res. Commun.*, **127**, 484–92.,
Riccio, P., Bobba, A., De Santis, A. and Quagliariello, E. (1985b) *Ital. J. Biochem.*, **34**, 412A–14A.
Riccio, P., Masotti, L., Cavatorta, P. *et al.* (1986) *Biochem. Biophys Res. Commun.*, **134**, 313–19.
Riccio, P., Jirillo, E., Caretto, G. *et al.* (1987) *EOS-Riv. Immunol. Immunofarm.*, **7**, 172–6.
Riccio, P., Jirillo, E., Bobba, A. and Liuzzi, G.M. (1988) in *An Update on Multiple Sclerosis*, (eds A.M. Battaglia and G. Crimi), Monduzzi, Bologna, pp. 21–4.
Riccio, P., Jirillo, E., Bobba, A. *et al.* (1989) *J. Clin. Lab. Anal.*, **4**, 2–4.
Riccio, P., Liuzzi, G.M. and Quagliariello, E. (1991) *Mol. Chem. Neuropathol.*
Smith, R. (1977) *Biochim. Biophys. Acta*, **470**, 170–84.
Smith, R., Cook, J. and Dickens, P.A. (1984) *J. Neurochem.*, **42**, 306–13.
Zehetbauer, B., Massacesi, L., Liuzzi, G.M. Vergelli, M., Olivotto, J., Grassi, L., Riccio, P. and Amaducci, L. (1991) *Acta Neurol.*

6 Regulation of immune cells by serine phospholipids

A. BRUNI,[1] L. MIETTO,[2] F. BELLINI,[2] D. PONZIN,[2]
E. CASELLI[2] and G. TOFFANO[2]
[1]Department of Pharmacology, University of Padova, Italy; [2]Fidia Research Laboratories, Abano Terme, Italy

6.1 INTRODUCTION

Following initial studies on the pharmacological action of exogenous phosphatidylserine (PS) and lysoPS (Bruni *et al.*, 1976; Bigon *et al.*, 1979a, 1979b) it has been demonstrated that lymphocytes, macrophages and mast cells are target cells for these phospholipids in vivo and in vitro. Analysis of phospholipid metabolism in mast cells shows that lysoPS influences the metabolic pathway yielding phospholipid-derived second messengers (Bellini *et al.*, 1988, 1990). These data are consistent with our hypothesis that PS and lysoPS induce a first messenger effect in immune cells. This action becomes manifest when cell damage triggers the exposure of PS to the extracellular environment and the production of lysoPS (Fig. 6.1). The regulatory action induced by serine phospholipids in lymphocytes and mast cells is briefly reviewed in this chapter.

6.2 LYMPHOCYTES

Early studies on the interaction of phospholipid vesicles with immune cells focused on macrophages, which were believed to be the only cells capable of binding and internalizing liposomes (reviewed by Poste, 1983). At variance, we have shown that rat lymphocytes are as active as macrophages in the binding and in the internalization of PS vesicles (Mietto *et al.*, 1989). Our observations stem from experiments designed to investigate the interaction of this phospholipid with rat peritoneal mononuclear cells, elicited by casein. In a comparison with phosphatidylcholine and phosphatidylinositol, we found that PS was taken up by these cells more efficiently that the other phospholipids. When the population of mononuclear cells was fractionated into the adherent

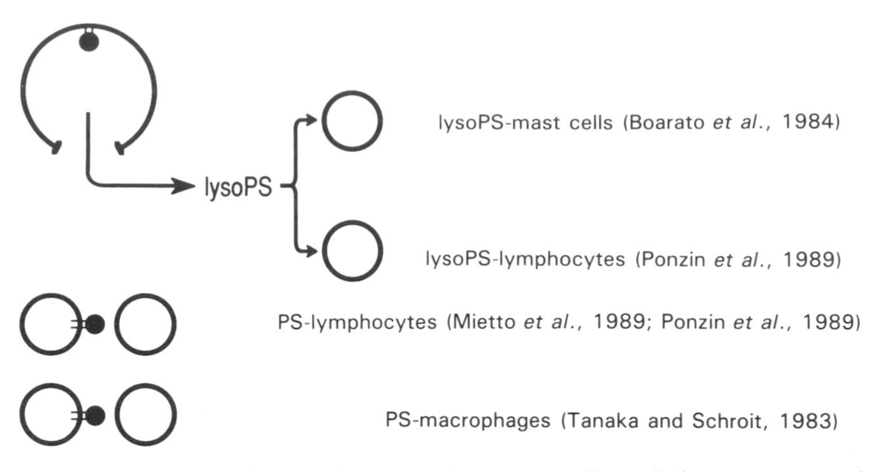

Fig. 6.1 Interaction of PS and lysoPS with immune cells. Cell damage triggers the uncontrolled activation of phospholipase A_2, producing lysoPS from PS. In addition, the depletion of high-energy compounds prevents lysoPS reacylation and promotes the exposure of PS to the external side of the plasma membrane. The two phospholipids may then interact with mast cells, lymphocytes and macrophages.

cells (macrophages) and the non-adherent cells (lymphocytes), PS uptake was detected in both fractions. Pre-treatment of lymphocytes with metabolic poisons partially prevented the PS uptake. Equal inhibition was induced by cytochalasin B. Furthermore, the effectiveness of lymphocytes decreased when the test were repeated with the PS enantiomer containing D-serine. Parallel experiments indicated that during the incubation with lymphocytes part of PS was deacylated to lyosPS which was subsequently incorporated and converted to PS. Thus, lymphocytes internalize PS by two mechanisms: endocytosis and the deacylation–reacylation cycle. The two pathways differ in the fate of the internalized phospholipid. In endocytosis, PS is directed into the intracellular sites of metabolism without alteration of plasma membrane phospholipid composition. In the deacylation–reacylation cycle the content of PS in plasma membrane increases for the short time required to ripristinate the physiological phospholipid composition.

The observation of a specific interaction between PS and lymphocytes prompted attempts to investigate changes in lymphocyte metabolic activity. We followed two different directions. In the first, the influence of PS on metabolism of those phospholipids involved in signalling mechanisms was examined. In the second we investigated the influence of PS on the mitogen-induced secretion of growth factors (e.g. interleukin-2). As shown in Table 6.1, when lymphocytes were incubated with

Table 6.1 Action of PS on the incorporation of soluble phospholipid precursors. Rat macrophages (390 μg of protein) and lymphocytes (280 μg of protein) were incubated for 60 min at 37 °C in the presence of PS vesicles (260 nmol/mg of protein, corresponding to 16 nmol/10^6 cells) and a trace amount of the following radiolabelled phospholipid precursors: [^{3}H]choline (1 μCi/sample), [^{3}H]serine (1 μCi/sample), [^{14}C]ethanolamine (5 μCi/sample), [^{3}H]inositol (10 μCi/sample). The cells were washed three times and extracted with chloroform–methanol. The organic phase was counted after washing with 0.2 vol of 10 mM $CaCl_2$. Means ± SE for three to four experiments unless indicated otherwise

Cells	*Precursors*	*Incorporation (DPM/mg of protein)*		
		None	*PS*	*% inhibition*
Macrophages	Choline[a]	46 450	49 530	-
	Serine (4)	151 634 ± 32 930	123 646 ± 24 639	18
	Ethanolamine (4)	38 514 ± 5 925	24 960 ± 5 348	35
	Inositol (4)	24 226 ± 2 704	16 761 ± 1 124	31
Lymphocytes	Choline[a]	154 150	136 450	11
	Serine (3)	116 060 ± 13 529	78 474 ± 10 475	32
	Ethanolamine (3)	82 332 ± 21 333	47 023 ± 17 153	43
	Inositol (4)	66 674 ± 4 825	24 966 ± 4 753	62

[a]Additional two experiments with 10 and 2.5 μCi [^{3}H]choline yielded the same results. Other experimental conditions as in Mietto *et al.* (1989).

radiolabelled phospholipid precursors in the presence of PS, their incorporation into the liposoluble radioactivity was variably changed. While PS did not affect choline incorporation, it inhibited the incorporation of serine, ethanolamine and inositol. Control experiments showed that PS did not affect the hydrosoluble radioactivity found inside the cells, thus excluding the influence on the transfer of precursors from the extracellular medium to the cytosol. In addition, resolution of

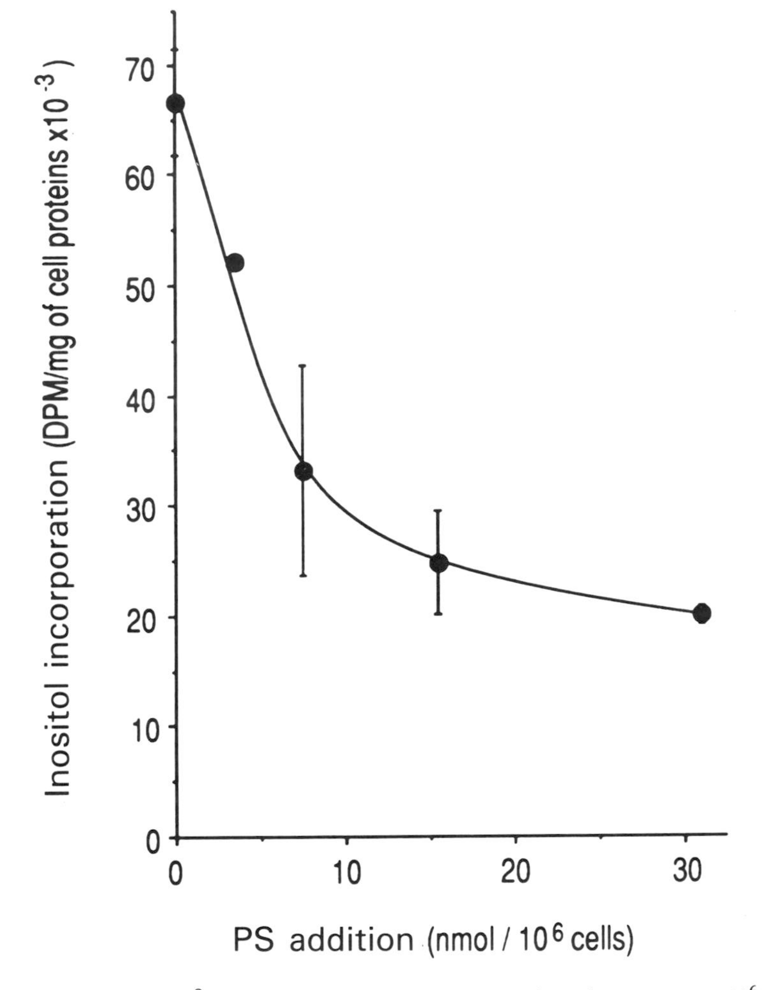

Fig. 6.2 Inhibition of [^{3}H]inositol incorporation in rat lymphocytes. 5 × 10^6 non-adherent mononuclear cells (lymphocytes) were incubated for 60 min at 37 °C with 10 μCi of [^{3}H]inositol and the indicated amount of PS in 1 ml of a buffered saline solution containing 1 mM Ca^{2+} and 1 mg/ml of bovine serum albumin. The cells were washed and extracted as described in Table 6.1. Experimental details are in Mietto *et al.* (1989).

liposoluble radioactivity by thin-layer chromatography confirmed that the step affected by PS was the incorporation of precursors into the respective phospholipids. Since it has been demonstrated that PS uptake decreases PS synthesis (Nishijima *et al.*, 1986), the effect on serine and ethanolamine incorporation was compatible with a down-regulation of the endogenous synthesis of PS and its main metabolite, phosphatidylethanolamine. However, the decreased incorporation of inositol was unexpected. Further experiments (Fig. 6.2) demonstrated that PS-induced inhibition was

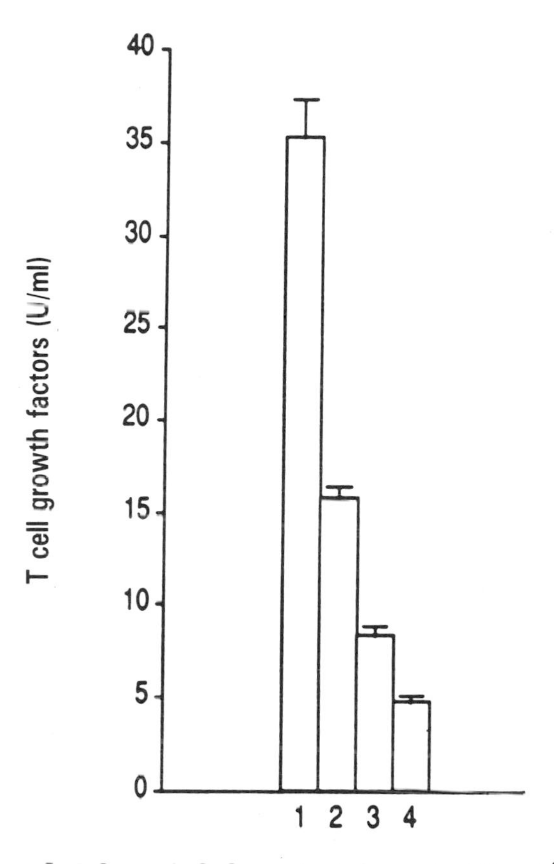

Fig. 6.3 Inhibition of interleukin 2 secretion in mouse splenocytes. 10^7 mouse splenocytes were incubated for 18 h in 1 ml of a medium containing 2 μg of concanavalin A and 10 ng of tetradecanoylphorbolacetate. The interleukin 2 secreted in the medium was measured in a line of interleukin 2-dependent cytoxtoxic T-lymphocytes. One unit is the amount producing half-maximal increase in thymidine incorporation. Interleukin 2 secretion was tested in the absence of PS (1) and in the presence of 1.2 nmol (2), 3 nmol (3) and 6 nmol of PS/10^6 cells (4). Drawn from the data of Ponzin *et al.* (1989).

manifest at a low phospholipid–cell ratio, suggesting a phospholipid effect on mechanisms regulating phosphatidylinositol turnover. Since this phospholipid is required for signal transduction, these results suggested that the lymphocyte response to stimuli transmitted by phosphoinositidase C could be influenced by PS. Experiments performed on mouse splenocytes showed this to be the case (Fig. 6.3). When spleen cells were activated by concanavalin A – a mitogen acting through the phosphoinositide-dependent transducing system – the subsequent secretion of interleukin 2 was decreased by PS. Appropriate controls described in detail elsewhere (Ponzin *et al.*, 1989) excluded that the action of PS was due to inhibition of the macrophage population of spleen cells. Indeed, macrophages were required in our system to secrete interleukin 1, a factor needed for full activation of lymphocytes.

6.3 MAST CELLS

The action of serine phospholipids on rodent mast cells has been fully characterized. Following the initial observation of Goth *et al.* (1971) showing the enhancing effect of PS on the antigen-induced histamine release in rat peritoneal mast cells, it has been demonstrated that lysoPS is several times more active. Thus, the deacylated derivative may be the active species (Martin and Lagunoff, 1979; Smith *et al.*, 1979). Further studies (Bruni *et al.*, 1982; Boarato *et al.*, 1984) showed that not only extraneous compounds such as antigens, lectins or dextran are potentiated by lysoPS, but also endogenous compounds acting on mast cells (e.g. nerve growth factor). In addition, lysoPS acts differently in rat, mouse and gerbil mast cells. While mast cell activation in rats requires the concomitant addition of lysoPS and additional ligands, in mouse and gerbil mast cells lysoPS acts alone. Studies on the structure–activity relationship demonstrated that the amino group and the carboxylate group of serine as well as an acyl chain of adequate length are required for the lysoPS-induced mast cell activation (Martin and Lagunoff, 1979; Smith *et al.*, 1979). Optimal activity requires the α-carbon atoms of serine to be in L configuration. In addition the carbon chain needs to be linked to the glycerol backbone by an ester bond (Horigome *et al.*, 1986; Tamori-Natori *et al.*, 1986). The same structure–activity relationship has been found in mast cell activation induced by lysoPS in vivo (Bruni *et al.*, 1984; Chang *et al.*, 1988; Monastra *et al.*, 1990).

Recent developments have elucidated the signalling mechanism activated by lysoPS. When mouse peritoneal mast cells were pre-labelled with $^{32}PO_4$ or [^{3}H] arachidonate, the increased labelling of phosphatidate and diacylglycerol induced by lysoPS indicated the activation of phosphoinositidase C. In agreement, lysoPS increased the production of inositol phosphates in [^{3}H] inositol-labelled mast cells (Bellini *et al.*, 1990). Using [^{3}H]arachidonate it was observed that the increased labelling of phosphatidate and diacylglycerol was coincident with the maximal rate of histamine secretion (2–5 min). It is interesting to note that the specific effect induced by lysoPS is not due to its reacylation into PS. Indeed, equal effects are produced by the analogue lacking the OH group required for reacylation (deoxy-lysoPS). Essentially the same results were obtained in rat peritoneal mast cells using the combined action of lysoPS and nerve growth factor (Bellini *et al.*, 1988).

6.4 CONCLUSIONS

The phospholipid composition of cell membranes is strictly controlled. Thus, the addition of exogenous phospholipids to cultured cells does not induce a change in their phospholipid content. Rather, inhibition of endogenous phospholipid synthesis or activation of their degradation is expected to occur. Our experiments with PS or lysoPS show the occurrence of these events in rodent lymphocytes and mast cells. In lymphocytes, the addition of PS causes a decrease of PS and phosphatidylethanolamine labelling when radioactive serine or ethanolamine is present in the incubation medium. At the same time the turnover of phosphatidylinositol is decreased. In mast cells, lysoPS enhances the hydrolysis of phosphoinositides by the activation of phosphoinositidase C. These effects result in a change of cell activity as shown by the decreased interleukin-2 secretion in lymphocytes and by degranulation in mast cells. The pharmacological action of serine phospholipids explains the need to maintain constant the phospholipid composition of circulating lipoproteins. Indeed, PS and lysoPS are excluded or kept very low in these plasma constituents.

REFERENCES

Bellini, F., Toffano, G. and Bruni, A. (1988) Activation of phosphoinositide hydrolysis by nerve growth factor and lysophosphatidylserine in rat peritoneal mast cells. *Biochim. Biophys. Acta*, **970**, 187–93.

Bellini, F., Viola, G., Menegus, A.M. *et al.* (1990) Signalling mechanism in the action of lysophosphatidylserine on mouse mast cells. *Biochim. Biophys. Acta*, **1052**, 216–20.

Bigon, E., Boarato, E., Bruni, A. *et al.* (1979a) Pharmacological effects of phosphatidylserine liposomes: regulation of glycolysis and energy level in brain. *Br. J. Pharmacol.*, **66**, 167–74.

Bigon, E., Boarato, E., Bruni, A. *et al.* (1979b) Pharmacological effects of phosphatidylserine liposomes: the role of lysophosphatidylserine. *Br. J. Pharmacol.*, **67**, 611–16.

Boarato, E., Mietto, L., Toffano, G. *et al.* (1984) Different responses of rodent mast cells to lysophosphatidylserine. *Agents Actions*, **14**, 613–18.

Bruni, A., Toffano, G., Leon, A. and Boarato, E. (1976) Pharmacological effects of phosphatidylserine liposomes. *Nature*, **260**, 331–3.

Bruni, A. Bigon, E., Boarato, E. *et al.* (1982) Interaction between nerve growth factor and lysophosphatidylserine on rat peritoneal mast cells. *FEBS Lett.*, **138**, 190–2.

Bruni, A., Bigon, E., Battistella, A. *et al.* (1984) Lysophosphatidylserine as histamine releaser in mice and rats. *Agents Actions*, **14**, 619–25.

Chang, H.W., Inoue, K., Bruni, A. *et al.* (1988) Stereoselective effects of lysophosphatidylserine in rodents. *Br. J. Pharmacol.*, **93**, 647–53.

Goth, A., Adams, H.R. and Knoohuizen, M. (1971) Phosphatidylserine: selective enhancer of histamine release. *Science*, **173**, 1034–5.

Horigome, K., Tamori-Natori, Y., Inoue, K. and Nojima, S. (1986) Effect of serine phospholipid structure on the enhancement of concanavalin A-induced degranulation in rat mast cells. *J. Biochem.*, **100**, 571–9.

Martin, T.W. and Lagunoff, D. (1979) Interactions of lysophospholipids and mast cells. *Nature*, **279**, 250–2.

Mietto, L., Boarato, E., Toffano, G. and Bruni, A. (1989) Internalization of phosphatidylserine by adherent and non-adherent rat mononuclear cells. *Biochim. Biophys. Acta*, **1013**, 1–6.

Monastra, G., Pege, G., Zanoni, R. *et al.* (1991) Lysophosphatidylserine-induced activation of mast cells in mice. *J. Lipid Mediators*, **3**, 39–50.

Nishijima, M., Kuge, O. and Akamatsu, Y. (1986) Phosphatidylserine biosynthesis in cultured Chinese hamster ovary cells. I. Inhibition of de novo phosphatidylserine biosynthesis by exogenous phosphatidylserine and its efficient incorporation. *J. Biol. Chem.*, **261**, 5784–9.

Ponzin, D., Mancini, C., Toffano, G. *et al.* (1989) Phosphatidylserine-induced modulation of the immune response in mice: effect of the intravenous administration. *Immunopharmacology*, **18**, 167–76.

Poste, G. (1983) Liposome targeting in vivo: problems and opportunities. *Biol. Cell*, **47**, 19–38.

Smith, G.A., Hesketh, T.R., Plumb, R.W. and Metcalfe, J.C. (1979) The exogenous lipid requirement for histamine release from rat peritoneal mast cells stimulated by concanavalin A. *FEBS Lett.*, **105**, 58–62.

Tamori-Natori, Y., Horigome, K., Inoue, K. and Nojima, S. (1986) Metabolism of lysophosphatidylserine, a potentiator of histamine release in rat mast cells. *J. Biochem.*, **100**, 581–90.

Tanaka, Y. and Schroit, A.J. (1983) Insertion of fluorescent phosphatidylserine into the plasma membrane of red blood cells: recognition by autologous macrophages. *J. Biol. Chem.*, **258**, 11335–43.

7 Cell adhesion molecules belonging to the immunoglobulin superfamily

G. GENNARINI, F. VITIELLO, P. CORSI, G. CIBELLI, M. BUTTIGLIONE AND C. DI BENEDETTA
Institute of Human Physiology, University of Bari, Italy

Cell–cell or cell–matrix interactions play a key role in several events involved in neural morphogenesis, such as cell migration, axon guidance and selection of the proper target cell (Schachner *et al.*, 1985; Edelman, 1986; Rutishauser, 1986; Harrelson and Goodman, 1988; Dodd *et al.* 1988). Over the last decade a number of molecules mediating such kinds of interactions have been identified and it became evident that most neural cells express more than one type of adhesion molecule (Jessell, 1988). Therefore, it has been proposed that the cell surface adhesive properties result from a combination of, or a competition between, different binding specificities (Rutishauser and Jessell, 1988).

According to structural criteria, cell–cell and cell–matrix adhesion molecules can be classified in distinct gene families. At the same time, a relationship has been demonstrated between their structure and the kind of adhesion mechanism they are involved in.

Molecules mediating the Ca^{2+}-dependent adhesion belong to the cadherins gene family, which includes E-, N- and P-cadherins, identified in epithelial, neural and placental tissues respectively (Takeichi, 1988). Adhesive interactions between cells and the extracellular matrix are mediated by the integral membrane glycoproteins which belong to the integrin family and recognize, besides extracellular matrix components like fibronectin and laminin, also cell surface glycoproteins of platelets, lymphoid and myeloid cells (Hynes, 1987; Rouslahti, 1988).

The immunoglobulin (Ig) supergene family (Williams, 1987; Williams and Barclay, 1988) includes a number of soluble or cell surface glycoproteins which share, as a structural feature, the homology with the constant and variable regions of the Igs and from a functional point of view serve a general role in interaction phenomena both at the molecular and cellular levels. Members of this family expressed in the immune system are, among others, the T-cell receptor, the CD8 and CD4 molecules, LFA3 and CD2 (Selvaray *et al.*, 1987); all are involved in adhesive interactions of the T-lymphocytes.

The structural criterion for inclusion in the Ig superfamily is the presence of one or several Ig homology units or domains (Edelman, 1970; Hood *et al.*, 1985). According to structural similarity, Ig domains have been divided into three classes: (1) V domains with nine β strands; (2) C1 domains with seven β strands; (3) C2 domains which, although folded like C domains, show greater sequence similarity to V domains (Williams, 1987).

In the nervous system several adhesive glycoproteins share homologies with the Ig variable and constant domains and therefore have been included in the Ig supergene family. Among them, N-CAM is up to now the most extensively characterized. It is involved in the formation of contacts between neural cells via a Ca^{2+}-independent homophilic binding (Cunningham *et al.*, 1987). Other components of the Ig superfamily are the myelin associated glycoprotein (MAG), mediating the neurone–oligodendrocyte and oligodendrocyte–oligodendrocyte interactions (Poltorak *et al.*, 1987), the mouse L1 antigen (Moos *et al.*, 1988) and its chicken homologue NgCAM (Grumet and Edelman, 1984), and axonal glycoproteins F11, G4 (Rathjen *et al.*, 1987a) and the related chicken brain glycoprotein neurofascin, involved in neurite fasciculation (Rathjen *et al.*, 1987b), as well as two recently identified insect proteins, fasciclin II (Harrelson and Goodman, 1988) and amalgam (Seeger *et al.*, 1988). With the exception of the P_0, a glycoprotein of the peripheral myelin (Lemke *et al.*, 1988), all of them have been grouped into the C2 set.

All the neural surface proteins with Ig domains of the C2 type seem to serve similar functions, since they have been shown to be involved in surface interactions between neural cells and between processes in different species. Therefore, a common phylogenetically ancient role of C2-type domains has been postulated in mediating cell recognition and adhesion phenomena during nervous system development (Lai *et al.*, 1987; Harrelson and Goodman, 1988).

Some of these molecules also share the L2/HNK-1 carbohydrate epitope, which contains sulphated glucuronic acid (Kruse *et al.*, 1984) This sugar moiety is also found on cell adhesion molecules not belonging to the Ig superfamily and may directly participate in cell–cell or cell–matrix adhesion (Künemund *et al.*, 1988).

In our laboratory we have recently identified and cloned a new member of the Ig superfamily, the mouse brain cell-surface glycoprotein F3, also bearing the L2/HNK-1 epitope (Gennarini *et al.*, 1989a, 1989b, 1989c).

F3 was originally isolated from the culture medium of 4- to 6-day-old mouse forebrain explants (Gennarini *et al.*, 1989a). Antibodies raised in rabbits against the purified molecule were used to isolate homologous cDNA sequences by screening a mouse brain cDNA library prepared in the λ gt11 expression vector. The translation product of one of these clones in *E. coli* has been used as an immunogen to obtain antibody probes with defined specificity for the characterization of the antigen molecular forms.

Two prominent polypeptides of 135 (F3.135) and 90 (F3.90) kDa were recognized in adult mouse brain. It has been shown that the 135 kDa chain carried the 12/HNK-1 epitope and its expression on the surface of cultured brain cell has been clearly demonstrated (Gennari *et al.*, 1989b).

Probing Northern blots of poly(A)$^+$RNA from adult and developing mouse forebrain with an F3 cDNA showed a single hybridizing mRNA of 6.3 kb, expressed at the highest levels between the first and the second postnatal week. A similar developmental pattern was observed for the F3 glycoprotein in mouse forebrain (Gennarini *et al.*, 1989b).

At the immunohistochemical level, in the postnatal day 10 cerebellar cortex, F3 antibodies labelled very faintly the surface of the proliferating cells in the external granular layer, while a very strong reaction was observed on the membrane of postmitotic migrating neurones; as differentiation proceeded the labelling became stronger on parallel fibres, thus suggesting that in mature nervous tissue the F3 glycoprotein was mainly expressed at the level of the axon tracts (Fig. 7.1(a) and (b); see also Gennarini *et al.*, 1989a, 1989b, 1989c).

Evidence of the association between F3 glycoprotein and axons was also obtained in vitro. In primary cultures of mouse forebrain cerebral cortex, F3 antibodies stained both the cell bodies and neurites of cells with neuronal morphology soon after seeding in culture. However, after maturation by culturing in a defined medium, F3 was found to be mainly localized at the axonal level (Gennarini *et al.*, 1989c).

Analysis of the sequence of overlapping cDNA clones allowed the prediction of a polypetide of 1020 amino acids (Gennarini *et al.*, 1989b). The N-terminal hydrophobic segment had the features of signal sequences common to secreted and membrane-bound proteins. The hydrophilic portion of F3 contained 14 cysteines, 12 of which were evenly spaced in the N-terminal half and allowed definition of the six repeating units of C2 type. There were nine potential sites of asparagine-linked glycosylation which would explain the large relative molecular mass shift of the F3 protein from 135 to 110 kDa upon sugar cleavage by endoglycosidase F (Gennarini *et al.*, 1989b).

The deduced protein sequence was organized into eight domains: the six N-terminal C2 repeats, the pre-membrane region of around 400 amino acids containing two regions of about 85 residues each, exhibiting

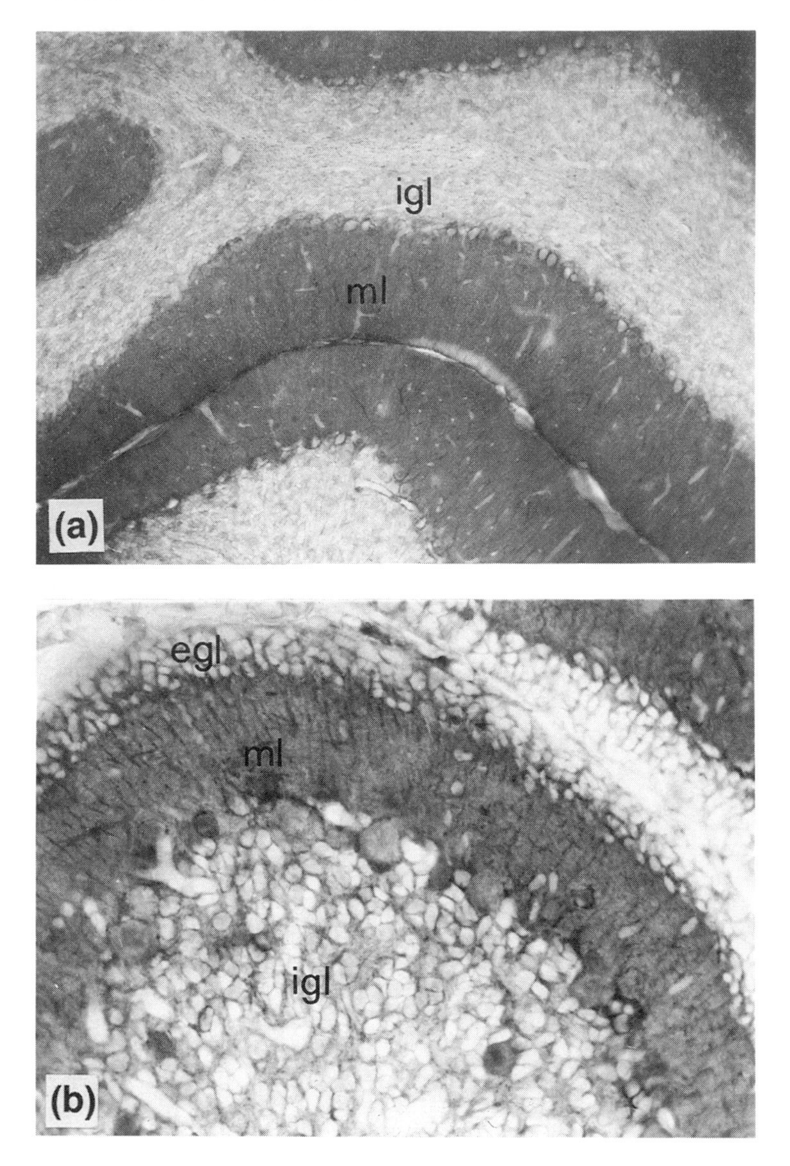

Fig. 7.1 Immunohistochemical localization of the F3 antigen in the adult (a) and developing (b) mouse cerebellar cortex. ml, molecular layer; igl, internal granular layer; egl, external granular layer.

similarity to fibronectin type III repeats, and the eighth domain consisting of a moderately, terminal, hydrophobic segment. A similar structure is usually found in proteins that are attached to the membrane via a covalent bond to PI-containing glycolipids (Ferguson and Williams, 1988). Accordingly, F3 could be released in soluble form by phospholipase C treatment of brain membranes (Gennarini *et al.*, 1989b). We cannot exclude, however, either that an isoform of F3 exists that spans the membrane, or that it is also genuinely secreted, as recently shown for NCAM (Gower *et al.*, 1988).

By in situ hybridization we could assign the F3 gene to band F of the mouse chromosone 15. No other genes of proteins belonging to the Ig superfamily have yet been localized in this region, thus underlining the fact that genes coding for these proteins are widely distributed in the mammalian genome (Gennarini *et al.*, 1989b).

When compared to other members of the Ig superfamily, F3 revealed, in its extracellular region, a 77% degree of positional identity with chicken F11 glycoprotein (Brümmendorf *et al.*, 1989); thus F3 may be considered the homologue of this protein in mouse. Another chicken protein showing a high degree of homology with F3 is contactin (Ranscht, 1988). In contrast to F3, contactin is a transmembrane protein and has a cytoplasmic domain. The difference in biochemical properties detected may be the consequence of different modes of membrane association. In fact, contactin is described as a highly insoluble, cytoskeletal-associated membrane glycoprotein, while most of F3 can be solubilized in non-ionic detergent and even in buffer.

Among the other proteins with C2-type Ig domains, F3 seems most closely related to L1 in overall structure, since both proteins have six Ig domains and a long pre-membrane region containing fibronectin type III repeats (Moos *et al.*, 1988); further, F3 seems to be related to L1 in terms of its expression: the staining pattern of young postnatal mouse cerebellum obtained with F3 and L1 antibodies is virtually indistinguishable. It seems reasonable to consider L1, contactin, F11 and F3 as prototypes of a subgroup of neural surface proteins differently distributed between cell bodies and processes and probably involved in the phenomenon of neurite fasciculation. From the date available, contactin and F3 seem to be most concentrated in fibre-rich areas of the brain (Gennarini *et al.*, 1989c) and might be mainly involved in fibre–fibre interactions. Moreover, the PI-mediated membrane attachment of F3 might be responsible for a distinctive role of the molecule since it could serve both as cell surface molecule and as extra cellular matrix component after cleavage of the phospholipid anchor.

ACKNOWLEDGEMENTS

This research was partially supported by MPI (40% and 60%) and

CNR grants. The work concerning cDNA cloning and sequencing has been carried out at the Centre d'Immunologie de Marseille–Luminy (France).

REFERENCES

Brümmendorf, P., Wolff, J.M., Frank, R. and Rathjen, F.G. (1989) *Neuron*, **2**, 1351–61.

Cunningham, B.A., Hemperly, J.J., Murray, B.A. *et al.* (1987) *Science*, **236**, 799–806.

Dodd, J., Morton, S.B., Karagogeos, D., Yamamoto, M. and Jessell, T.M. (1988) *Neuron*, **1**, 105–16.

Edelman, G.M. (1970) *Biochemistry*, **9**, 3197–205.

Edelman, G.M. (1986) *Annu. Rev. Cell Biol.*, **2**, 81–116.

Ferguson, M.A.J. and Williams, A.F. (1988) *Annu. Rev. Biochem.*, **57**, 285–320.

Gennarini, G., Rougon, G., Vitiello, F. *et al.* (1989a) *J. Neurosci. Res.*, **22**, 1–12.

Gennarini, G., Cibelli, G., Rougon, G. *et al.* (1989b). *J. Cell Biol.*, **109**, 775–88.

Gennarini, F., Rougon, G. and Goridis, C. (1989c) *Acta Histochemica*, band XXXVIII (suppl.), S. 65–9.

Gower, H.J., Barton, C.H., Elsom, V.L. *et al.* (1988) *Cell*, **55**, 955–64.

Grumet, M. and Edelman, G.M. (1984) *J. Cell Biol.*, **98**, 1746–56.

Harrelson, A.L. and Goodman, C.S. (1988) *Science*, **242**, 700–8.

Hood, L., Kronenberg, M. and Hunkapiller, T. (1985) *Cell*, **40**, 225–9.

Hynes, R.O. (1987) *Cell*, **48**, 549–54.

Jessell, T.M. (1988) *Neuron*, **1**, 3–13.

Kronenberg, M., Siu, G., Hood, L.E. and Shastri, N. (1986) *Annu. Rev. Immunol.*, **4**, 529–591.

Kruse, J., Mailhammer, R., Wernecke, H. *et al.* (1984) *Nature*, **311**, 153–5.

Künemund, V., Jungalwala, F.B., Fischer, G. *et al.* (1988) *J. Cell Biol.*, **106**, 213–23.

Lai, C., Watson, J.B., Bloom, F.E. *et al.* (1987) *Immunol. Rev.*, **100**, 129–51.

Lemke, G., Lamar, E. and Patterson, J. (1988) *Neuron*, **1**, 73–83.

Littman, D.R. (1987) *Annu. Rev. Immunol.*, **5**, 561–84.

Moos, M., Tacke, R., Scherer, H. *et al.* (1988) *Nature (Lond.)*, **334**, 701–3.

Poltorak, M., Sadoul, R., Keilhauer, G. *et al.* (1987) *J. Cell Biol.*, **105**, 1893–9.

Ranscht, B. (1988) *J. Cell Biol.*, **107**, 1561–73.

Ranscht, B., Moss, D.J. and Thomas, C. (1984) *J. Cell Biol.*, **99**, 1803–13.

Rathjen, F.G., Wolff, J.M., Frank, R. *et al.* (1987a) *J. Cell Biol.*, **104**, 343–53.

Rathjen, F.G., Wolff, M., Chang, S. *et al.* (1987b) *Cell*, **51**, 841–9.

Rouslahti, E. (1988) *Annu. Rev. Biochem.*, **57**, 375–413.

Rutishauser, U. (1986) *Trends Neurosci.*, **9**, 374–8.

Rutishauser, U. and Jessell, T.M. (1988) *Physiol. Rev.*, **68**, 819–57.

Schachner, M., Faissner, A., Fischer, G. *et al.* (1985) *The Cell in Contact*, Wiley, New York.

Seeger, M.A., Haffley, L. and Kaufman, T.C. (1988) *Cell*, **55**, 589–600.

Selvaray, P., Plunkett, M.L., Dustin, M. *et al.* (1987) *Nature*, **326**, 400–3.
Springer, T.A., Dustin, M.L., Kishimoto, T.K. and Marlin, S.D. (1987) *Annu. Rev. Immunol.*, **5**, 223–52.
Takeichi, M. (1988) *Development*, **102**, 639–55.
Unkeless, J., Scigliano, E. and Freedman, V.H. (1988) *Annu. Rev. Immunol.*, **6**, 251–81.
Williams, A.F. (1987) *Immunol. Today*, **8**, 298–303.
Williams, A.F. and Barclay, A.N. (1988) *Annu. Rev. Immunol.*, **6**, 381–405.

8 Neurogenic control of cerebral blood vessels

C. ALAFACI, F.M. SALPIETRO AND F. TOMASELLO
Department of Neurosurgery, University of Messina, Italy

8.1 INTRODUCTION

For more than half a century it was thought that cerebral blood vessels were innervated only by noradrenergic and cholinergic nerve fibres. In the last few years, with the development of increasingly selective and sensitive immunocytochemical methods, several neuropeptides have been identified within the cerebral perivascular nerve fibres. The complexity of the neurogenic mechanisms in the cerebrovascular bed has thus become increasingly obvious. A defined role has been ascribed to sympathetic nerves in modulating cerebral capacitance and extending the upper limit of autoregulation. We speculate that these new transmitter candidates may play a role in maintaining cerebral homeostasis by inducing rapid modifications of the vessel tone. In addition to perivascular nerves, it has become evident also that the endothelium may modulate cerebral vascular smooth muscle tone.

8.2 PERIVASCULAR NERVES

It is now well established that the cerebral vasculature receives a sympathetic, parasympathetic and sensory nerve supply.

8.2.1 Sympathetic innervation

(a) Noradrenaline

Cerebral vessels are richly supplied by perivascular noradrenergic nerve fibres which originate mainly from the superior cervical ganglia (SCG) (Nielsen and Owman, 1967; Edvinsson *et al.*, 1972), with additional inputs from the stellate ganglia to the vertebrobasilar system (Arbab *et al.*, 1988). Small pial and intraparenchymal vessels receive central projections from the locus coeruleus and related cell groups in the lower brain stem (Edvinsson *et al.*, 1973; Hartman *et al.*, 1972). These nerve fibres are suggested to play an important role in blood–brain barrier functions possibly via direct influence on vascular permeability.

Noradrenaline (NA), whether applied directly, released by tyramine, or upon sympathetic nerve stimulation, leads to α-adrenoceptor-mediated vasoconstriction of cerebral vessels (Edvinsson, 1982). This contractile response can be specifically blocked by α-adrenoceptor blockers. While the large cerebral arteries of most species contract in response to NA, pial arteries of the adult pig and the rat basilar artery, which lack functional α-adrenoceptors, relax via the activation of β-adrenoceptors (Winquist and Bohr, 1982; Winquist *et al.*, 1982). Such regional and inter-species variations in the size and duration of NA-induced vascular responses (Auer *et al.*, 1981; Hamel *et al.*, 1985) most probably reflect the different pattern and distribution of perivascular noradrenergic nerves and vascular adrenoceptors (Bevan *et al.*, 1987; Edvinsson, 1982; Owman *et al.*, 1984). Brain vessels of some species (e.g. rat, monkey and man) contain post-junctional α_1-adrenoceptors (Toda, 1983) whereas in other species, such as cat, dog and cow, the α-adrenoceptors are of the α_2-subtype (Skärby *et al.*, 1983). In the canine basilar artery it has recently been shown that the size of the NA-induced contraction is diminished upon endothelium removal, possibly via the release of an endothelium-derived contractile factor such as arachidonic acid metabolites.

Besides the extensive nature of its cerebrovascular innervation, the noradrenergic system does not appear to play an important role in the regulation of resting cerebral blood flow (CBF) (Heistad and Marcus, 1978). There is evidence to suggest that this system has a protective role during episodes of pronounced hypertension by raising the upper limit of autoregulation (Edvinsson *et al.*, 1976) and a trophic role in controlling vascular smooth muscle thickness (Bevan, 1984; Lee *et al.*, 1987).

(b) Neuropeptide Y

Neuropeptide Y (NPY) is a 36-amino-acid peptide originally isolated from porcine brain (Tatemoto *et al.*, 1982). A rich plexus of perivascular nerves containing NPY has been demonstrated in the cerebral walls of many species, including man, where it is distributed in a manner similar to that of noradrenergic fibres (Alafaci *et al.*, 1985; Edvinsson *et al.*, 1988a). This peptide has been shown to coexist with NA in the majority of the sympathetic nerves supplying blood vessels and in cell bodies of the superior cervical ganglion. NA is present alone in the small synaptic vesicles, whereas large-sized dense-core vesicles contain both NA and NPY. NPY-containing nerve fibres do not completely disappear following superior cervical ganglionectomy, the number of NPY fibres decreasing by only 40% after sympathectomy. This finding indicates that only a limited population of sympathetic noradrenergic neurones contain NPY, which is also located in non-adrenergic neurones. Recently, NPY has been localized in a population of non-adrenergic cerebral perivascular nerves containing vasoactive intestinal polypeptide (VIP) (Gibbins and Morris, 1988) and also colocalized with VIP in cranial parasympathetic neurones (Leblanc *et al.*, 1987). The existence of this population of non-sympathetic NPY explains the persistence of NPY-positive nerve fibres in cerebral vessels after sympathetic denervation (Schon *et al.*, 1985a).

The primary action of NPY on cerebrovascular smooth muscle is a potent, dose-dependent and direct vasoconstrictor action via post-junctional smooth muscle receptors, which is not modified by adrenoceptor blockade or by 5-hydroxytryptamine (5-HT) antagonists, but is blocked by calcium entry blockers (Edvinsson *et al.*, 1983c; Edvinsson, 1985). NPY-induced contraction is of a similar magnitude to that produced by NA but is significantly more sustained, and is greater in pial arteries than veins. In contrast, in most peripheral vessels, in view of its coexistence with NA, NPY has revealed both pre- and post-junctional neuromodulatory mechanisms (Edvinsson *et al.*, 1987a, 1988). There is little evidence for a post-junctional potentiating action of NPY on NA-induced vasoconstriction in the cerebral vasculature (Edvinsson *et al.*, 1988a; Hanko *et al.*, 1986). An inhibitory action of NPY-induced vasoconstriction of canine cerebral arteries has recently been reported (Suzuki *et al.*, 1988)., reflecting a pre-junctional inhibition of NA release. Regarding the effect of NPY on CBF, a decrease in vertebral blood flow in dogs and cortical blood flow in rats has been observed following intravertebral (Suzuki *et al.*, 1988) and intracarotid (Allen *et al.*, 1984) administrations, respectively.

(c) Serotonin (5-hydroxytryptamine)

Immunohistochemical and ultrastructural studies have shown that 5-HT is localized within perivascular nerves supplying the cerebral vasculature (Alafaci *et al.*, 1986; Dhital and Burnstock, 1987; Jackowski *et al.*, 1988). Controversy still surrounds the origin of these fibres, and as to whether or not they are capable of 5-HT synthesis. Evidence has been presented for both a central (Edvinsson *et al.*, 1983a; Reinhard *et al.*, 1979) and a peripheral (Alafaci *et al.*, 1986) origin for these nerve fibres. There is now considerable agreement that in large vessels of brain 5-HT acts as a false neurotransmitter by being actively taken up and stored together with NA in sympathetic nerves, and that it is released upon transmural nerve stimulation to produce vasoconstriction (Chang *et al.*, 1988; Saito and Lee, 1987). This is supported by the immunohistochemical colocalization of 5-HT with dopamine β-hydroxylase (DBH) in cerebrovascular nerve fibres, and that visualization of both substances is lost following surgical sympathectomy (Chang *et al.*, 1988). In contrast, the small pial vessels and intraparenchymal microvessels receive true 5-HT-containing nerve fibres originating centrally from neurones in the raphe nuclei. Electolytic lesions of these central nuclei and intracerebral administration of 5,7-DHT leads to a significant decrease in the 5-HT content of cerebral microvessels (Edvinsson *et al.*, 1983a; Reinhard *et al.*, 1979). This system offers the unique possibility of a coupling between blood flow and metabolism in the brain.

5-HT induces both contractile and dilatory responses in the cerebral circulation, depending on vessel diameter and tone (Edvinsson *et al.*, 1985; Harper and MacKenzie, 1977). Small vessels (resting diameter less than 70 μm) dilate while large vessels (diameter greater than 200 μm) tend to constrict in response to 5-HT. The vasoconstriction appears to be mediated via direct action on 5-HT_2 receptors in the smooth muscle, while the vasodilatation is mediated via β-adrenoceptors (Edvinsson *et al.*, 1985). Intravenous administration of 5-HT increases blood–brain barrier permeability of cerebral microvessels via activation of endothelial 5-HT_2 receptors (Olesen, 1985), raising the possibility that endothelium-dependent 5-HT-induced vasodilatation may be present in cerebral vessels. Because of its constrictor effect, 5-HT has ben implicated in the pathogenesis of cerebral vasospasm after subarachnoid haemorrhage (SAH) and migraine.

8.2.2 Parasympathetic innervation

(a) Acetylcholine

The presence of cerebrovascular cholinergic nerves, based on the histochemical localization of acetylcholinesterase (AChE), is well documented in several species (Edvinsson *et al.*, 1972). This enzyme is not a specific marker for cholinergic neurones but is also present in sympathetic nerves and may have several other functions besides cleaving acetylcholine (ACh); for example, it is able to degrade peptides and therefore may also label sensory nerves (Chubb *et al.*, 1980). In view of the lack of specificity of AChE for cholinergic nerves, an improvement of this method involves the use of recently developed antibodies against choline acetyltransferase (ChAT) (Eckenstein and Thoenen, 1982). ChAT immunoreactivity has subsequently been localized in endothelial cells and nerve fibres of rat cortical microvessels (Arneric *et al.*, 1987; Parnavelas *et al.*, 1985).

Denervation studies suggest a parasympathetic origin for some of these nerve fibres from the sphenopalatine and otic ganglia (Hara *et al.*, 1985). Local nerve cells positive for AChE have also been observed within the cholinergic perivascular plexus of the dog circle of Willis (Borodulya and Pletchkova, 1976). Ultrastructural studies have shown a very close association between cholinergic and adrenergic nerve varicosities in the cerebral perivascular nerves, often within the same Schwann cell sheath (Edvinsson *et al.*, 1972; Iwayama *et al.*, 1970). This raises the possibility that ACh, in addition to its action on vascular smooth muscle and endothelium, may modulate NA release by acting on muscarinic receptors in adrenergic nerve terminals. NA may have a similarly reciprocal action on cholinergic terminals to modulate ACh release.

ACh causes vasodilatation of cerebral vessels and leads to an increase in CBF (Heistad *et al.*, 1980; Lee *et al.*, 1978). This vasodilatory response is mediated through muscarinic receptors via the release of an endothelium-derived relaxing factor (EDRF); in the absence of the endothelium, ACh causes contraction of cerebral vessels (Lee, 1980). The relaxation is competitively antagonized by the classical muscarinic antagonist atropine (Edvinsson *et al.*, 1977).

(b) Vasoactive intestinal polypeptide

The presence of VIP in cerebral vessels and within cerebrovascular nerve fibres in several species has been documented by radioimmunoassay and immunohistochemical techniques (Dhital and Burnstock, 1987; Edvinsson, 1985). A study using a retrograde axonal tracing technique has shown that VIP is colocalized with AChE in perivascular nerves of cerebral vessels and in neurones within the sphenopalatine and otic ganglia (Hara *et al.*, 1985). Peptide histidine isoleucine (PHI), which shares considerable sequence homology with VIP, has also been shown to coexist with VIP in a population of cerebrovascular nerves (Edvinsson and McCulloch, 1985). More recently, double immunostaining has revealed a population of NPY-containing nerve fibres colocalized with VIP (Gibbins and Morris, 1988). Non-sympathetic NPY coexisting with VIP and arising from the sphenopalatine ganglia comprises 25% of the total cerebrovascular NPY nerve fibres. The functional significance of this co-localization is not yet known; however, it has been shown that NPY, at low doses, may antagonize ACh- and SP-induced vasodilatation (Fallgren *et al.*, 1989).

In vivo experiments indicate that VIP mediates vasodilatation of cerebral vessels and increases CBF via specific VIP receptors (Edvinsson, 1985); this response is independent of the endothelium (Lee *et al.*, 1984) and occurs in parallel with activation of cerebrovascular adenylate cyclase. VIP binding sites have been demonstrated in the medial layer of bovine cerebral vessels (Poulin *et al.*, 1986).

Intracerebral injections of VIP have shown potent but complex effects on glucose utilization and metabolism (Edvinsson, 1986). PHI is less potent than VIP, both in vitro and in situ, in mediating relaxation of cerebral vessels, but studies so far indicate that both substances act at the same receptor (Edvinsson and McCulloch, 1985).

Despite their localization within perivascular nerve fibres and characterization of their potent pharmacological actions in vitro and in vivo, a specific role for VIP and PHI in the regulation of local resting CBF remains speculative.

8.2.3 Sensory innervation

(a) Calcitonin gene-related peptide

Calcitonin gene-related peptide (CGRP) is a 37-amino-acid peptide arising from the alternative processing of RNA, transcribed from the calcitonin gene in neural tissue (Rosenfeld *et al.*, 1983). Immuno-histochemical studies have shown that cerebral arteries and veins from different mammalian species, including man, are richly surrounded by fine varicose nerve fibres which display CGRP-like immunoreactivity (Dhital and Burnstock, 1987; Edvinsson *et al.*, 1987). CGRP-containing cell bodies have been demonstrated in the trigeminal ganglia, lesion of which leads to a decrease of CGRP content and CGRP-immunoreactive nerve fibres in the cerebral circulation (Uddman *et al.*, 1985). CGRP coexists with both SP and neurokinin A in a population of trigeminal neurones (Edvinsson *et al.*, 1988). Activation of perivascular sensory nerves causes the release of multiple neuropeptides (CGRP and tachykinins) which cooperate in inducing vascular responses. Following capsaicin treatment, there is a significant loss of perivascular CGRP-immunoreactive nerve fibres, indicating a sensory origin for the majority of these fibres (Wharton *et al.*, 1986). A significant rise in the CGRP content of pial vessels following chronic sympathectomy has been reported in the rat (Schon *et al.*, 1985b).

CGRP is the most potent of all the vasodilator neuropeptides so far identified in the perivascular nerve plexus of cerebral vessels, as evidenced by both in vitro and in situ studies (McCulloch *et al.*, 1986). This relaxation is independent of adrenergic, cholinergic, histaminergic or tachykinin receptors (Edvinsson *et al.*, 1988a; McCulloch *et al.*, 1986; Hanko *et al.*, 1985) and is not dependent on vascular endothelium (Hanko *et al.*, 1985).

CGRP may act as an emergency neurogenic system capable of an immediate and local reflex response in conditions where excessive cerebrovascular constriction occurs, such as in subarachnoid haemorrhage and migraine (McCulloch *et al.*, 1986). Chronic lesions of the trigeminal ganglia do not alter CBF or local cerebral glucose utilization but markedly prolong the duration of cerebrovascular constriction (Edvinsson *et al.*, 1986). Unilateral lesion of the ophthalmic division of the trigeminal nerve is associated with a decreased luminal diameter of cerebral vessels, the restoration of which is markedly delayed following excessive vasoconstriction (Tsai *et al.*, 1988). This finding suggests that CGRP may be a candidate as a nerve-derived trophic factor.

(b) Tachykinins

Tachykinins are a family of peptides which share a similar C-terminal amino acid sequence (Erspamer, 1981). Until a few years ago, SP was the only tachykinin known to be present in the mammalian nervous system. Recently, two new mammalian tachykinins have been described: neurokinin A (NKA) and neurokinin B (NKB). They are both decapeptides isolated from porcine spinal cord (Kimura *et al.*, 1983). SP and NKA arise from the same precursor and this finding has been supported by the recent evidence of cotransmission of these peptides in trigeminal cell bodies and around cerebral vessels (Edvinsson *et al.*, 1988a). Both SP and NKA also coexist with CGRP in cerebrovascular nerves and in cell bodies of the trigeminal ganglia (Edvinsson *et al.*, 1988a). Lesions experiments also indicate an alternative source for SP-containing nerve fibres supplying the caudal part of the vertebrobasilar system, possibly from dorsal root ganglia or brain stem neurones where SP is colocalized with 5-HT (Yamamoto *et al.*, 1983). The biological effects of tachykinins are mediated via receptor subtypes classified as NK-1, NK-2 and NK-3, according to the rank order of potency of the agonists. At the NK-1 receptor SP is the most potent tachykinin while NKA and NKB show high affinity for the NK-2 and NK-3 subtypes, respectively.

The identification of these new peptides suggests that more than one neuropeptide is involved in both central and peripheral sensory transmission and also raises the question of whether these new tachykinins may cause the SP-mediated responses so far observed.

All three peptides induce endothelium-dependent relaxation of cerebral vessels; SP- and NKA-induced dilatation is blocked by tachykinin antagonists (Edvinsson *et al.*, 1988a; Edvinsson and Jansen,

1987). It is suggested that both tachykinins act via the NK-1 type of receptors.

The release of SP from cerebral vessels has been examined in a number of animals. Potassium, capsaicin or electrical stimulation of cerebral vessels causes a calcium-dependent increase in SP release (Edvinsson *et al.*, 1983b; Moskowitz *et al.*, 1983). The activation of the trigeminal ganglion by thermocoagulation in man, or by electrical stimulation in the cat, also leads to a local increase of SP and CGRP (Edvinsson *et al.*, 1987a). Stimulation of sensory nerves may also lead to 'axon reflex'-mediated antidromic release of SP from perivascular nerve fibres (Burnstock, 1977).

Intracarotid infusion of SP in man does not significantly alter the internal carotid blood flow (Samnegard *et al.* 1978). In addition, an ultrastructural study has shown that SP-positive terminals are located in the outer adventitia of cerebral vessels (Itakura *et al.*, 1984).

These studies indicate that while tachykinins have a potent vasodilatory action, they probably play only a minor role in the regulation of resting cerebrovascular tone. Their origin from the trigeminal ganglia might explain an involvement in the neurogenic inflammatory responses in blood vessels during the headache phase of migraine.

In addition to the neurotransmitters discussed above, there are several other substances which are present in cerebral perivascular nerve fibres and which may play a role in the regulation of cerebrovascular tone, such as dynorphin B (Moskowitz *et al.*, 1987), vasopressin (Jojart *et al.*, 1984), gastrin-releasing peptide (Uddman *et al.*, 1983) and atrial natriuretic peptide (ANP) which, although not identified within perivascular nerves, may cause cerebral vasodilatation (Kawai and Ohhashi, 1987).

8.3 ENDOTHELIUM

Following the original discovery by Furchgott and Zawadski (1980) of endothelium-dependent vasodilatation, vascular endothelium has been recognized to have a key role in the regulation of the cerebral circulation. It is now widely accepted that it is able to modulate vascular smooth muscle function. It has been shown that the endothelium releases endothelium-derived relaxing factor(s) (EDRF) in response to ACh and a number of other vasoactive substances (e.g. bradykinin, SP, thrombin, CGRP, histamine, adenosine nucleotides) (Furchgott, 1983). The capacity to produce EDRF has been demonstrated in numerous species, including humans, and the mediator has been identified as nitric oxide (Palmer *et al.*, 1987).

Endothelial cells may also produce vasoconstrictor substances (endothelium-derived contracting factor(s) (EDCF) (Vanhoutte, 1987). The first evidence that the endothelium may release substances inducing cerebral contraction came from experiments in which a decrease in PO_2 produced endothelium-dependent contractions of peripheral, coronary and cerebral arteries (Katusic and Vanhoutte, 1986; Rubanyi and Vanhoutte, 1985). One of these agents has been identified as endothelin (ET). This is a 21-amino-acid peptide, with two intra-chain disulphide bridges, originally isolated from cultures of porcine aortic endothelial cells (Yanagisawa *et al.*, 1988a). A family of vasoactive isopeptides has recently been described: ET-1, ET-2, and ET-3 (Inoue *et al.*, 1989), ET-1 being the most potent member of this family. Endothelin has been reported to have a potent contractile effect both in vitro and in situ. The vasoconstriction is concentration dependent, long-lasting and difficult to wash out, and has been observed in coronary, renal and pulmonary vessels from several mammalian species (Brain *et al.*, 1988; Tomobe *et al.*, 1988; Yanagisawa *et al.*, 1988a, 1988b). The strong and very potent vasoconstriction induced by endothelin may suggest a role in the pathogenesis of cerebrovascular disorders such as migraine and subarachnoid haemorrhage. There is evidence supporting the involvement of the peptide in the development of delayed vasospasm after subarachnoid haemorrhage.

1. Endothelin is one of the most potent vasoconstrictors known in the cerebral circulation (Jansen *et al.*, 1989).
2. Specific, high-affinity ET-1 binding sites have been localized in small arteries (Power *et al.*, 1989).
3. Prolonged basilar artery vasospasm follows intracisternal administration of the peptide (Asano *et al.*, 1989; Mima *et al.*, 1989).
4. Endothelin possesses a potent growth-promoting activity for cultured vascular smooth muscle cells (Kumoro *et al.*, 1988), which may explain the medial hypertrophy during the chronic stage of cerebral vasospasm.
5. Thrombin, which may be present in the subarachnoid clots, can induce pre-proendothelin mRNA in endothelial cells (Yanagisawa *et al.*, 1988a).
6. An enhanced vasoconstrictor effect of endothelin on basilar artery in vitro has been described in rats with subarachnoid haemorrhage (Alafaci *et al.*, 1990).
7. Raised plasma endothelin in aneurysmal subarachnoid haemorrhage has recently been reported (Masaoka *et al.*, 1989).

Regarding the mechanism of endothelin-induced contraction there is agreement that ET stimulates phosphatidylinositol turnover (Simonson *et al.*, 1989) and its effect partially relies on the passage of extracellular calcium (Kai *et al.*, 1988).

In addition to a vasoconstrictor activity, a vasodilator action in isolated rat carotid (Warner *et al.*, 1989) and feline systemic vasculature (Lippton *et al.*, 1988) has been reported.

REFERENCES

Alafaci, C., Cowen, T., Crockard, H.A. and Burnstock, G. (1985) *J. Cereb. Blood Flow Metab.*, **5** (suppl. 1), S543.

Alafaci, C., Cowen, T., Crockard, H.A. and Burnstock, G. (1986) *Brain Res. Bull.*, **16**, 303–4.

Alafaci, C., Jansen, I., Arbab, M.A.R. *et al.* (1990) *Acta Physiol. Scand.*, **138**, 317–19.

Allen, J.M., Schon, F., Todd, N. *et al.* (1984) *Lancet*, **ii**, 550–2.

Arbab, M.A.R., Wiklund, L., Delgado, T. and Svendgaard, N.A. (1988) *Brain Res.*, **445**, 175–80.

Arneric, S.P., Honig, M.A., Milner, T.A. *et al.* (1987) *J. Cereb. Blood Flow Metab.*, **7** (suppl. 1), S330.

Asano, T., Ikegaki, I., Suzuki, Y. *et al.* (1989) *Biochem. Biophys. Res. Commun.*, **159**, 1345–51.

Auer, L., Johansson, B.B. and Lund, S. (1981) *Stroke*, **12**, 528–31.

Bevan, R.D. (1984) *Hypertension*, **6** (suppl. III), 19–26.

Bevan, A.J., Duckworth, J., Laher, I. *et al.* (1987) *FASEB J.*, **1**, 193–8.

Borodulya, A.V. and Pletchkova, E.K. (1976) *Acta Anat.*, **96**, 135–47.

Brain, S.D., Tippins, J.R. and Williams, T.J. (1988) *Br. J. Pharmacol.*, **95**, 1005–7.

Burnstock, G. (1977) *J. Invest. Dermatol.*, **69**, 47–57.

Chang, J.Y., Owman, C. and Steinbusch, H.W.M. (1988) *Brain Res.*, **438**, 237–46.

Chubb, I.W., Hodgson, A.J. and White, G.H. (1980) *Neuroscience*, **5**, 2065–72.

Dhital, K.K. and Burnstock, G. (1987) *Les Maladies de la Paroi Arterielle*, Flammarion, Paris, pp. 77–105.

Eckenstein, F. and Thoenen, H. (1982) *EMBO J.*, **1**, 363–8.

Edvinsson, L. (1982) *TINS*, **5**, 425–8.

Edvinsson, L. (1985) *Acta Physiol. Scand.*, **125**, 33–41.

Edvinsson, L. and Jansen, I. (1987) *J. Pharmacol.*, **90**, 553–9.

Edvinsson, L. and McCulloch, J. (1985) *Regul. Pept.*, **10**, 345–56.

Edvinsson, L., Nielsen, K.C., Owman, C. and Sporrong, B. (1972) *Z. Zellforsch.*, **134**, 311–25.

Edvinsson, L., Lindvall, M., Nielsen, K.C. and Owman, C. (1973) *Brain Res.*, **117**, 519–23.

Edvinsson, L., Owman, C. and Siesjo, B.(1976) *Brain Res.*, **117**, 519–23.

Edvinsson, L., Falck, B. and Owman, C. (1977) *J. Pharmacol. Exp. Ther.*, **200**, 117–26.

Edvinsson, L., Degueurce, A., Duverger, D. *et al.* (1983a) *Nature*, **306**, 55–56.

Edvinsson, L., Rosendahl-Helgesen, S. and Uddman, R. (1983b) *Cell. Tissue Res.*, **234**, 1–7.

Edvinsson, L., Emson, P., McCulloch, J. *et al.* (1983c) *Neurosci. Lett.*, **43**, 79–84.

Edvinsson, L., MacKenzie, E.T. and Scatton, B. (1985) *Prog. Appl. Microcirc.*, **8**, 213–24.

Edvinsson, L., McCulloch, J., Kingman, T.A. and Uddman, R. (1986) *Neural Regulation of Brain Circulation*, Elsevier, Amsterdam, pp. 407–18.
Edvinsson, L., Ekman, R., Jansen, I. *et al.* (1987a) *J. Cereb. Blood Flow Metab.*, **7**, 720–8.
Edvinsson, L., Håkanson, R., Wahlestedt, C. and Uddman, R. (1987b) *TIPS*, **8**, 231–5.
Edvinsson, L., Brodin, E., Jansen, I. and Uddman, R. (1988) *Regul. Pept.* **20**, 181–97.
Edvinsson, L., McCulloch, J., Kelly, P.A.T. *et al.* (1988b) *Ann. N.V. Acad. Sci.*, **527**, 378–92.
Erspamer, V. (1981) *TINS*, **4**, 267–9.
Fallgren, B., Ekblad, E. and Edvinsson, L. (1989) *Neurosci. Lett.*, **100**, 71–6.
Furchgott, R.F. (1983) *Circ. Res.*, **53**, 557–73.
Furchgott, R.F. and Zawadski, J.V. (1980) *Nature*, **288**, 373–6.
Gibbins, I.L. and Morris, J.L. (1988) *Brain Res.*, **444**, 402–6.
Hamel, E., Edvinsson, L. and MacKenzie, E.T. (1985) *J. Cereb. Blood Flow Metab.*, **5**, S553–4.
Hanko, J.H., Hardebo, J.E., Kåhrström, J. *et al.* (1985) *Neurosci. Lett.* , 91–5.
Hanko, J.H., Tornebrandt, K., Hardebo, J.E. *et al.* (1986) *J. Auton. Pharmacol.*, **6**, 117–24.
Hara, H., Hamill, G.S. and Jacobowitz, D.M. (1985) *Brain Res. Bull.*, **14**, 179–88.
Harper, A.M. and MacKenzie, E.T. (1977) *J. Physiol.*, **271**, 735–46.
Hartman, B.K., Zide, D. and Udenfriend, S. (1972) *Proc. Natl Acad. Sci. USA*, **69**, 2722–6.
Heistad, D.D. and Marcus M.L. (1978) *Circ. Res.*, **42**, 295–302.
Heistad, D.D., Marcus, M.L., Said, S.I. and Gross, P.M. (1980) *Am. J. Physiol.*, **239**, H73–H80.
Inoue, A., Yanagisawa, M., Kimura, S. *et al.* (1989) *Proc. Natl Acad. Sci. USA*, **86**, 2863–7.
Itakura, T., Okuno, T. and Nakakita, K. (1984) *J. Cereb. Blood Flow Metab.*, **4**, 407–14.
Iwayama, T., Furness, J.B. and Burnstock, G. (1970) *Circ. Res.*, **26**, 635–46.
Jackowski, A., Crockard, H.A. and Burnstock, G. (1988) *Brain Res.*, **443**, 159–65.
Jansen, I., Fallgren, B. and Edvinsson, L. (1989) *J. Cereb. Blood Flow Metab.*, **9**, 743–7.
Jojart, I., Joo, F., Siklos, L. and Laszlo, F.A. (1984) *Neurosci. Lett.*, **51**, 259–64.
Kai, H., Kanaide, H. and Nakamura, M. (1988) *Biochem. Biophys. Res. Commun.*, **158**, 235–43.
Katusic, Z.S. and Vanhoutte, P.M. (1986) *J. Cardiovasc. Pharmacol.*, **2**, S97–S101.
Kawai, Y. and Ohhashi, T. (1987) *Experientia*, **43**, 568–70.
Kimura, S., Okada, M., Sugita, W. *et al.* (1983) *Proc. Jpn. Acad. Sci.*, **59**, 101–4.
Kumoro, I., Kurihara, H., Sugiyama, T. *et al.* (1988) *FEBS Lett.*, **238**, 249–52.
Leblanc, G.G., Trjmmer, B.A. and Landis, S.C. (1987) *Proc. Natl Acad. Sci, USA*, **84**, 3571–5.
Lee, T.J.-F. (1980) *Eur. J. Pharmacol.*, **68**, 393–4.
Lee, T.J.-F., Hume, W.R., Su, C. and Bevan, J.A. (1978) *Circ. Res.*, **42**, 535–42.
Lee, T.J.-F., Saito, A. and Berezin, I. (1984) *Science*, **224**, 898–901.

Lee, R.M.K.W., Triggle, C.R., Cheung, D.W.T. and Coughlin, M.D. (1987) *Hypertension*, **10**, 328–38.
Lippton, H., Goff, J. and Hyman, A. (1988) *Eur. J. Pharmacol.*, **155**, 197–9.
Masaoka, H., Suzuki, R., Hizata, Y. *et al.* (1989) *Lancet*, **ii**, 1402.
McCulloch, J., Uddman, R., Kingman, T. and Edvinsson, L. (1986) *Proc. Natl Acad. Sci. USA*, **83**, 5731–5.
Mima, T., Yanagisawa, M., Shigeno, T. *et al.* (1989) *Stroke*, **20**, 1553–6.
Moskowitz, M.A., Brody, M. and Liu-Chen, L.-Y. (1983) *Neuroscience*, **9**, 809–14.
Moskowitz, M.A. , Saito, K., Brezina, L. and Dickson, J. (1987) *Neuroscience*, **23**, 731–7.
Nielsen, K.C. and Owman, C. (1967) *Brain Res.*, **6**, 773–6.
Olesen, S.P. (1985) *J. Physiol.*, **361**, 103–13.
Owman, C., Andersson, J., Hanko, J. and Hardebo, E. (1984) *Neurotransmitters and the Cerebral Circulation*, Raven Press, New York, pp. 11–38.
Palmer, R.M.J., Ferrige, A.G. and Moncada, S. (1987) *Nature*, **327**, 524–6.
Parnavelas, J., Kelly, W. and Burnstock, G. (1985) *Nature*, **316**, 724–5.
Poulin, P., Suzuki, Y., Lederis, K. and Rorstad, O.P. (1986) *Brain Res.*, **381**, 382–4.
Power, R.F., Wharton, J., Salas, S.P. *et al.* (1989) *Eur. J. Pharmacol.*, **160**, 199–200.
Reinhard, J.F., Liebmann, J.E. and Schlosberg, A.J. (1979) *Science*, **206**, 85–6.
Rosenfeld, M.G., Mermod, J.-J., Amara, S.G. *et al.* (1983) *Nature*, **304**, 129–35.
Rubanyi, G.M. and Vanhoutte, P.M. (1985) *J. Physiol.*, **364**, 45–56.
Saito, A. and Lee, T.J.F. (1987) *Circ. Res.*, **60**, 220–8.
Samnegard, H., Thulin, L., Tyden, G. *et al.* (1978) *Acta Physiol. Scand.*, **104**, 491–5.
Schon, F., Allen, J.M., Yeats, J.C. *et al.* (1985a) *Neurosci. Lett.*, **57**, 65–71.
Schon, F., Ghatei, M., Allen, J.M. *et al.* (1985b) *Brain Res.*, **348**, 197–200.
Simonson, M.S., Wann, S., Mene, P. *et al.* (1989) *J. Clin. Invest.*, **83**, 708–12.
Skärby, T., Andersson, K.-E. and Edvinsson, L. (1983) *Acta Physiol. Scand.*, **117**, 63–73.
Suzuki, Y., Shibuya, M., Ikegaki, I. *et al.* (1988) *Eur. J. Pharmacol.*, **146**, 271–7.
Tatemoto, K., Carlquist, M. and Mutt, V. (1982) *Nature*, **296**, 659–60.
Tsai, S.-H., Tew, J.M., McLean, J.H. and Shipley, M.T. (1988) *J. Comp. Neurol.*, **271**, 435–44.
Toda, N. (1983) *J. Pharmacol. Exp. Ther.*, **226**, 861–8.
Tomobe, Y., Migauchi, T., Saito, A. *et al.* (1988) *Eur. J. Pharmacol.*, **152**, 373–4.
Uddman, R., Edvinsson, L., Owman, C. and Sundler, F. (1983) *J. Cereb. Blood Flow Metab.*, **3**, 386–90.
Uddman, R., Edvinsson, L., Ekman, R. *et al.* (1985) *Neurosci. Lett.*, **62**, 131–6.
Vanhoutte, P.M. (1987) *Blood Vessels*, **24**, 141–4.
Warner, T.D., de Nucci, G. and Vane, J.R. (1989) *Eur. J. Pharmacol.*,**159**, 325–6.
Wharton, J., Gulbenkian, S., Mulderry, P.K. *et al.* (1986) *J. Autonom. Nerv. Syst.*, **16**, 289–309.

Winquist, R.J. and Bohr, D.F. (1982) *Experientia*, **38**, 1187–8.
Winquist, R.J., Webb, C. and Bohr, D.F. (1982) *Circ. Res.*, **51**, 769–76.
Yamamoto, K., Matsuyama, T., Shiosaka, S. *et al.* (1983) *J. Comp. Neurol.*, **215**, 421–6.
Yanagisawa, M., Kurihara, H., Kimura, S. *et al.* (1988a) *Nature*, **332**, 411–15.
Yanagisawa, M., Inoue, A., Ishikawa, T. *et al.* (1988b) *Proc. Natl Acad. Sci. USA*, **85**, 6964–7.

9 The fine structural aspects of brain oedema and associated microvascular and glial changes

A. HIRANO AND R. WAKI
Division of Neuropathology, Department of Pathology, Montefiore Medical Center, Albert Einstein College of Medicine, Bronx, New York, 10467, USA

9.1 INTRODUCTION

The brain is unique among the various organs of the body in several ways. Perhaps the most obvious is its almost total confinement within the rigid, bony calvarium. Another well-known difference between the mature brain and many other organs is the blood–brain barrier, which somehow results in the poor penetration of certain molecules into the brain parenchyma that readily permeate most other tissues. The purpose of this presentation is to outline the fine structural alterations observed under various oedematous conditions. Particular attention will be paid to the endothelium of the small vessels of the brain and the surrounding cerebral parenchyma.

9.2 ALTERATIONS OF CEREBRAL ENDOTHELIUM ASSOCIATED WITH BRAIN OEDEMA

Under normal conditions, the endothelium of the capillaries of the central nervous system has certain characteristic features which are different from those of other organs. First, cerebral endothelial cells are markedly attenuated and usually not fenestrated. The peripheral capillary wall is only 0.1 μm thick. Areas containing fenestrated blood vessels under normal conditions lack a blood–brain barrier. They include the choroid plexus, the pineal and pituitary glands, the area postrema, the tuber cinereum and the median eminence. The 50 nm diameter pores in fenestrated vessels are formed by the fusion of the luminael and basal plasma membranes. The pore is apparently sealed by a continuous diaphragm about 5 nm thick. Fine structural analysis of endothelial fenestral diaphragms has revealed a radial arrangement of wedge-shaped channels of 5 nm diameter separated by fibrillar septa (Bearer and Orci, 1985).

Second, adjacent endothelial cells are bound by elaborate continuous tight junctions. As revealed by freeze-fracture technique these junctions are associated with anastomosing strands (Brightman, 1989). The composition of the strands is still unclear: they may consist of micellar lipids rather than protein (Kachar and Reese, 1982).

Third, there is a paucity of pinocytotic vesicles in the cerebral endothelial cells.

These features are most clearly demonstrated when tracers such as horseradish peroxidase are administered intravenously (Hirano and Kato, 1988). The tracer is confined to the lumen. It seems to be stopped at the tight junction and is not internalized in the endothelial cell. Thus, it seems clear that the anatomical site of the blood–brain barrier, at least for horseradish peroxidase, is at the cerebral endothelium (Reese and Karnovsky, 1967).

In brain oedema, abnormal accumulations of fluid appear within the brain parenchyma. In haematogenous oedema, the blood–brain barrier is broken at the level of the endothelium. The possible routes of penetration of the fluid through the endothelium has been the subject of numerous studies and several routes have been suggested based on the results of tracer studies (Hirano *et al.*, 1969; Hirano and Llena, 1983).

9.2.1 Tight junctions and vesicles

Opening of the tight junction is one of the possible routes (Hirano, 1974; Hirano and Llena, 1983). This possibility has been most extensively studied in cerebral vessels flushed with hyperosmotic fluid, which causes shrinkage of the endothelial cells (Brightman *et al.*, 1973; Dorovini-Zis *et al.*, 1983; Nagy *et al.*, 1984; Rapoport and Robinson, 1986; Brightman, 1989). Under these conditions the tracer can be found between, and subluminal to, endothelial cells, suggesting a discontinuity in the tight junction which is not apparent in thin section.

Under various pathological conditions resulting in brain oedema, many vesicles may be visualized in the endothelium of the cerebral vessel, in contrast to their paucity in normal conditions (Hirano and Llena, 1983). The possibility of vesicular transport across capillary endothelia has been considered. An alternative interpretation is the formation of channels rather than their actual movement (Noguchi *et al.*, 1987). Some material transversing the endothelium via vesicles may be deposited in lysosomes (Broadwell *et al.*, 1988). The role of receptor-mediated endocytosis by coated vesicles in the uptake of specific substances and their transport has also been reviewed (Brightman, 1989).

9.2.2 Surface infolding, extravasation of haematogenous cells, diffusion through damaged endothelium, and fenestrations

Under certain pathological conditions, the surface of the altered endothelium may show prominent and irregular infolding. The infolding conceivably produces large pockets within the endothelium, facilitating massive transfer of haematogenous material (Hirano and Llena, 1983). Various haematogenous cells also penetrate the vascular wall and appear in the perivascular space, under various pathological conditions, especially in the acute stage of inflammation.

Focal permeation of tracer, in the endothelial ground substance, may be confined to part of the endothelial cell or may permeate the entire cytoplasm of the cell. If such permeation is seen the adjacent perivascular space may also contain the tracer material. This phenomenon may be interpreted as artefact but may also be considered to be diffusion of the tracer through damaged endothelium. Necrosis of endothelium is a common pathological phenomenon, and often causes actual haemorrhage.

Occasionally, fenestrated blood vessels have been demonstrated in areas of the central nervous system where pores are normally characteristically absent. This has been observed in both neoplastic and non-neoplastic pathological conditions. A list of such conditions has been previously reported (Hirano and Llena, 1983).

9.2.3 Other alterations of the endothelium in brain oedema

Another alteration of cerebral endothelium sometimes associated with brain oedema is endothelial proliferation. This is often associated with the appearance of Weibel–Palade bodies (Hirano and Matsui, 1975) and rarely tubular arrays (Hirano *et al.*, 1975).

The electric charge on the cell membrane of the endothelial cells and its role in the uptake of various substances has been studied in both normal and pathological conditions (Simionescu *et al.*, 1981; Brightman, 1989; Nag, 1988).

The role and localization of enzymes associated with the endothelium and its plasma membrane have also been the subject of many investigations. These include α-glutamyltranspeptidase (Betz *et al.*, 1980), Na^+–K^+-activated adenosine triphosphatase (ATPase) (Vorbrodt *et al.*, 1982), Ca^{2+}-activated ATPase (Nag, 1988) and K^+-dependent *p*-nitrophenylphosphatase (K^+-PNPPase) (Kato and Nakamura, 1989), among others (Brightman, 1989).

In addition, various other immunohistochemical markers for endothelium have been reported. These include MB1 for all endothelial and haematopoietic cells except mature erythrocytes (Labastie *et al.*, 1986) and the monoclonal antibody PAL-E, specific for endothelium of rat brain microvessels which does not stain the endothelium of other organs, large vessels or the choroid plexus, but also reacts with the brush border of kidney proximal tubules and liver bile canaliculi (Mickalak *et al.*, 1986). Application of these markers in the study of altered cerebral endothelium in brain oedema will be interesting. The 'blood–brain barrier protein' recognized by a monoclonal antibody reported by Sternberger and Sternberger in 1987 will be especially fascinating. Non-endothelial cells, and even endothelial cells in other organs, are generally negative for this antibody and much less positive or absent in the endothelium of the area postrema and choroid plexus (Sternberger and Sternberger, 1987).

9.3 BRAIN OEDEMA

The most common type of brain oedema is haematogenous. The source of the oedema fluid is the plasma, which transverses the endothelial cells, underlying basement membrane and spreads into the perivascular space. It then penetrates between, or sometimes through, disrupted perivascular astrocytic cell processes and appears in the extracellular space of the brain parenchyma. The punctate adhesions and gap junctions between astrocytic cell processes do not seem to impede the penetration of macromolecules. Tight junctions are absent (Hirano, 1987).

The increase of water content associated with oedema results in an increase in brain volume and weight. In vasogenic oedema fluid accumulates within the white matter much more than the grey matter, because the stream of oedema fluid generally follows the anatomical pathway of least resistance. The white matter is composed of many bundles of myelinated and unmyelinated axons, interspersed with the interfascicular oligodendrocytes and astrocytes as well as the blood vessels. These elements are arranged in a highly organized manner with remarkably narrow extracellular spaces in normal conditions. When oedema fluid appears in the parenchymatous extracellular space, it extends to the nearby white matter. Once there it separates individual nerve fibres from one another and fluid-filled spaces are readily demonstrated in properly fixed tissue (Hirano *et al.*, 1964). An expanded extracellular space becomes more conspicuous when an electron-dense tracer substance is used. The absence of demonstrable junctions between myelinated axons and the formation of tracts by bundles of nerve fibres facilitate the passage of oedema fluid along the fibre tracts. However, adjacent cell bodies among the interfascicular oligodendroglia and adjacent astrocytes are connected by a junctional apparatus which resists separation.

Infiltration of oedema fluid within the grey matter is limited because of the presence of numerous junctions between the cells in these regions. Most numerous of these junctions are the synapses between neurones. Junctions are also present between glial cells. In addition, dendrites of neurones generally extend in an individual way rather than in bundles. These many dendritic trees are decorated by numerous spines. All of these anatomical features may well contribute to the limited expansion of the extracellular space in the grey matter of the cerebral cortex.

The location of the extracellular fluid not only shifts but also changes in volume with the passage of time. It can be visualized by the computed tomography (CT) scan, magnetic resonance imaging (MRI), and on gross examination. Eventually the fluid is absorbed by various types of cells.

The astrocytes are also actively involved in oedema. Dramatic swelling is often associated with the accumulation of large amounts of glycogen granules in acute oedematous lesions (Hirano *et al.*, 1965). Actual rupture of astrocytic plasma membrane results in the penetration of oedema fluid into the cells. An increase in glial fibrils within the astrocytes and phagocytic activity take place during the oedematous process. Mitotic division of astrocytes as well as movement of immature astrocytes along the stream of oedema fluid have been documented in the oedema produced by cold injury (Ikuta *et al.*, 1983).

9.4 EXTRACELLULAR COMPARTMENTS ASSOCIATED WITH MYELINATED AXONS

White matter is the main site of fluid accumulation in brain oedema and the myelinated axon is the major element in this tissue. Five extracellular compartments, which normally are not easily recognizable, are associated with the myelinated axon which can become clearly demonstrable in various oedematous conditions. They are:

1. the extracellular space surrounding the myelin sheath, the major site of accumulation of haematogenous oedema fluid associated with the breakdown of the blood–brain barrier;
2. intramyelinic splits at the level of the intraperiod line; a characteristic feature of certain conditions such as triethyltin intoxication (Hirano, 1983);
3. enlargement of the periaxonal space (Hirano, 1983);
4. infiltration of dense fluid in the region of Schmidt–Lanterman cleft (Hirano, 1982); and
5. narrow channels of extracellular spaces between the transverse bands at the periaxonal interface of the paranode (Hirano and Dembitzer, 1982).

Special junctional devices generally separate these compartments (Hirano, 1985).

REFERENCES

Bearer, E.L., and Orci, L. (1985) *J. Cell. Biol.*, **100**, 418–28.
Betz, A.L., Firth, J.A. and Goldstein, G.W. (1980) *Brain Res.*, **192**, 17–28.
Brightman, M.W. (1989) in *Implications of the Blood–Brain Barrier and its Manipulation,* Vol. 1 (ed. E. A. Neuwelt), Plenum Medical. New York, pp. 53–83.
Brightman, M.W., Hori, M., Rapoport, S.I. *et al.* (1973) *J. Comp. Neurol.*, **152**, 317–26.
Broadwell, R.D., Balin, B. and Salcman, M. (1988) *Proc. Natl Acad. Sci. USA*, **80**, 7352–6.
Dorovini-Zis, K., Sato, M., Goping, G. *et al.* (1983) *Acta Neuropathol. (Berl.)*, **60**, 49–60.
Hirano, A. (1974) in *Pathology of Cerebral Microcirculation* (ed. J. Cervos-Navarro), de Gruyter, Berlin, pp. 203–17.
Hirano, A. (1982) *Acta Neuropathol. (Berl,)*, **58**, 34–8.
Hirano, A. (1983) in *Progress in Neuropathology,* Vol. 5 (ed. H.M. Zimmerman), Raven Press, New York, pp. 99–112.
Hirano, A. (1985) in *Proceedings of the 6th International Symposium on Brain Edema* (ed, Y. Inaba), Springer-Verlag, New York, pp. 6–13.
Hirano, A. (1987) in *Stroke and Cerebrospinal Microcirculation* (eds. J. Cervos-Navarro and R. Ferszt), Raven Press, New York, pp. 219–22.
Hirano, A. and Dembitzer, H.M. (1982) *J. Neurocytol.*, **11**, 861–6.
Hirano, A. and Kato, T. (1988) in *Neuromethod Vol. 9: Neuronal Microenvironment* (eds A.A. Boulton, G.B. Baker and W. Walz), Human Press, Clifton, NJ, pp. 105–26.
Hirano, A. and Llena, J.F (1983) in *Advances in Cellular Neurobiology*, Vol. 4 (eds S. Federoff, S. and L. Herz), Academic Press, New York, pp. 223–47.
Hirano, A. and Matsui, T. (1975) *Human Pathol.*, **6**, 611–21.
Hirano, A., Zimmerman, H.M. and Levine, S. (1964) *Am. J. Pathol.*, **45**, 1–19.
Hirano, A., Zimmerman, H.M. and Levine, S. (1965) *J. Neuropathol. Exp. Neurol.*, **24**, 386–97.
Hirano, A., Zimmerman, H.M. and Levine, S. (1968) *J. Neuropathol. Exp. Neurol.*, **27**, 571–80.
Hirano, A., Becker, N.H. and Zimmerman, H.M. (1969) *Arch. Neurol.*, **20**, 300–8.
Hirano, A., Llena, J.F. and Chung, H.D. (1975) *Acta Neuropathol. (Berl.)*, **32**, 103–13.
Ikuta, F., Ohama, E., Yoshida, Y. *et al.* (1983) *Adv. Neurol. Sci. (Tokyo)*, **27**, 839–56.
Kachar, B. and Reese, T.S. (1982) *Nature (Lond.)*, **296**, 464–6.
Kato, S. and Nakamura, H. (1989) *Acta Neuropathol. (Berl.)*, **77**, 455–64.
Labastie, M.C., Poole, T.J., Peault, B.M. and Le Douarin, N.M. (1986) *Proc. Natl Acad, Sci. USA*, **83**, 9016–20.
Mickalak, T., White, F.P., Gard, A.L. and Dutton, G.R. (1986) *Brain Res.*, **379**, 320–8.

Nag, S. (1988) *Acta Neuropathol. (Berl.)*, **75**, 547–53.
Nagy, Z., Peters, H. and Huttner, I. (1984) *Lab. Invest.*, **50**, 313–22.
Noguchi, Y., Shibata, Y. and Yamamoto, T. (1987) *Anat. Rec.*, **217** 355–60.
Rapoport, S. and Robinson, P.J. (1986) *Ann. NY Acad. Sci.*, **481**, 250–67.
Reese, T.S. and Karnovsky, M.J. (1967) *J. Cell. Biol.*, **34**, 207–17.
Schlingemann, R.O., Dingjan, G.M., Emeis, J.J. *et al.* (1985) *Lab. Invest.*, **52**, 71–6.
Simionescu, N., Simionescu, M. and Palade, G.E. (1981) *J. Cell. Biol.*, **90**, 605–13.
Sternberger, N.H. and Sternberger, L.A. (1987) *Proc. Natl. Acad. Sci. USA*, **84**, 8169–73.
Vorbrodt, A.W., Lossinsky, A.S. and Wisniewski, H.K.M. (1982) *Brain Res.*, **243**, 225–34.

10 Image-guided localized nuclear magnetic resonance spectroscopy of human brain tumours

W. HEINDEL, W. STEINBRICH, J. BUNKE AND G. FRIEDMANN
Department of Diagnostic Radiology, University of Cologne Medical School, Köln, Germany

10.1 INTRODUCTION

Complementary to morphological studies by magnetic resonance imaging (MRI), localized in vivo nuclear magnetic resonance spectroscopy (MRS) opens the way to a biochemical characterization of brain tumours. This study was designed to evaluate the diagnostic value of image-selected localized proton and phosphorus MRS in patients suffering from intracranial tumours.

10.2 EQUIPMENT AND METHODS

A 1.5 T whole-body MR system (Philips, Gyroscan S15) operating at 64 MHz for ^{1}H and 25.9 MHz for ^{31}P was used. ^{1}H MRI preceded spectroscopy in order to judge the tumour macroscopically and to define the volume of interest (VOI) within the lesion.

Localized water-suppressed ^{1}H MR spectra were obtained using a regular 30 cm imaging head coil (send/receive). An orthogonal ^{31}P closely coupled head coil was inserted into the proton coil. Thus, imaging and the spectroscopic studies could be done alternatively, without system modification or repositioning of the patient.

Volume selection to obtain water-suppressed ^{1}H spectra was achieved by a spatially selective 90°–180°–180° spin echo sequence. Typical echo times were 272 (or 136) ms, repetition time 2 s. For ^{31}P a modification (Luyten *et al.*, 1989) of the 'Image-selected in vivo spectroscopy (ISIS)' sequence (Ordidge *et al.*, 1986) utilizing adiabatic inversion and excitation pulses was used (repetition time 3 s).

The spectra were uniformly post-processed. Tissue pH was determined from the PCr–Pi chemical shift difference using the formula proposed by Petroff *et al.* (1985).

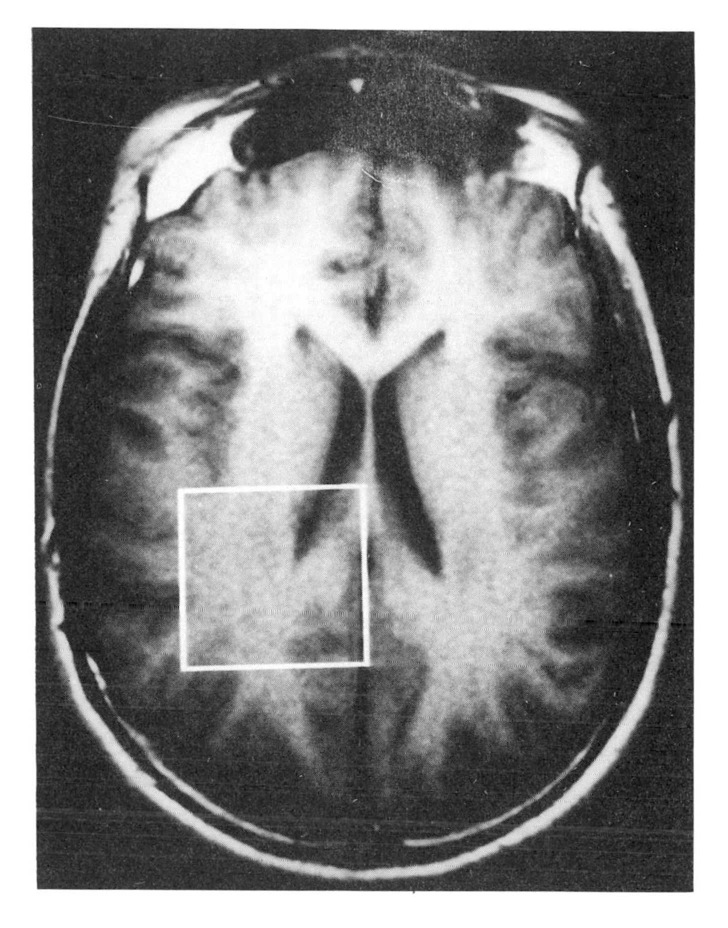

Fig. 10.1 A 21-year-old volunteer. (a) ^{1}H MRI image (SE 300/15) for localization of the cube-shaped VOI (40 × 40 × 40 mm) in the parieto-occipital region. (b) Volume-selective ^{31}P MR spectrum of healthy brain tissue. For assignments see text. Intracellular pH calculated from the chemical shift of inorganic phosphate versus PCr is 7.03. (c) Localized, water-suppressed MR spectrum of the identical volume. For assignments see text.

10.3 RESULTS AND DISCUSSION

In order to establish the examination protocol, healthy volunteers of different age and sex were studied. Fig. 10.1 demonstrates the results in a 21-year-old man: MRI was used to centre the 64 cm^3 VOI in the parieto-occipital region for spectroscopy of predominantly white matter. The proton spectrum shows resonances from *N*-acetyl aspartate (NAA), creatine and phosphocreratine ((P)Cr), choline-containing compounds

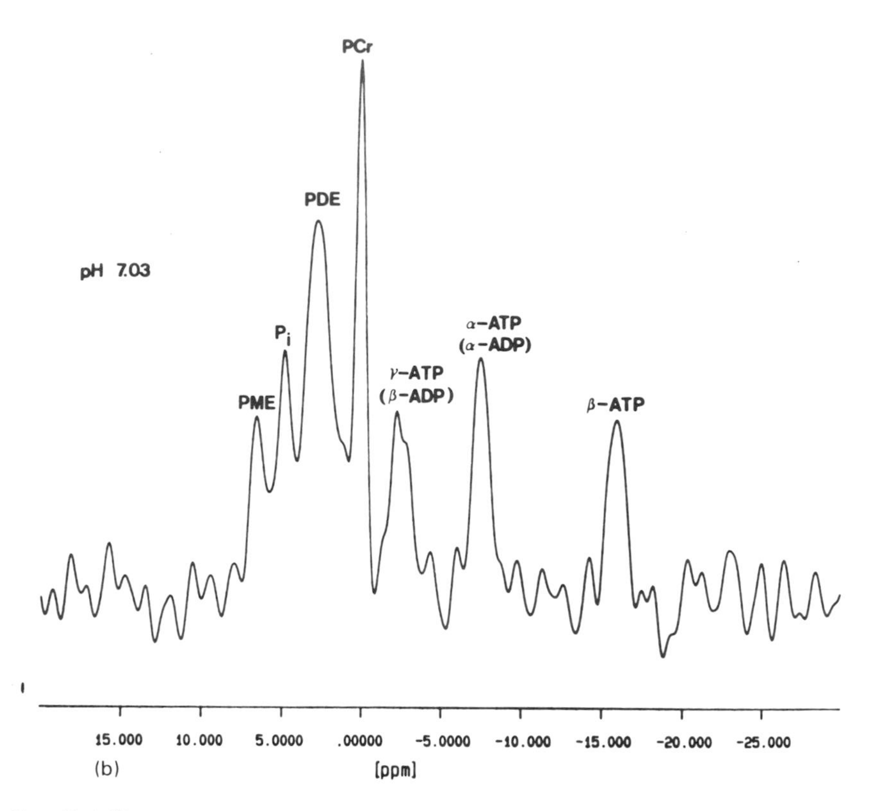

Fig. 10.1 (b)

(Cho) and glutamine (Gln), the phosphorus spectrum signals from adenosine triphosphate (ATP), PCr, inorganic phosphate (Pi) and phosphomono- and diesters (PME and PDE). The evaluation of the ^{31}P spectra yielded an intracellular pH of 7.01 ± 0.06 for normal brain tissue (Heindel *et al.*, 1988). Comparison of the spectra from the same anatomical region confirmed reproducibility of the results using this approach; however, spectra from different regions in the brain revealed more pronounced differences in ^{1}H MRS that in ^{31}P MRS. For example, the relative concentration of NAA in the cerebellum was shown to be significantly lower than in Fig. 10.1.

The aim of this study was to examine identical volumes, though the high intrinsic MR sensitivity and the biological abundance of protons proved advantageous as compared to phosphorus. All presented ^{1}H spectra needed only 128 averages compared to 512 excitations in ^{31}P. In addition, in several patients a bad signal-to-noise ratio resulted with

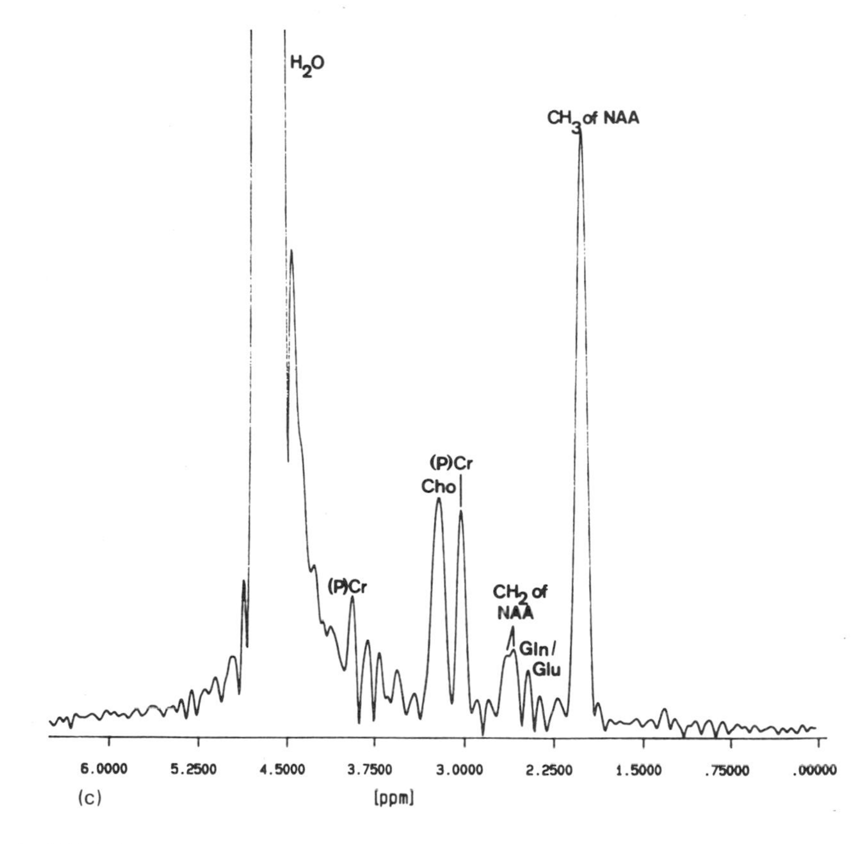

Fig 10.1 (c)

^{31}P MRS, which was either due to tumour necrosis (e.g. in a few glioblastomas) or due to a too small VOI as determined from the images.

In gliomas ^{31}P MRS typically demonstrated decreased PDE and PCr signals as well as normal or alkaline pH (Heindel *et al.*, 1988). The ^{1}H spectra showed an increased Cho/(P)Cr ratio, while NAA was reduced. Lactate concentration was characteristically elevated in astrocytomas and glioblastomas. In an oligodendroglioma only increased lipid signals were observed.

Fig. 10.2 depicts the results in a 25-year-old man with a primary neuroectodermal tumour. ^{31}P shows a reduced PDE signal in particular, but also a diminished PCr resonance in this malignant glioma. In ^{1}H considerably reduced signals from NAA and Cr/PCr can be observed, while the choline concentration is unchanged. Calculation of tissue pH from ^{31}P indicates an alkaline environment in the tumorous tissue,

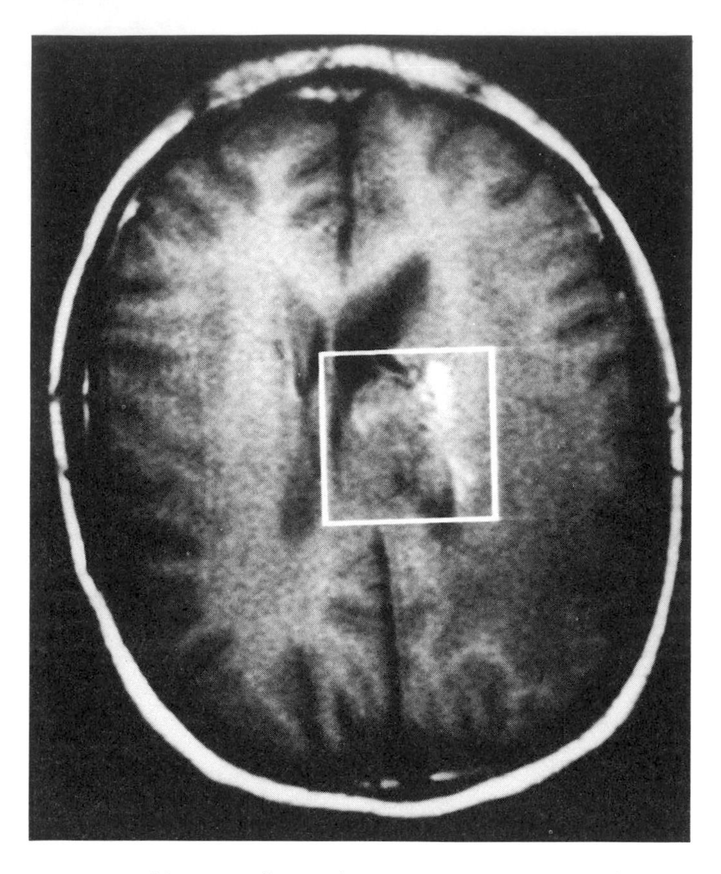

Fig. 10.2 A 25-year-old man suffering from a primary neuroectodermal tumour. (a) ^{1}H MRI (SE 300/15) documents the tumour. The 40 × 40 × 40 mm VOI selected for spectroscopy is indicated.

though lactic acid is demonstrated by ^{1}H. Similar observations in animal studies (Hossmann *et al.*, 1986) have been interpreted as evidence for an active regulation of intracellular pH.

Compared with healthy brain tissue, spectra from meningiomas demonstrated the most obvious differences: In ^{31}P the PCr peak decreased below the level of ATP, and the phosphodiester signal was reduced. Tumour pH values were often within the alkaline range. In ^{1}H neither NAA nor (P)Cr was observed in one meningioma, while the Cho/(P)Cr ratio was increased in another. In both tumours a doublet at 1.47 p.p.m. was observed, which could be assigned to the amino acid alanine. Up to now, no combined study of an identical volume

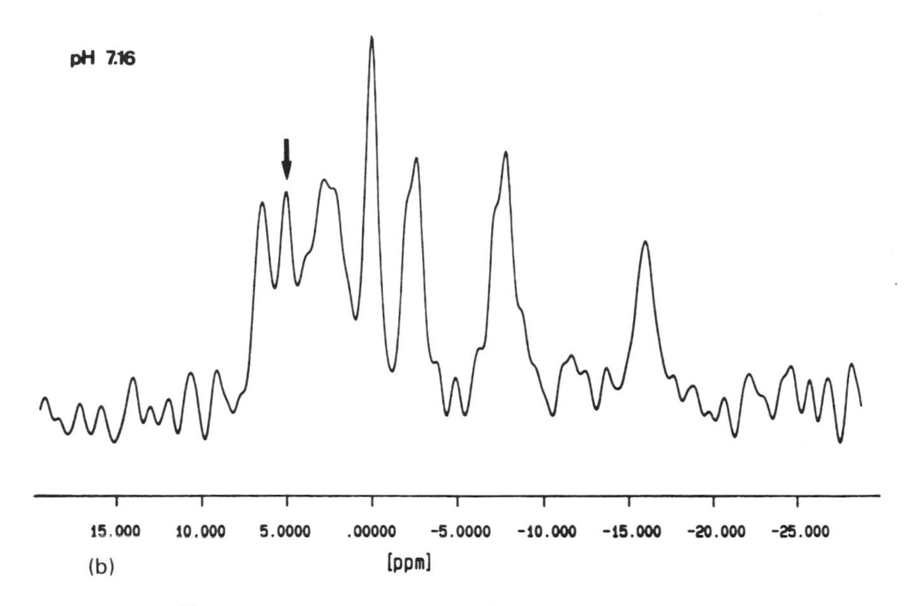

Fig. 10.2 (b) ^{31}P MR spectrum shows in particular the reduction of phosphodiesters, but also some decrease of the phosphocreatine signal as compared with healthy brain tissue. The tumorous pH (7.16) indicates alkaline environment.

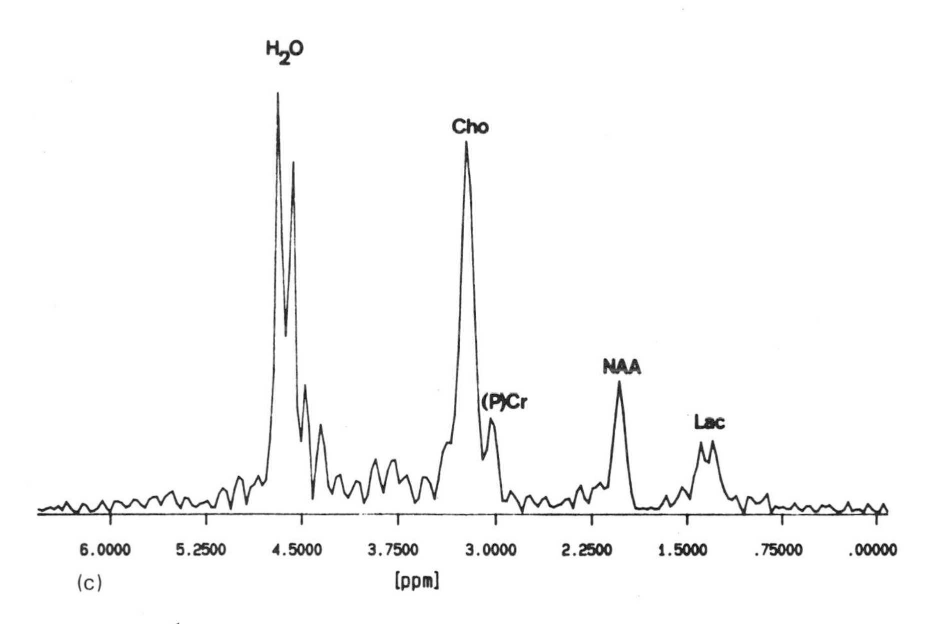

Fig. 10.2 (c) ^{1}H MR spectrum reveals an increase in the Cho/(P)Cr ratio, the decrease of the NAA signal and the appearance of a strong signal from lactic acid (1.3 p.p.m.).

localized in all dimensions within a meningioma has been performed. Therefore further studies are necessary to clarify if the appearance of a creatine signal only in the ^{31}P spectrum is due to partial volume effects of the volume selection technique. Other effects may be involved as well, as spectra localized in a large cystic component within a haemangioblastoma showed no observable resonances both with ^{1}H and ^{31}P.

10.4 CONCLUSIONS

The procedure for a combined MR imaging and ^{1}H and ^{31}P spectroscopic examination of the brain on a clinical MR system could be established. Spectra from normal human brain tissue demonstrated good reproducibility, though further investigations to determine the physiological variability appear to be necessary. Spectra from human brain tumours revealed unambiguous differences. Volume-selective ^{1}H MRS can demonstrate spectral patterns for brain tumours of different histology, as it has been shown for localized ^{31}P MRS in a large number of patients (Heindel *et al.*, 1988; Shoubridge *et al.*, 1988). Combination of ^{31}P MRS and ^{1}H MRS revealed complementary metabolic information on tumorous tissue (e.g. alkaline pH in spite of increased lactic acid). However, it cannot be deduced from the preliminary data of the present study whether the diagnostic specificity of MR can be further improved by the spectroscopic criteria.

ACKNOWLEDGEMENTS

We thank Professor Dr W.-D Heiss (Department of Neurology), Professor Dr N. Klug (Department of Neurosurgery and coworkers for their kind cooperation. This study was supported by the Deutsche Forschungsgemeinschaft (DFG).

REFERENCES

Heindel, W., Bünke, J., Glethe, *et al.* (1988) *J. Comput. Assist. Tomogr.*, **12**, 907.

Hossmann, K.A., Miess, G., Paschen, W., *et al.* (1986) *Acta Neuropathol.* **69**, 139.

Luyten, P.R., Groen, J.P., Vermeulen, J.W.A.H., *et al.* (1989) *Magn. Res. Med.*, **11**, 1.

Ordidge, R.J., Connelly, A., Lohmann, J.A.B., *et al.* (1986) *J. Magn. Res.*, **66**, 283.

Petroff, O.A.C., Prichard, J.W., Behar, K.L., *et al.* (1985) *Neurology*, **35**, 781.

Shoubridge, E., Arnold, D.L. Emrich, J.F., *et al.* (1988) *SMRM (Abstr.)*, **2**, 616.

11 Neuroimaging study of frontal lobes

G.L. LENZI, A. PADOVANI, P. PANTANO, M. IACOBONI, M. RICCI, P. FRANCO AND V. DI PIERO
Neurological Clinic, Department of Neurological Sciences, University 'La Sapienza', Rome, Italy

It has been known for a long time that the functional activity of brain structures and the blood flow to them are closely and directly related to one another. However, only recently has this relationship been well substantiated. Plausible hypotheses can now be explored concerning changes in one variable on the basis of detectable changes in another, even within small regions of the brain.

By the application of neuroimaging techniques, it is in fact possible to assess either cerebral blood flow (CBF) or metabolism and, through these, levels of neuronal activity in the depths of the human brain. The familiarity of neurologists with 'sliced' representation of the brain, as achieved with computed tomography (CT) scan, allows us to minimize the description of the single-photon (SPECT) and positron emission tomography (PET) techniques.

PET is the tomographic representation of tissue radioactivity due to positron-emitting isotopes which decay by emitting a positively charged electron. On encountering an electron, this decays, releasing two photons which travel in precisely opposite directions. The two photons will impact detectors appropriately set in opposed banks around the brain. The detectors may be positioned differently around the source in order to achieve the necessary angular data for the computer reconstruction which, through processes of back projection and filtering, provides the trans-axial distribution of radioactivity responsible for the recorded events. The inherent spatial resolution of the detector system defines the details with which the distribution of radioactivity is measured and represented.

SPECT utilizes, on the other hand, a signal produced by the excitation of a γ-ray detector by photons emitted by radioisotopes during their natural decay. The event is thus the arrival of the γ-ray over the detecting crystal. The localization of the photon source is achieved through a lead collimator placed in front of the crystal. The single-photon emitter radioisotopes may be natural or artificially produced by accelerators or cyclotrons. In contrast, all the positron emitter isotopes are cyclotron produced. Different types of radioisotope are now available for clinical studies.

The characteristics of the radioisotope in action determines the explored function. The temporal profile of the radioisotope distribution in the different brain areas, however, is fundamental to understanding how a specific cerebral function is explored.

The tomography reconstruction procedures utilized for these technologies require that an adequate number of 'events' is recorded for each tomographic slice, otherwise the statistical variability of the reconstructions might affect the significance of the results. A worthwhile number, in terms of evidence, is in the order of a hundred thousand 'events' for each slice, although evidence which would be useful statistically would require millions of 'events' for each slice.

Because of this indispensable requirement, two problems arise. First of all, we have to deal with the radioisotope dosage, i.e. with the issue of dosimetry, target organ and ethics, in particular when rest–activation–control protocols are performed on humans.

In addition, time is an important issue, i.e. the examination period required in order to record a sufficient number of 'events'.

All through this period, the 'to be explored' cerebral function needs to maintain a steady state regarding a specific region, since the final neuroimages will represent the sum of all the recorded events in that specific region.

These limiting aspects of the emission tomography techniques as well as of the physiological models employed in the study of cerebral functions with specific radioisotopes were certainly substantial until the early 1980s. Recently, however, 'fast' emission tomography techniques regarding CBF have been developed and activation studies have been resumed, as the examination period has been shortened to a few seconds.

In addition, the utilization of brief emission isotopes, as at PET, thus allows one to subtract a rest image recorded earlier from the image 'in activation', during the same examination session.

There are correlation models in progress by which the activity rate changes of specific pixels (brain areas) can be correlated to changes of more distant pixels. The model in question, although not yet fully tested, may represent a helpful tool in order to investigate the neuropsychological functions of the brain, in particular of the frontal lobes.

As previously stated, the cerebral functions that may be investigated by emission tomography techniques depend on the employed isotopes and models. The functions investigated by SPECT consist basically of CBF through isotopes like xenon-133 and technetium-99m, whereas the functions investigated by PET are theoretically endless. In fact, positron emitting isotopes like carbon (^{11}C), oxygen (^{15}O) and nitrogen (^{13}N) are normally present naturally. These isotopes can be utilized either by themselves or linked to a more complex molecule like [^{11}C]glucose, [^{11}C]methyl-tyroxine, and so on, the latter maintaining its own metabolic pathways.

This, in addition to the fact that each neurone consumes moment by moment an amount of glucose and oxygen in accordance with its functional level of activity, is largely responsible for attracting the interest of the neuroscientist. In fact, the higher the neurone discharge frequency, the higher is the substrate and blood flow demand.

The utilization of such a physiological relationship which links function to metabolism and metabolism to blood flow at a microregional level enables us thus to derive a neuronal functional activity from flow or metabolic rate changes. This was utilized recently to define sensorial pathways like the auditory and visual ones using very simple stimuli (Mazziotta *et al.*, 1982).

More rcently, more integrated functions like those of the frontal lobes have been explored. With regard to this, the collected data are so far interesting but merely descriptive. Typical is the case report (EVR) described by Eslinger and Damasio (1985) utilizing SPECT. The neuroimages show a wide lesion affecting frontal lobes bilaterally, except the left dorsolateral prefrontal cortex. The patient, despite the lesion, was cognitively adequate with regard to formal neuropsychological disturbances. Interestingly the patient was completely unable to cope with daily life situations, suggesting that EVR 'was unable either to analyse or to integrate the premises of a real-life problem as a consequence of 'impaired evocation of contextual (episodic) memories for social behaviour'.

Such a subject is without doubt very interesting at the neuropsychological level. However, the SPECT study performed actually confirms the nuclear magnetic resonance findings showing only that a CBF reduction occurred in those areas affected by anatomical damage. This descriptive approach, although impressive, simplifies emission tomography applications.

More interesting is the question of whether or not CBF and/or metabolism changes occur in morphologically intact areas, and even more interesting is the question of whether specific CBF and/or metabolism changes occur as a consequence of specific function activation. It is not surprising therefore that in diffuse degenerative diseases the atrophy rate has been correlated with a cortical CBF and metabolic reduction, and severity of clinical signs with the CBF and metabolic changes (Frackowiak *et al.*, 1981).

On the same line an expected pattern of reduced CBF and metabolism, in particular of the frontal lobes, has been observed in Pick's disease, which cause a marked frontal–temporal cortical atrophy. However, Alzheimer's-type dementia and Pick's disease represent typical diffuse cortical morphological alterations which are directly related to the metabolic changes observed.

The questions described earlier originated after the demonstration of a diaschisis phenomenon of reduced CBF and metabolism occurring in non-contiguous areas which are, however, functionally connected to the primary lesion and in particular to frontal areas (Lenzi, 1982). With regard to this, an important requirement is that the areas affected by diaschisis have to be anatomically normal.

Interestingly, data collected in patients with subcortical dementias cognitively characterized by a 'frontal syndrome' have shown significant reduction of CBF and metabolism at the frontal cortex level. A frontal lobe reduction of metabolism was observed in fact in progressive supranuclear palsy and recently significant correlations have been demonstrated between neuropsychological deficit severity and frontal metabolic levels (Leenders *et al.*, 1988).

Although less consistently, frontal lobe metabolic and CBF decrease was observed in cognitively impaired Parkinson's patients: in particular, a typical pattern at the prefrontal level was demonstrated in Parkinson's patients with depression. Activation studies utilizing tasks involving specifically the prefrontal cortex have confirmed the prefrontal dysfunctions in Parkinson's disease.

Huntington's disease is also understood to be a classical subcortical dementia. Among others, neuropsychological deficits suggesting frontal lobe dysfunction have been documented. In the past, many authors have failed to demonstrate any changes of CBF and metabolism at the frontal lobe level but a typical dorsolateral prefrontal CBF pattern was observed in Huntington's patients during activation. A frontal CBF reduction at rest was recently demonstrated (Hasselbach *et al.*, 1989).

Such findings obtained in patients affected by diseases not directly involving cortical structures are in accordance with the diaschisis–disconnection theory and confirm previous observations collected in patients with deep cerebrovascular lesions.

The correlation between clinical course, as assessed by neuropsychological evaluation, and neuroimaging data in patients with frontal diaschisis has been studied by many groups using SPECT and PET. Patients with deep cerebrovascular lesions showed a marked CBF and metabolism reduction in areas anatomically intact: a clear correlation between site of diaschisis effect and symptoms was demonstrated, in particular in patients with Broca's aphasia (Metter *et al.*, 1989).

Typical prefrontal CBF and metabolic patterns have been observed in mental disorders characterized by frontal signs even if there was no anatomical evidence of lesion.

Depression, either unipolar or bipolar, has been repeatedly correlated to a dorsolateral prefrontal metabolism reduction. A correlation between depression severity and metabolic level at the dorsolateral prefrontal has been also demonstrated as well as metabolic improvements in the recovery phase (Baxter *et al.*, 1989).

CBF and metabolism have been observed in schizophrenia as characterized by a hypofrontality pattern. Recently, activation procedures have been carried out in schizophrenics. In comparison with the controls, there was no increase of CBF in the dorsolateral prefrontal cortex during activation using a test widely adopted as selective of frontal function. The performance on this test was found to correlate with the CBF level (Weinberger *et al.*, 1986).

In conclusion, although the progress and the application of functional neuroimaging do not enable us to understand the aetiology of the presented diseases, we can now utilize non-invasive measures which reflect regional functions and adopt procedures which are either neuropsychological or, more suggestively, pharmacological, which may demonstrate changes of those functions explored. Unfortunately, most of these data are very recent and obtained with 'first'-generation technical devices with significant methodological limitations. Therefore, further studies are needed to confirm and to complete the present observations in this field.

REFERENCES

Baxter, L.R., Schwartz, J.M., Phelps, M.E. *et al.* (1989) *Arch. Gen. Psychiat.*, **46**, 243–50.

Eslinger, P.J. and Damasio, A.R. (1985) *Neurology*, **35**, 1731–41.

Frackowiak, R.S.J., Pozzilli, C., Legg, N.J. *et al.* (1981) *Brain*, **104**, 753–78.

Hasselbach, S., Andersen, A., Sorensen, S., *et al.* (1989) *J. Cereb. Blood Flow Metab.*, **9** (suppl. 1), S350

Leenders, K.L., Frackowiak, R.J.S. and Lees, A.J. (1988) *Brain*, **111**, 615–30.

Lenzi, G.L. (1982) *Arch. Ital. Biol.*, **120**, 189–200.

Mazziotta, J.C., Phelps, M.E., Carson, R.E. and Kuhl, D.E. (1982) *Neurology*, **32**, 921–37.

Metter, E.J., Kempler, D., Jackson, C. *et al.* (1989) *Arch. Neurol.*, **46**, 27–34.

Weinberger, D.R., Berman, K.F., and Zec, R.F. (1986) *Arch. Gen. Psychiat.*, **43**, 114–24.

PART TWO

Neuropathies, Myopathies and other Dysfunctions

12 Biochemical indicators of ischaemic brain damage

H. SOMER AND R.O. ROINE
Department of Neurology, University of Helsinki, Finland

12.1 INTRODUCTION

Laboratory methods such as lactate and pyruvate measurements and various serum enzyme tests have been in routine clinical use to reveal acute ischaemic injury of the heart or the skeletal muscle. The same enzymes, such as aspartate aminotransferase (ASAT), lactate dehydrogenase (LD) and creatine kinase (CK), occur abundantly also in the brain, and there should be a clear analogy for their use in the clinical research of ischaemic brain damage. Various problems, like insensitivity in the methods in cerebrospinal fluid (CSF) analysis, non-specificity of the markers regarding the tissue of origin (brain, erythrocytes, leucocytes), problems concerning sample taking and its timing, and the historical inclination of neuroscientists to use neurophysiological, radiological and morphological methods rather than biochemical ones, have hampered widespread use of biochemical markers for this purpose.

In this review we summarize some of the recent findings and solutions to methodological problems in clinical research on ischaemic brain damage (Table 12.1).

12.2 POSSIBLE INDICATORS

12.1.1 Lactate

Because glucose and oxygen are the main substrates for energy production in brain, hypoxaemia is accompanied by changes in cerebral carbohydrate and energy metabolism. Under anaerobic conditions, glycolysis is the only source of adenosine triphosphate (ATP) production, with lactate as the main product. Glycolysis is accelerated, which results in increased

Table 12.1 Biochemical indicators of ischaemic brain damage

Indicators	*Methods*	*Source of analysis*	*Time scale*	*Comments*
Lactate	Chemical	CSF	A few hours to a few days	Correlation with global ischaemia Does not predict prognosis Non-specific
Hypoxanthine	Fluorometry HPLC	CSF	1–20 h	Correlation with ischaemia and brain damage
Creatine kinase (CK)	Enzymatic (spectrophotometry, fluorometry, bioluminescence)	CSF	24–72 h	Good correlation with ischaemic damage in experimental and clinical studies
Creatine kinase BB isoenzyme (CK-BB)	Electrophoresis + Fluorometry chromatography, RIA	serum	A few hours	Suggestive evidence in some studies
		CSF	24–72 h	Good correlation in experimental and clinical studies
Aspartate amino-transferase (ASAT)	Spectrophotometry	CSF	24–72 h	Some correlation with ischaemic damage in experimental and clinical studies
Lactate dehydrogenase (LD)	Spectrophotometry	CSF	24–72 h	Some correlation with ischaemic damage in experimental and clinical studies
Neurone-specific enolase (NSE)	RIA	serum	A few days	Some correlation with ischaemic damage, moderate correlation in severe global ischaemic damage
		CSF	24–72 h	Good correlation in experimental and clinical ischaemic brain damage
S-100 protein	RIA EIA	CSF	24–72 h	Good correlation in experimental and clinical ischaemic brain damage

HPLC, high-performance liquid chromatography; RIA, radioimmunoassay; EIA, enzyme immunoassay.

glucose utilization and lactate accumulation. Measurement of lactate is a standard laboratory procedure and can be applied to CSF samples from clinical material as well. Increased CSF lactate does not necessarily reflect the situation in the brain, but may be derived from leucocytes present in inflammatory conditions, or it may be caused by generalized tissue hypoxia due to poor ventilation or asphyxia. In severe brain ischaemia, CSF lactate increases within hours (Siesjö, 1981; Bøhmer *et al.*, 1983).

12.2.2 Hypoxanthine

In hypoxic brain ATP stores are rapidly depleted and adenosine monophosphate (AMP) concentration is increased. Its degradation products are converted to hypoxanthine, which is rapidly released into the CSF (Saugstad *et al.*, 1988). In severe brain ischaemia hypoxanthine levels increase up to tenfold within a few hours, but usually return to normal within 24 h (Bøhmer *et al.*, 1983).

12.2.3 Creatine kinase

CK, (EC 2.7.3.2) is an enzyme which occurs mainly in skeletal muscle, heart and brain. It has the molecular weight of 81 kDa. The enzyme is structurally a dimer composed of either B-subunits (brain), or M-subunits (muscle); heart tissue has a substantial proportion of its CK activity as the MB form (Dawson and Fine, 1967). These isoenzymes occur exclusively in the cytoplasm. A separate isoenzyme occurs in the mitochondria (Wevers *et al.*, 1981). In the brain most of the CK exists as the CK-BB isoenzyme and is localized mainly in the astrocytes (Thompson *et al.*, 1980).

Herschkowitz and Cumings (1964) reported increased CK activities in the CSF in some neurological conditions, but their diagnostic value remained unknown for several years. CK measurements of CSF were problematic since the enzyme activity was usually rather low, and the enzyme activity was easily inactivated, possibly through several mechanisms (Urdal and Strømme, 1985). Traditional spectrophotometric assays for CSF total CK activity have, however, been successfully used in clinical studies (Vaagenes *et al.*, 1980, 1986). A greater sensitivity is obtained by bioluminescent assay (Halonen *et al.*, 1982), which, however, needs special equipment. Several radioimmunoassays have been developed for CK-BB measurement (Bell *et al.*, 1978; Jackson *et al.*, 1984). They usually provide sufficient sensitivity and escape the problem of enzyme inactivation. The antibodies usually recognize the B-subunit also present in the MB-isoenzyme, at least to some extent. This is not a problem in CSF analysis since most of the CK is composed of

CK-BB isoenzyme, but causes substantial limitations if serum samples are analysed with this methodology.

12.2.4 Neurone-specific enolase

Enolase (EC 4.2.1.11) is an enzyme involved in glycolytic metabolism. There are multiple isoenzymes of enolase, which are dimers with molecular weights from 77 to 87 kDa, formed from three subunits (α, β and γ). The γ-subunit of enolase is considerably enriched in whole brain, and α–γ and γ–γ enolase isoenzymes represent about 2% of the total soluble protein in the whole brain. Neurone-specific enolase (NSE) is defined as the γ-subunit of enolase. It is localized in the cytoplasm of the neurones. No other tissue contains more than 3% of the brain levels of γ–γ enolase or 13% of the levels of the dimer α–γ. NSE can be measured by a sensitive radioimmunoassay (Mokuno *et al.*, 1983).

12.2.5 S-100 protein

S-100 is an acidic calcium-binding protein (Calissano *et al.*, 1969), which is functionally related to a family of calcium-binding proteins. It is a dimer with a molecular weight of 21 kDa. The α–α is present exclusively in the neurones, while β–β and α–β forms occur mainly in the glial cells (Isobe *et al.*, 1984). The intracellular location is mainly cytoplasmic. S-100 protein can be measured in the CSF by radioimmunoassay (Persson *et al.*, 1987) or by enzyme immunoassay (Kato *et al.*, 1982).

12.3 CEREBROSPINAL FLUID OR SERUM?

CSF represents an extension of brain extracellular fluid and supposedly reflects brain metabolism better than blood does. Experimental and clinical studies have shown that CSF lactate and hypoxanthine are increased within hours of the onset of hypoxia (Bøhmer *et al.*, 1983; Vaagenes *et al.*, 1988), but may return to normal within 24 h. Blood lactate or hypoxanthine values usually illustrate general tissue hypoxia and cannot be taken as specific measures of cerebral hypoxia.

CK, neurone-specific enolase and S-100 protein reach their peak values in 24–72 h in experimental studies (Steinberg *et al.*, 1984; Vaagenes *et al.* 1988; Hårdemark, 1988). Two or three serial samples give a better estimate of the extent of hypoxic brain injury (Hårdemark, 1988) but this is of course difficult to obtain except when a ventricular shunt is installed.

Lumbar puncture is not a routine diagnostic procedure in ischaemic brain disorder and it cannot be performed repeatedly in the same patient.

A biochemical indicator measurable from serum samples would be extremely useful. The blood–brain barrier is normally impermeable to molecules of the size of CK. Several enzymes, including CK, have been reported to be increased in the serum after ischaemic brain disorder (Dubo *et al.*, 1967; Wolintz *et al.*, 1969), although questions remained concerning the origin of these serum CK elevations. The development of new, more specific isoenzyme methods (Somer and Konttinen, 1972) created new possibilities for examining the problem. Subsequent experimental and clinical observation confirmed that the CK elevations were derived mainly from extracerebral sources, but some CK-BB was indeed released from brain to peripheral blood (Somer *et al.*, 1975). The appearance of CK-BB isoenzyme in blood was found to correlate with severe prognosis (Kaste *et al.*, 1977). Unfortunately, CK-BB isoenzyme disappears during the first 24 h from the serum in almost all cases. This is apparently due to a much shorter half-life of the CK-BB isoenzyme, compared to the other isoenzymes (Roberts and Sobel, 1978). If the CK-B subunit is measured by radioimmunoassay, it can usually be detected for a longer period in the serum and shows some correlation to the prognosis (Bell *et al.*, 1978). There are, however, some unexplained variations (Pfeiffer *et al.*, 1983), which are probably due to the fact that especially subarachnoid haemorrhage, but also other acute cerebral catastrophes, are associated with obvious myocardial lesions (Kaste *et al.*, 1978), causing elevation of CK-MB isoenzyme in the serum.

Increased values of serum NSE have been reported in some studies concerning ischaemic brain damage (Persson *et al.*, 1987; Roine *et al.*, 1989). Reports on S-100 protein are limited in this respect (Persson *et al.*, 1987).

The present data suggest that biochemical consequences of ischaemic brain damage can be monitored better from CSF than from serum samples. There is, however, substantial evidence to show that the blood–brain barrier is disrupted in severe ischaemic brain damage, at least for a short period. Intracellular soluble compounds are then released into the circulation. At the moment, we do not have satisfactory methods for monitoring brain damage from serum samples in clinical trials. The situation may change if new brain-specific markers are found, particularly if they can be measured in serum for suitable periods with sensitive methods.

12.4 EXPERIMENTAL STUDIES

Dynamics of intracellular enzyme release from brain to CSF and serum have been clarified repeatedly (Wakim and Fleisher, 1956; Somer *et al.*, 1975; Maas, 1977a, 1977b). Steinberg *et al.* (1984) applied the four-vessel occlusion model to study the consequences of 10, 20 and 30 min total ischaemia on CSF NSE levels. There was a correlation between the duration of the ischaemia and the degree of NSE elevation. Maximal elevations, averaging ninefold compared to basal values, were observed at 6–24 h. NSE levels remained elevated up to 192 h. The authors conclude that NSE release is a continuous process lasting several days, with a clearance rate of 4 h. Onset of ischaemia was followed by transient changes in exploration behaviour and neurological state of the animals that were no longer visible 24 h later, Histological observations 3 days after the ischaemia showed neuronal loss as well as neuronal damage in several forebrain areas.

Vaagenes *et al.*, (1988) applied two models where cardiac arrest was caused in dogs either by ventricular fibrillation or by asphyxiation. Several enzymes, CK, ASAT and LD were measured in the CSF. Peak activity of enzymes in CSF at 48–72 h post-arrest correlated with outcome, and CK was the best predictor. Brain histopathological damage score at autopsy 96 h post-arrest correlated with CK level in CSF (r = 0.79), and with neurological deficit (r = 0.70). Enzymatic analysis of the brain autopsied at 6 h showed that the corresponding enzymes were decreased in areas with worst histological damage (grey matter of neocortex, hippocampus, nucleus caudatus, cerebellum). Enzymatic analysis at later observation points showed a further decline in brain CK activity. The authors conclude CSF enzyme analysis to be 'a chemical brain biopsy method' that is useful in predicting outcome from brain damage caused by cardiac arrest.

Enzyme release from local infarct area has been recently studied in the dog by producing embolic stroke and measuring CSF CK-BB, and in the rat by occluding the middle cerebral artery and measuring NSE in the CSF obtained either through a subdural or an extradural implantation technique (Hårdemark, *et al.*, 1988). Both studies showed a severalfold increase of the enzyme as compared to the basal values with the same time scale (1–3 days). Significant correlations were detected between the relative size of the infarct and the markers (CK-BB, r = 0.94; NSE, r = 0.97). The same model was used to demonstrate the release of S-100 protein to CSF. Both markers showed a fairly similar release pattern in ischaemic lesions. The release was much slower in ischaemic as compared to experimental cortical contusion (Hårdemark, 1988).

12.5 CLINICAL STUDIES IN GLOBAL CEREBRAL ISCHAEMIA

Kjekshus *et al.* (1980) found increased CK activity in the CSF in patients after cardiac resuscitation. They also tried to correlate the CSF CK values to clinical outcome and to histopathological changes of those 40 patients who died more than 30 days after successful resuscitation (Vaagenes *et al.*, 1980). A curvilinear relationship was found between maximum CSF CK activity and the extent and magnitude of the defined brain damage. No cerebral damage was found when maximum CK activity remained below 4 U/l. The frontal cortex was less severely damaged than other regions examined. CSF CK increased to a range of 4–10 U/l in patients without damage to the frontal cortex but with slight to severe damage in the other regions. In deeply comatose patients CK activity always exceeded 10 U/l and neuronal necrosis was found in all regions. Longstreth *et al.* (1981) confirmed the finding of increased CK activity after cardiac arrest. In a prospective study Longstreth *et al.* (1984) found that patients who never awoke had significantly higher CK values than those awakening. Using a cut-off of 25 U/l only 5 out of 29 patients were misclassified.

We have studied 75 consecutive patients who were resuscitated from out-of-hospital ventricular fibrillation by a mobile intensive care unit. The mean duration of the circulatory arrest was 5.2 min and the mean duration of cardiopulmonary resuscitation was 15.2 min. All patients were treated according to a standard protocol and neurological examination, including Glasgow coma scale, was carried out by the same neurologist at 1 h, 24 h and 1 week after cardiac arrest. Patients were divided into two groups. Those who recovered consciousness as defined by the ability to obey command (40) constituted group 1, and those who remained unconscious (26) constituted group 2. CSF was obtained by lumbar puncture approximately 24 h (range 20–26 h) after cardiac arrest. CK-BB and NSE were analysed from CSF and serum samples by radioimmunoassay methods.

Patients who remained unconscious had a mean CSF NSE value of 99.7 ± 39.7 ng/ml, whereas those who recovered consciousness had NSE of 10.7 ± 0.7 ng/ml (Fig. 12.1), which is also higher than that of the controls (6.5 ± 0.5 ng/ml; mean ± SEM). All patients with NSE concentration higher than 24 ng/ml remained unconscious and died. When this level is taken as a cut-off value, the test has a sensitivity of 74%, specificity and positive predictive value of 100%, and negative predictive value of 89% in detecting patients who did not recover. CK-BB showed very similar results (Fig. 12.1). Patients who never recovered consciousness had CK-BB concentration of 96.8 ± 24.4 ng/ml, as compared to those who recovered, 6.7 ± 1.2 ng/ml, and controls 0.7 ± 0.1 ng/ml. The correlation between NSE and CK-BB CSF values was 0.84.

Serum NSE was usually increased in patients who never recovered consciousness, whereas those patients who recovered consciousness had significantly lower NSE levels, usually within the normal range

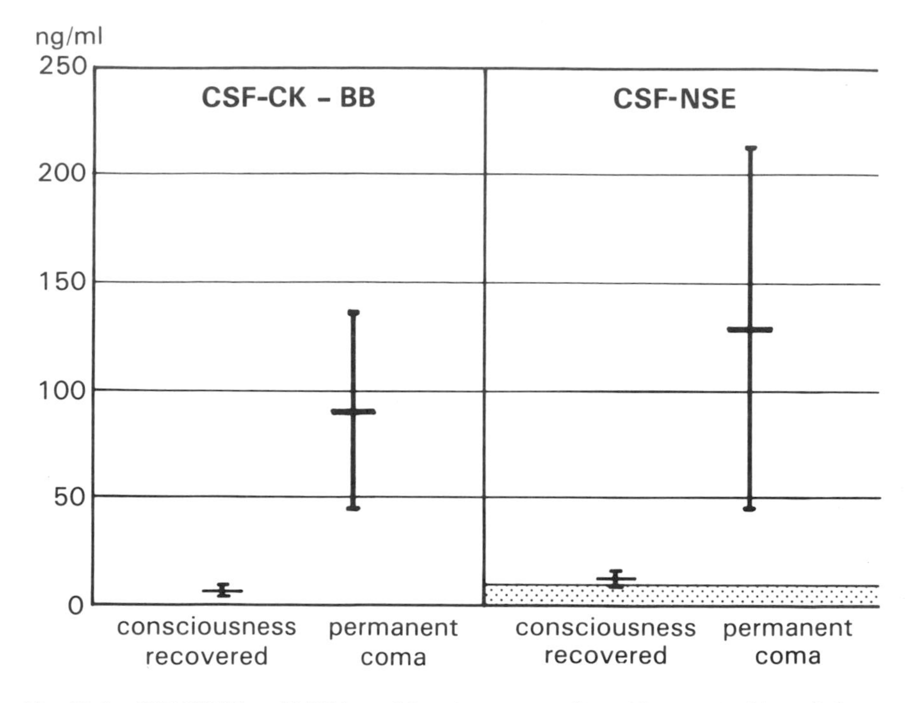

Fig. 12.1 CSF CK-BB and NSE from 66 patients resuscitated from out-of-hospital ventricular fibrillation. Forty patients recovered consciousness; 26 remained in permanent coma and died. Results are expressed as 95% confidence intervals of the means for both groups. Shaded area illustrates the reference range. Results are based on the work of Roine *et al.* (1989).

(Fig. 12.2). If an arbitrary cut-off value of 17 ng/ml is chosen, the test has a sensitivity of 40%, specificity of 98%, a positive predictive value of 98% and a negative predictive value of 79% in detecting patients who did not recover consciousness. Serum NSE correlated with CSF NSE concentration (Fig. 12.3). The results suggest that serum NSE measurement has predictive value, although it does not have the same discriminatory value as CSF NSE assay.

Serum CK-BB was increased in group 1 (20.5 ± 3.5 ng/ml, controls 10 ng/ml), and even more so in group 2 (39.1 ± 8.2 ng/ml, but no cut-off value could be established between the two groups. Serum CK-BB did not correlate with CSF CK-BB level. The negative results obtained by serum CK-BB measurement are easily explained by the fact that all the patients underwent resuscitation; myocardial infarction was verified in the majority. This was responsible for the release of CK-BB isoenzyme into the serum.

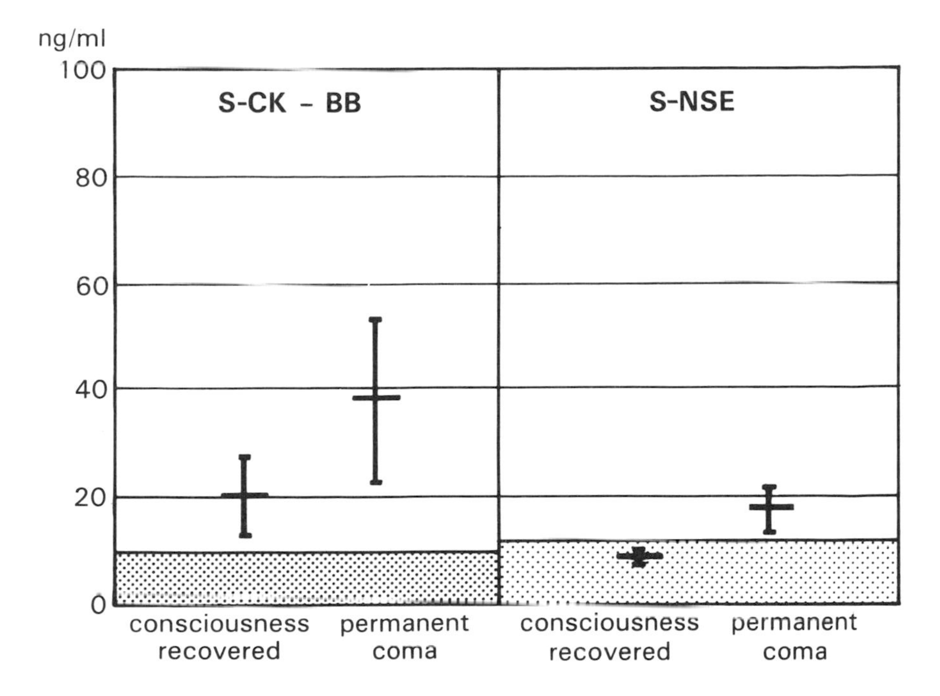

Fig 12.2 Serum CK-BB and NSE from 66 patients resuscitated from out-of-hospital ventricular fibrillation. Forty patients recovered consciousness; 26 remained in permanent coma and died. Results are expressed as 95% confidence intervals of the means for both groups. Shaded area illustrates the reference range. Results are based on the work of Roine *et al.* (1989).

Clinical correlation between Glasgow coma scale at 1 day and 7 days and CSF results with both NSE and CK-BB showed correlations from 0.63 to 0.75 (Spearman rank correlation coefficients). Similar correlations were obtained with Glasgow outcome scale. Serum NSE showed a correlation of 0.40 at the maximum, but serum CK-BB showed no correlation. The results do not justify the use of serum NSE or CK-BB in the clinical decision-making process. The recovery of consciousness and survival of cardiac arrest victims can be predicted reliably by the measurement of NSE or CK-BB in the CSF 24 h after resuscitation.

12.6 CLINICAL STUDIES IN ISCHAEMIC STROKE

Various enzymes have been measured in the CSF in patients with brain infarction (Green *et al.*, 1957; Wolintz *et al.*, 1969; Mokuno *et al.*, 1983; Royds *et al.*, 1983). The results usually show a wide scattering of results.

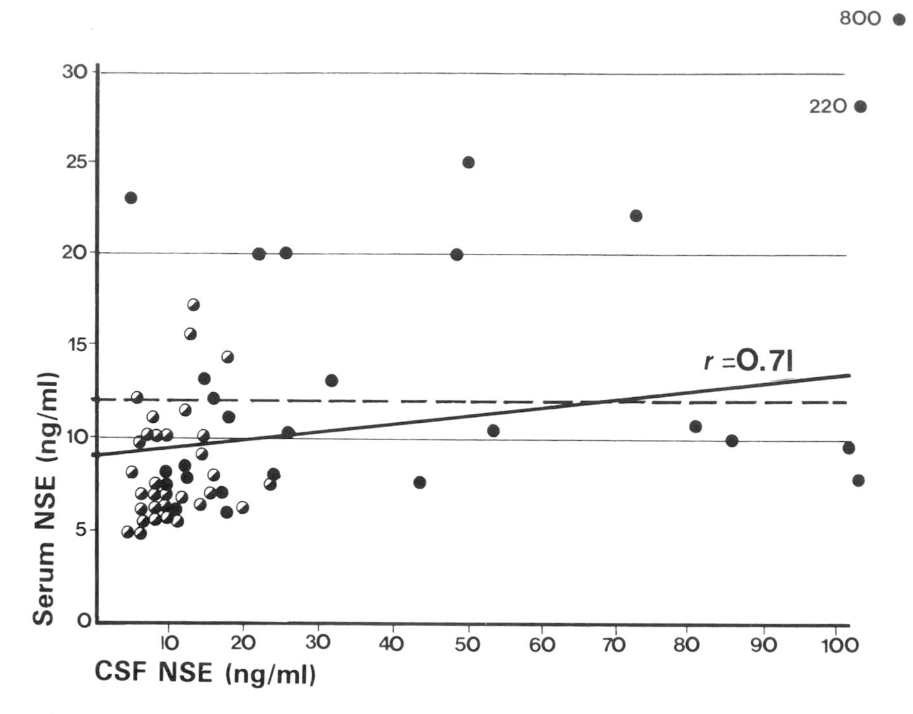

Fig 12.3 Correlation between serum and CSF NSE concentration (Pearson's correlation coefficient, r = 0.71). Results are based on the work of Roine *et al.* (1989).

If a panel of neurological diseases have been studied, highest values have usually been detected in the group of stroke patients, but the majority of stroke patients have similar levels to those with other neurological disorders. As computed tomography (CT) and nuclear magnetic resonance imaging became the first methods of choice in diagnosis of focal cerebrovascular diseases, the clinical values of CSF enzyme measurements has remained unknown.

Vaagenes *et al.* (1986) have re-evaluated the issue by measuring CK, CK-BB, LD and ASAT activities in the CSF of 35 patients with acute stroke. Patients were divided into three groups: deeply comatose (8), patients with residual motor defects (20), and patients with transient minor dysfunction (7). The level of CK-BB seemed to discriminate best between the various neurological outcome groups, whereas LD level seemed slightly more sensitive in reflecting minor transient disturbances. Total CK activity was mainly of CK-BB activity, but some CK-MM activity was occasionally detected, reflecting possible disruption of the blood–brain barrier. There was some overlap between the first two groups with these best markers, but the results clearly show a correlation between CSF results and the clinical outcome.

Hay *et al.* (1984) studied CSF enolase in 26 patients with stroke. CSF enolase level correlated with infarct volume (measured by CT of brain) and with clinical outcome. High CSF enolase was always accompanied with a poor prognosis. Persson *et al.* (1987) found increased concentrations of both NSE and S-100 in the CSF after large infarcts, whereas in small infarcts and transient ischaemic attacks only NSE was increased.

The value of serum enzyme measurements is still unsolved. CK-BB is relatively brain specific and could theoretically be an ideal marker, especially as CK isoenzyme methods are available in practically all hospitals, Although the appearance of CK-BB isoenzyme in serum was associated with grave prognosis, the isoenzyme was detectable for a few hours only. It is therefore difficult to organize proper sampling for diagnostic purposes or for research projects. NSE has not been studied systematically in serum samples of patients with stroke, but since only 40% of our patients with global ischaemic brain damage showed elevated values it appears likely that also this test is too insensitive in patients with stroke.

12.7 PERINATAL ASPHYXIA

Asphyxia is known to occur in a small percentage of deliveries and is recognized as an important cause of neurological impairment. For obvious reasons it is of utmost interest to recognize such cases and to develop methods to monitor the situation during delivery. This is even more important since traditional clinical methods – Apgar score and metabolic acidosis – have relatively poor predictive value, as shown again recently (Ruth and Raivio, 1988).

Traditional parameters like blood lactate have been used widely, but it is obvious that increased lactate level in umbilical cord blood correlates well with the duration of intrapartum asphyxia, but does not necessarily reflect the situation within the brain. CSF lactate is obviously a better candidate. Mathew *et al.* (1980) found increased CSF lactate in 55% of asphyxiated infants, provided the sample was taken within 8 h after birth, but not afterwards. CK-BB was therefore considered a potential biochemical indicator (Becker and Menzel, 1978), but recent results with 31 asphyxiated infants show that although the test predicts neonatal death it does not predict neurological damage among survivors (Ruth, 1989). Various other indicators of hypoxia, like arginine vasopressin, erythropoietin and hypoxanthine, have been shown to be good indicators of asphyxia at birth, but they do not predict neurological outcome (Ruth *et al.*, 1988).

REFERENCES

Becker, M. and Menzel, K. (1978) Brain-typical creatine kinase in the serum of newborn infants with perinatal brain damage. *Acta Paediatr. Scand.*, **67**, 177–80.

Bell, R.D., Rosenberg, R.N., Ting, R. *et al.* (1978) Creatine kinase BB isoenzyme levels by radioimmunoassay in patients with neurological disease. *Ann. Neurol.*, **3**, 52–9.

Bell, R.D., Alexander, G.M., Nguyen, T. and Albin, M.S. (1986) Quantification of cerebral infarct size by creatine kinase BB isoenzyme. *Stroke*, **17**, 254–60.

Bøhmer, T., Kjekshus, J. and Vaagenes, P. (1983) Biochemical indices of cerebral ischemic injury. *Scand. J. Clin. Lab. Invest.*, **43**, 261–5.

Calissano, P., Moore, B. and Friesen, A. (1969) Effect of calcium on S-100, a protein of the nervous system. *Biochemistry*, **8**, 4318–26.

Dawson, D.M. and Fine, I.H. (1967) Creatine kinase in human tissues. *Arch. Neurol.*, **16**, 175–80.

Dubo, H., Park, D.C., Pennington, R.J.T. *et al.* (1967) Serum creatine-kinase in cases of stroke, head injury, and meningitis. *Lancet*, **ii**, 743–8.

Green, J.B., Oldwurtel, H.A., O'Doherty, D.S. *et al.* (1957) Cerebrospinal fluid glutamic oxalacetic transaminase activity in neurologic disease. *Neurology*, **7**, 312–22.

Halonen, T., Näntö, V., Frey, H. and Lövgren, T. (1982) Bioluminescent assay of total and brain-specific creatine kinase activity in cerebrospinal fluid. *J. Neurochem.*, **39**, 36–43.

Hårdemark, H.-G. (1988) S-100 protein and neuron-specific enolase cerebrospinal fluid markers of brain damage: clinical and experimental studies. *Acta Univ. Upsaliensis.* Comprehensive summaries of Uppsala Dissertations from the Faculty of Medicine.

Hay, E., Royds, J., Davies-Jones, G. *et al.* (1984) Cerebrospinal fluid enolase in stroke. *J. Neurol. Neurosurg. Psychiat.*, **47**, 724–9.

Herschkowitz, N. and Cumings, J.N. (1964) Creatine kinase in cerebrospinal fluid. *J. Neurol. Neurosurg. Psychiat.* **27**, 247–50.

Isobe, T., Takahshi, K. and Okuyama, T. (1984) S-100 protein is present in neurons of central and peripheral nervous system. *J. Neurochem.*, **43**, 1494–6.

Jackson, A.P., Siddle, K. and Thompson, R.J. (1984) Two-site monoclonal antibody assays for human heart- and brain-type creatine kinase. *Clin. Chem.*, **30**, 1157–62.

Kaste, M., Somer, H. and Konttinen, A. (1977) Brain-type creatine kinase isoenzyme: occurrence in serum in acute cerebral disorders. *Arch. Neurol.*, **34**, 142–4.

Kaste, M., Somer, H. and Konttinen, H. (1978) Heart type creatine kinase isoenzyme (CK MB) in acute cerebral disorders. *Br. Heart J.*, **40**, 802–5.

Kato, K., Ishiguro, Y., Suzuki, F. *et al.* (1982) Distribution of nervous system specific forms of enolase in peripheral tissues. *Brain Res.*, **237**, 441–8.

Kjekshus, J.K., Vaagenes, P. and Hetland, O. (1980) Assessment of cerebral injury with spinal fluid creatine kinase (CSF-CK) in patients after cardiac resuscitation. *Scand. J. Clin. Lab. Invest.*, **40**, 437–44.

Longstreth, W.T. Jr, Clayson, K.J. and Sumi, M.S. (1981) Cerebrospinal

fluid and serum creatine kinase BB activity after out-of-hospital cardiac arrest. *Neurology*, **31**, 455–8.
Longstreth, W.T., Jr, Clauson, K.J., Chandler, W.L. and Sumi, S.M. (1984) Cerebrospinal fluid creatine kinase activity and neurologic recovery after cardiac arrest. *Neurology*, **34**, 834–7.
Maas, A.I.R. (1977a) Cerebrospinal fluid enzymes in acute brain injury. 1. Dynamics of changes in CSF enzyme activity after acute experimental brain injury. *J. Neurol. Neurosurg. Psychiat.*, **40**, 655–65.
Maas, A.I.R. (1977b) Cerebrospinal fluid enzymes in acute brain injury. 2. Relation of CSF enzyme activity to extent of brain injury. *J. Neurol. Neurosurg. Psychiat.*, **40**, 666–74.
Mathew, O.P., Bland, H., Boxerman, S.B. and James, E. (1980) CSF lactate levels in high risk neonates with and without asphyxia. *Pediatrics*, **66**, 224–7.
Mokuno, K., Kato, K., Kawai, K. *et al.* (1983) neuron-specific and S-100 protein levels in cerebrospinal fluid of patients with various neurological diseases. *J. Neurol. Sci.*, **60**, 443–51.
Persson, L., Hårdemark, H.-G., Gustafsson, J. *et al.* (1987) S-100 protein and neuron-specific enolase in cerebrospinal fluid and serum: markers of cell damage in human central nervous system. *Stroke*, **18**, 911–18.
Persson, L., Hårdemark, H.-G., Edner, G. *et al.* (1988) S-100 protein in cerebrospinal fluid of patients with subarachnoid hemorrhage: a potential marker of brain damage. *Acta Neurochir.*, **93**, 116–22.
Pfeiffer, F.E., Homburger, H.A. and Yanagihara, T. (1983) Creatine kinase BB isoenzyme in CSF in neurologic diseases: measurement by radioimmunoassay. *Arch. Neurol.*, **40**, 169–72.
Roberts, R. and Sobel, B.E. (1978) Creatine kinase isoenzymes in the assessment of heart disease. *Am. Heart J.*, **95**, 521–8.
Roine, R.O., Somer, H., Kaste, M. *et al.* (1989) Neurological outcome after out-of-hospital cardiac arrest: prediction by cerebrospinal fluid enzyme analysis. *Arch. Neurol.*, **46**, 753–6.
Royds, J.A., Davies-Jones, G.A.B., Lewtas, N.A. *et al.* (1983) Enolase isoenzymes in the cerebrospinal fluid of patients with diseases of the nervous system. *J. Neurol. Neurosurg. Psychiat.*, **46**, 1031–6.
Ruth, V.J. (1989) Prognostic value of creatine kinase BB-isoenzyme in high risk newborn infants. *Arch. Dis. Child.*, **64**, 563–8.
Ruth, V.J. and Raivio, K.O. (1988) Perinatal brain damage: predictive value of metabolic acidosis and the Apgar score. *Br. Med. J.*, **297**, 24–7.
Ruth, V.J., Autti-Rämö, I., Granström, M.-L. *et al.* (1988) Prediction of perinatal brain damage by cord plasma vasopressin, erythropoietin, and hypoxanthine. *J. Pediatr.*, **113**, 693–5.
Saugstad, O.D. (1988) Hypoxanthine as an indicator of hypoxia: its role in health and disease through free radical production. *Pediatr. Res.*, **23**, 143–50.
Siesjö, B.K. (1981) Cell damage in brain: a speculative synthesis. *J. Cereb. Blood Flow Metab.*, **1**, 155–82.
Somer, H. and Konttinen, A. (1972) Demonstration of serum creatine kinase isoenzymes by fluorescence technique. *Clin. Chim. Acta*, **40**, 133–8.
Somer, H., Kaste, M., Troupp, H. and Konttinen, A. (1975) Brain creatine kinase in blood after acute brain injury. *J. Neurol. Neurosurg. Psychiat.*, **38**, 572–6.

Steinberg, R., Gueniau, C., Scarna, H. *et al.* (1984) Experimental brain ischaemia: neuronspecific enolase level in cerebrospinal fluid as an index of neuronal damage. *J. Neurochem.*, **43**, 19–24.

Thompson, R.J., Kynoch, P.A.M. and Sarjant, J. (1980) Immunohistochemical localization of creative kinase-BB isoenzyme to astrocytes in human brain. *Brain Res.*, **201**, 723–6.

Urdal, P. and Strømme, J.H. (1985) Creatine kinase BB in cerebrospinal fluid and blood: methodology and possible clinical application. *Scand. J. Clin. Lab. Invest.*, **45**, 481–7.

Vaagenes, P., Kjekshus, J. and Torvik, A. (1980) The relationship between cerebrospinal fluid creatine kinase and morphologic changes in the brain after transient cardiac arrest. *Circulation*, **61**, 1194–9.

Vaagenes, P., Urdal, P. Melvoll, R. and Valnes, K. (1986) Enzyme level changes in the cerebrospinal fluid of patients with acute stroke. *Arch. Neurol.*, **43**, 357–62.

Vaagenes, P., Safar, P., Diven, W. *et al.* (1988) Brain enzyme levels in CSF after cardiac arrest and resuscitation in dogs: markers of damage and predictors of outcome. *J. Cereb. Blood Flow Metab.*, **8**, 262–75.

Wakim, K.G. and Fleisher, G.A. (1956) The effect of experimental cerebral infarction on transaminase activity in serum, cerebrospinal fluid and infarcted tissue. *Proc. Mayo Clin.*, **31**, 391–9.

Wevers, R.A., Reutelingsperger, C.P.M., Dam, B. and Soons, J.B.J. (1981) Mitochondrial creatine kinase (EC 2.7.3.2) in the brain. *Clin. Chim. Acta*, **119**, 209–23.

Wolintz, A.H., Jacobs, L.D., Cristoff, N. *et al.* (1969) Serum and cerebrospinal fluid enzymes in cerebrovascular disease: creatine phosphokinase, aldolase and lactic dehydrogenase. *Arch. Neurol.*, **20**, 54–61.

13 Molecular basis of astroglial reaction to brain injury: regulation of glial fibrillary acidic protein mRNA in rat cerebral cortex and in primary astroglial cell cultures

D.F. CONDORELLI AND A.M. GIUFFRIDA STELLA
Institute of Biochemistry, Faculty of Medicine, University of Catania, Viale A. Doria 6, 95125 Catania, Italy

13.1 INTRODUCTION

In response to CNS injury astrocytes increase in size and show a substantial increase in the number of intermediate glial filaments (reactive astrocytes).

The glial fibrillary acidic protein (GFAP) is the major protein of intermediate filaments of astroglial cells (Eng *et al.*, 1971) and the rapid increase in GFAP immunoreactivity which follows neuronal damage has been employed as an index of reactive astrogliosis (Bignami and Dahl, 1976; Mathewson and Berry, 1985; Schiffer *et al.*, 1986; Takamiya *et al.*, 1988).

Several mechanisms have been suggested to explain this increase in GFAP immunoreactivity: an increase in antigenic epitopes caused by post-translational modifications, an increased accessibility of antibodies due to a disassembly of glial filaments, an accumulation of GFAP due to a decreased degradation rate, an increased translation of pre-existing GFAP mRNA, an increased transciption of the GFAP gene or a stabilization of GFAP mRNA.

In order to provide information about the molecular events underlying the increase in GFAP immunoreactivity we used a cDNA probe encoding the mouse GFAP (Lewis *et al.*, 1984) to measure the level of GFAP mRNA following a mechanical lesion in the rat cerebral cortex. Our results show that a rapid increase in GFAP mRNA level is one of the early events after brain injury.

In an attempt to elucidate the regulatory mechanism of GFAP expression we have used cell culture models consisting of cultures highly enriched in astroglial cells prepared from cerebral hemispheres of newborn rats. Different factors, which have been suggested as extracellular signals able to regulate GFAP expression during reactive astrogliosis, have been tested in rat glial cell cultures. Among the chemical factors, some growth factors, such as fibroblast growth factor (FGF), which is contained in neurones (Pettmann *et al.*, 1986), are believed to be involved in reactive gliosis.

Interleukin-1 (IL-1) is a mitogen for cultured astroglia and the intracerebral injection of IL-1 induces a striking increase in GFAP positive cell number and GFAP staining intensity (Giulian, 1987). These findings suggest that IL-1, released by activated macrophages and microglia at the lesion site, may promote the formation of astroglial scars in the damaged mammalian brain. Therefore IL-1 is a possible candidate as extracellular signal able to regulate GFAP expression during reactive astrogliosis.

In order to test the hypothesis that a factor, such as IL-1, released by activated macrophages might be able to induce rapidly the expression of the GFAP gene, we examined the effect of a medium conditioned by activated macrophages on GFAP mRNA level in cultured astrocytes. Moreover the possible role in the regulation of GFAP expression of two different signal transduction pathways, those dependent on cAMP and those mediated by protein kinase C (PKC), was also investigated.

13.2 MATERIALS AND METHODS

13.2.1 Brain injury

Male Wistar albino rats weighing 200–240 g were anaesthetized with diethyl ether and held in a stereotaxic apparatus. A midline skin incision was made and a small hole was opened in the skull (2 mm right of the midline and 4.5 mm posterior to the bregma). The right cerebral cortex was lesioned by pushing down a rotating dental drill (1 mm in diameter) 2.5 mm below the brain surface.

The following cortical areas were dissected: (1) a cortical zone surrounding the lesion site; (2) a cortical zone far from the lesion site in the ipsilateral hemisphere; (3) a cortical zone in the unlesioned contralateral hemisphere adjacent to the sagittal scissure.

13.2.2 Primary rat glial cell cultures

Primary glial cell cultures were obtained from cerebral hemispheres of newborn rats (Booher and Sensenbrenner, 1972; Avola *et al.*, 1988). The meninges were removed, and the hemispheres were passed through a sterile nylon sieve (82 μm pore size) into nutrient medium. The basal nutrient medium consisted of Dulbecco's modified Eagle's medium (DMEM), containing 10% heat-inactivated fetal calf serum (FCS),

2 mM glutamine, penicillin (50 U/ml) and streptomycin (0.05 mg/ml). Cells were seeded into Falcon Petri dishes (100 mm diameter) at a plating density of 0.5 × 10^5 cells/cm^2. The cultures were incubated at 37 °C in a humidified 5% CO_2/95% air atmosphere. The culture medium was changed after 6 days and then twice a week.

At 12 days in vitro (DIV), 48 h after the last medium change, cells were washed with DMEM without serum and were incubated in the same medium. Cells were deprived of serum for 24 h and incubated with various factors, as described in the legends of figures.

13.2.3 RNA isolation, electrophoresis and hybridization

Total RNA from cerebral tissue and cultured cells was extracted as described by Chomczynski and Sacchi (1987). Thirty micrograms of total cellular RNA of each sample was next subjected to electrophoresis through 1.1% agarose/2.2 M formaldehyde and afterwards blotted onto nitrocellulose. Pre-hybridization (4–6 h) and hybridization (18–24 h) were carried out at 42°C in a mixture containing 50% of formamide, 5 × SSC, 5 × Denhardt's solution, 20 mM sodium phosphate buffer pH 6.5, and 100 μg of denatured, sheared salmon sperm DNA per millilitre. In the case of hybridization, the mixture was supplemented with 10% of dextran sulphate and labelled denatured probe. GFAP cDNA probe (Lewis *et al.*, 1984) or β-actin cDNA probe (Gunning *et al.*, 1983) were ^{32}P-labelled by the random primed DNA labelling method developed by Feinberg and Vogelstein (1984).

Nitrocellulose membranes were washed three times (5 min each) with a mixture containing 2 × SSC (standard saline citrate: 0.15M NaCl, 0.015M trisodium citrate) and 0.1% sodium dodecyl sulphate (SDS) at room temperature and three times (30 min each washing) with a mixture of 0.1 × SSC and 0.1% SDS at 52 °C and then exposed to X-ray film at − 70 °C using intensifying screens.

13.2.4 Preparation of macrophages

Macrophages were prepared by peritonal lavage from Wistar albino male rats after injection of thioglycolate broth. Macrophages were plated in 75 cm^2 flasks and incubated for 1–2 h at 37 °C in a humidified 5% CO_2/95% air atmosphere and then washed three times to remove non-adherent cells. The macrophages were cultured for 24 h in DMEM supplemented with 10% FCS, 2 mM glutamine, penicillin and streptomycin and then lipopolysaccharide (LPS, 50 ng/ml) was added. After 24 h the medium containing serum and LPS was removed and replaced by serum-free DMEM. Incubation was continued for an additional 24 h and medium was collected and used as conditioned medium.

13.3 RESULTS

13.3.1 GFAP mRNA after brain injury

In the lesioned cortex (zone a) an increase in GFAP mRNA content was observed as early as 6 h after injury (Fig. 13.1). This astroglial response reached a peak 1–3 days after injury and then decreased. A similar temporal pattern was observed in cortical zones far from the injury site (zone b) or in the contralateral hemisphere (zone c); however, the absolute levels of GFAP mRNA were lower than in the lesioned cortex (Fig. 13.1).

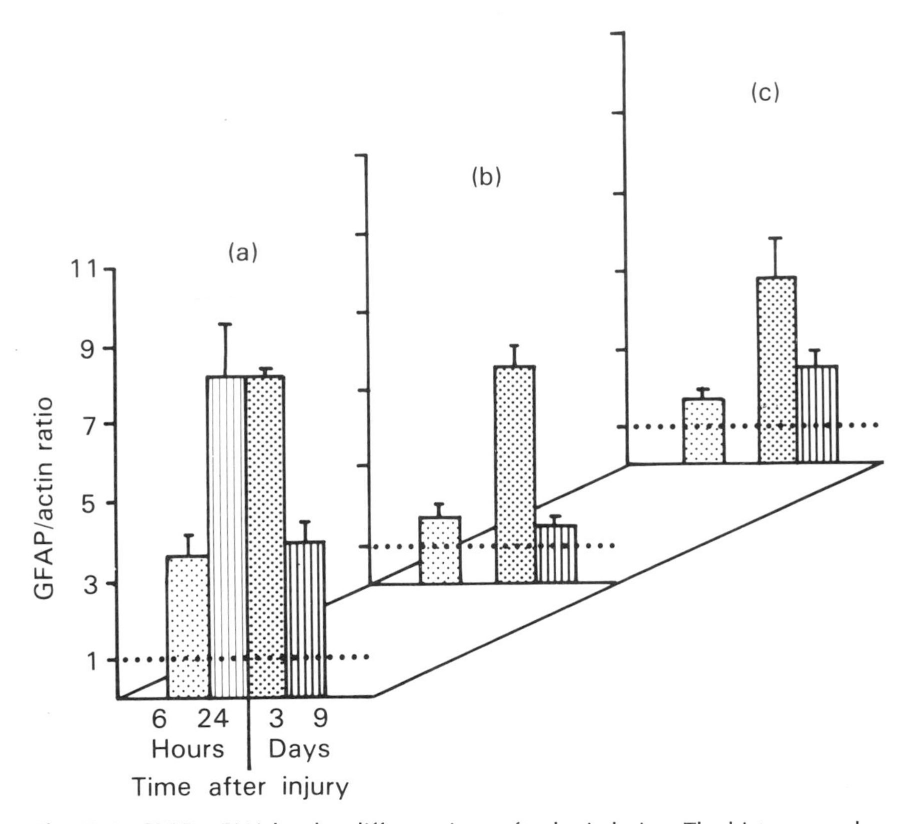

Fig. 13.1 GFAP mRNA level at different times after brain lesion. The histograms show the ratio between GFAP and ß-actin mRNA levels (obtained by densitometric scanning of the area of GFAP mRNA hybridization signal divided by the corresponding area of the ß-actin mRNA signal). Results are expressed in arbitrary units (control = 1) and are average of three independent experiments ± SEM: (a) cortical zone surrounding the lesion site in the ipsilateral hemisphere; (b) cortical zone far from the lesion site in the ipsilateral hemisphere; (c) cortical zone in the unlesioned contralateral hemisphere.

13.3.2 GFAP mRNA in glial cell cultures

In order to obtain more information about extracellular stimuli able to induce an increase in GFAP mRNA level, primary astroglial cell cultures were used. The effects of pituitary fibroblast growth factor (FGF) and a medium conditioned by activated macrophages were investigated. No effect on GFAP mRNA level was observed after incubation up to 10 h with FGF (10 ng/ml) or activated macrophage-conditioned medium (Fig. 13.2).

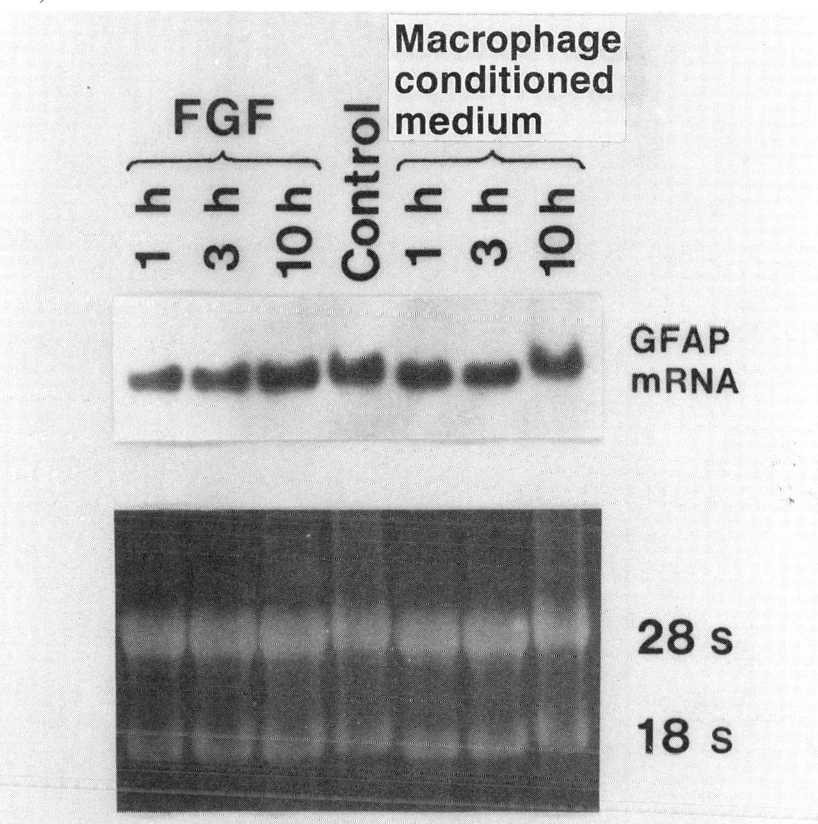

Fig. 13.2 GFAP mRNA level in primary rat astroglial cell cultures after treatment with pituitary FGF or activated macrophage-conditioned medium.

To determine whether the intracellular cAMP may regulate GFAP mRNA level, we examined the effects of 3-isobutyl-1-methylxanthine (IBMX) and of isoproterenol (IPR). IBMX is a phosphodiesterase inhibitor which blocks cAMP breakdown within the cell, while isoproterenol is a β-adrenergic receptor agonist which activates the adenylate cyclase

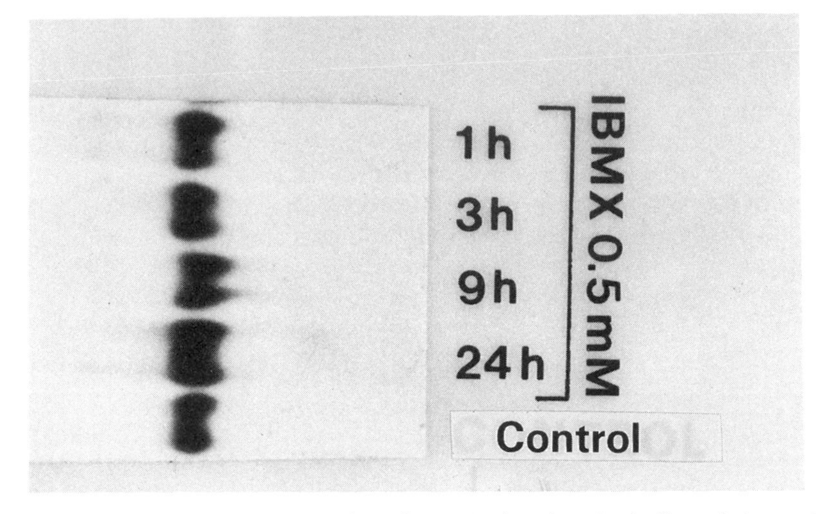

Fig. 13.3 Northern blot analysis of total RNA isolated at the indicated times after addition of 0.5 mM IBMX.

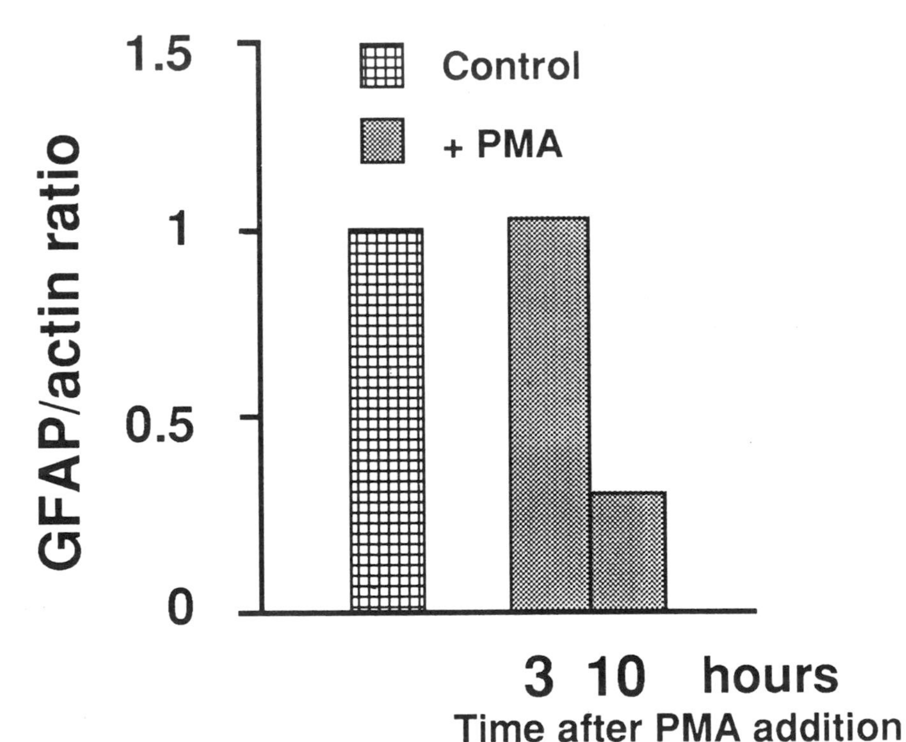

Fig. 13.4 Ratio between GFAP mRNA and ß-actin mRNA level 3 h and 10 h after addition of PMA (50 ng/ml) in primary astroglial cell cultures.

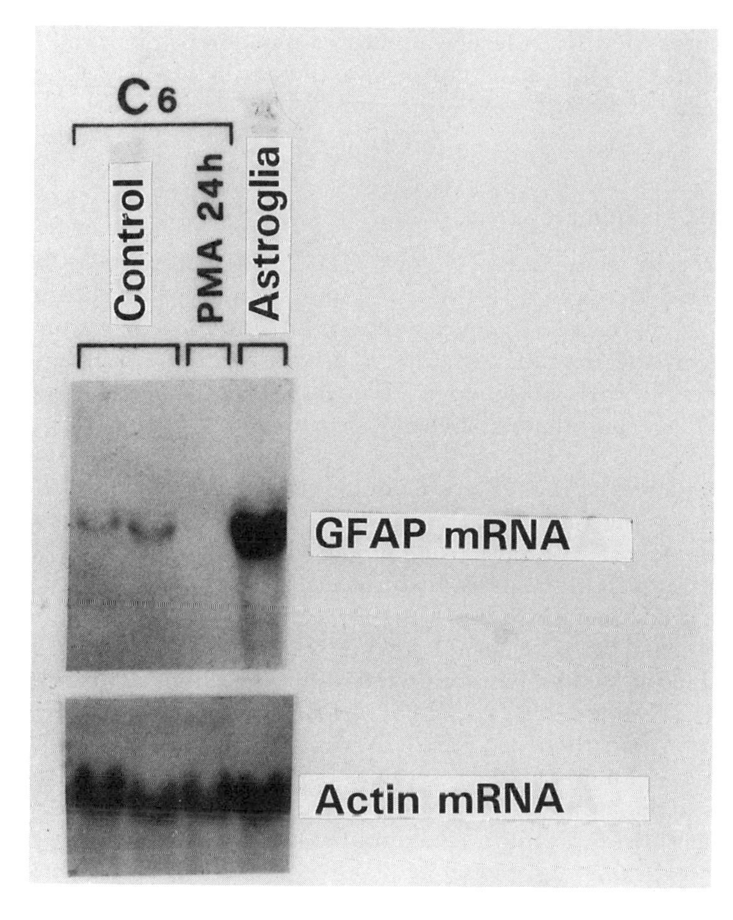

Fig. 13.5 Representative autoradiogram obtained by Northern blot analysis of total RNA extracted from C6 glioma cell cultures treated with PMA for 24 h. The hybridization signal obtained in primary astroglial cultures (Astroglia) is shown for comparison.

system. In primary astroglial cultures the exposure to IPR induces a rapid increase in intracellular cAMP concentration. This effect is concentration dependent (maximal stimulation at 10 μM) and reaches a peak after 5 min of incubation (Condorelli *et al.*, 1989). IBMX induces a two–threefold increase in cAMP level which remains constant for at least 12 h.

An increase in steady-rate GFAP mRNA level was observed after 24 h treatment with 0.5 mM IBMX (Fig. 13.3) in primary astroglial cell cultures. No effect was observed with IPR (10 μM) treatment up to 24 h. The effect of phorbol-12-myristate-13-acetate (PMA), an activator of

PKC, on GFAP mRNA level was also examined in primary rat astroglial cells and in C6 glioma cell culture. As shown in Figs. 13.4 and 13.5, PMA reduced GFAP steady-rate mRNA levels in both cell culture systems.

13.4 DISCUSSION

The observed remarkable rise in GFAP mRNA suggests that the well-known increase in GFAP immunoreactivity after brain injury reflects an increased de novo synthesis of this protein rather than a decrease in the degradation rate. This observation is in agreement with recent results obtained by Tetzlaff *et al.*, (1988), showing an increased incorporation of labelled precursor into GFAP during reactive astrogliosis.

The analysis of the temporal course of the glial response to injury, as shown by the quantification of GFAP mRNA level, reveals some interesting features of the astroglial reativity, such as the rapidity of the response. Indeed in the present study a clear increase in GFAP mRNA was observed 6 h post-lesion.

A very early increase in GFAP mRNA (24 h after lesion) was also observed in the rat basal forebrain following a mechanical fimbria–fornix transection (Cavicchioli *et al.*, 1988). GFAP mRNA level reached a peak at 1–3 days and then decreased. A similar increase of mRNA level, followed by a long-term decrease, has been reported in the striatum in response to dopaminergic deafferentation induced by 6-OHDA infusion in the rat substantia nigra (Rataboul *et al.*, 1988).

In the experimental model of reactive gliosis, obtained performing a stab wound with a needle in the rat cerebral cortex, GFAP-positive cells increase in number around the lesion at 2 days and then spread to the entire ipsilateral cortex (Mathewson and Berry, 1985; Takamiya *et al.*, 1988). According to Mathewson and Berry (1985), reactivity subsequently regressed leaving at 20 days post-lesion only a thin band of neural tissue surrounding the lesion, containing reactive astrocytes.

Our results show an astrocytic activation in cortical zones far from the injury site and in the contralateral hemisphere. In addition to the local astrocyte response in the immediate vicinity of the lesion there are reports of a reaction over a wide area following injury. Several authors have shown a spreading of GFAP immunoreactivity to the ipsilateral cortex and, in some cases, also an involvement of the contralateral hemisphere (Amaducci *et al.*, 1981; Ludwin, 1985). However, Miyake *et al.*, (1988) observed GFAP-positive reactive astrocyte just around a stab wound in rat cerebral cortex, but not further distantly. These contradictory observations might be explained by the fact that the widespread astrocytic response is evoked only when a lesion exceeds a threshold of intensity.

Schiffer *et al.* (1986) have shown that after a brain injury induced by laser, if the lesion is less than 300 μm in diameter, hypertrophy and hyperplasia of astrocytes is present only around the necrotic area, while in the case of larger lesions reactive astrocytes are observed in the whole homolateral hemisphere and, to a lesser degree, in the contralateral one.

In the present work, pituitary FGF did not influence GFAP mRNA level in astroglial culture. Our data are in agreement with recent observations showing that in cultured astrocytes maintained in chemically defined media supplemented with basic FGF intermediate filaments are present but rare (Perraud *et al.*, 1988). Moreover no effect of the activated macrophage-conditioned medium on GFAP mRNA level was observed in astroglial cells after incubation up to 10 h. This result is not in favour of the hypothesis that IL-1, released by activated macrophages in the lesioned site, might be one of the factors responsible for the rapid increase in GFAP mRNA level.

Previous studies have shown that GFAP level is increased by treatment with cAMP-stimulatory agents (Goldman and Chiu, 1984; Hertz *et al.*, 1978; Sensenbrenner *et al.*, 1980). In the present study a slight increase of GFAP mRNA level was observed after 24 h treatment with IBMX. A larger and earlier (6 h) increase in GFAP mRNA level induced by cAMP has been recently reported by Shafit-Zagardo *et al.*, (1988). These authors tested the effect of dibutyryl cAMP, a cAMP analogue, and of forskolin (a potent stimulatory of adenylate cyclase) plus IBMX in primary rat astrocyte cultures and in the human astrocytoma line U-373 MG. The quantitative differences with the results reported in the present paper might be explained by the different pharmacological properties of the cAMP-stimulating agents used in these studies. Treatment with forskolin plus IBMX produces a rapid increase in cAMP level, which remained elevated and constant throughout the 12 h time course (Shafit-Zagardo *et al.*, 1988), while IPR induced a rapid increase in cAMP level followed by a desensitization phenomenon and a return to basal cAMP level. The response elicited by IBMX alone is constant for 12 h but is much lower than that observed in the presence of forskolin plus IBMX.

A prolonged treatment with PMA, a phorbol ester which activates PKC, produced a dramatic decrease in GFAP mRNA in primary rat astroglial cultures and C6 cells. A similar result has been reported by Shafit-Zagardo *et al.*, (1988) in a human astrocytoma cell line.

Further studies are necessary to establish the physiological relevance of the regulation by cAMP or PKC and clarify the exact molecular mechanism of the modification of GFAP mRNA level as shown in the present paper.

In conclusion, it is worthwhile emphasizing that GFAP is a cell-specific protein whose mRNA level is highly modulated by pathological conditions and extracellular signals. The understanding of the molecular basis of its tissue-specific expression and its regulation by extracellular factors is a task of considerable interest.

REFERENCES

Amaducci, L.A., Forno, K.I. and Eng, L.F. (1981) *Neurosci. Lett.*, **21**, 27–32.
Avola, R., Condorelli, D.F., Surrentino, S. *et al.* (1988) *J. Neurosci. Res.*, **19**, 230–8.
Bignami, A. and Dahl, D. (1976) *Neuropathol. Appl. Neurobiol.*, **2**, 99–110.
Booher, J. and Sensenbrenner, M. (1972) *Neurobiology.*, **2**, 97–105.
Cavicchioli, L., Dickson, G., Prentice, H. *et al.* (1988) *Pharmacol. Res. Commun.*, **20**, 609–10.
Chomczynski, P. and Sacchi, N. (1987) *Anal. Biochem.*, **162**, 156–9.
Condorelli, D.F., Kaczmarek, L., Nicoletti, F. *et al.* (1989) *J. Neurosci. Res.*, **23**, 234–9.
Eng, L.F., Vanderheagen, J.J., Bignami, A. and Gerstl, B. (1971) *Brain Res.*, **28**, 351–4.
Feinberg, A.P. and Vogelstein, B. (1984) *Anal. Biochem.*, **137**, 266–7.
Giulian, D. (1987) *J. Neurosci. Res.*, **18**, 155–71.
Goldman, J.E. and Chiu, F.C. (1984) *Brain Res.*, **306**, 85–95.
Gunning, P., Ponte, P., Okayama, H. *et al.* (1983) *Mol. Cell Biol.*, **3**, 787–95.
Hertz, L., Bock, E. and Schousboe, A. (1978) *Dev. Neurosci.*, **1**, 226–38.
Lewis, S.A., Balcarek, J.M., Krek, V. *et al.* (1984) *Proc. Natl Acad. Sci USA*, **81**, 2743–6.
Ludwin, S.K. (1985) *Lab. Invest.*, **52**, 20–30.
Mathewson, A.J. and Berry, M. (1985) *Brain Res.*, **327**, 61–9.
Miyake, T., Hattori, T., Fukuda, M. *et al.* (1988) *Brain Res.*, **451**, 133–8.
Perraud, F., Labourdette, G., Miehe, M. *et al.* (1988) *J. Neurosci. Res.*, **20**, 1–11.
Pettman, B., Labourdette, G., Weibel, M. and Sensenbrenner, M. (1986) *Neurosci. Lett.*, **68**, 175–80.
Rataboul, P., Biguet, N.F., Vernier, P. *et al.* (1988) *J. Neurosci. Res.*, **20**, 165–75.
Schiffer, D., Giordana, M.T., Mighelli, A. *et al.* (1986) *Brain Res.*, **374**, 110–18.
Sensenbrenner, M., Labourdette, G., Delaunov, J.P. *et al.* (1980) *Tissue Cultures in Neurobiology* (eds Giacobini *et al.*), Raven Press, New York, pp. 385–95.
Shafit-Zagardo, B., Kume-Iwaki, and A. Goldman, J.E. (1988) *Glia*, **1**, 346–54.
Takamiya, Y., Kohsaka, S., Toya, S. *et al.* (1988) *Dev. Brain. Res.*, **38**, 201–10.
Tetzlaff, W., Graeber, M.B., Bisby, M.A. and Kreutzberg, G.W. (1988) *Glia*, **1**, 90–5.

14 Tissue plasminogen activator: acute thrombotic stroke*

GREGORY J. DEL ZOPPO
Department of Molecular and Experimental Medicine and Division of Hematology/Medical Oncology, Scripps Clinic and Research Foundation, La Jolla, California USA

14.1 INTRODUCTION

Ischaemic stroke is a primarily vascular disorder for which no uniformly successful treatment exists. The socioeconomic consequences of this disorder with the clinical approaches currently available are staggering in the negative: it is the second most common vascular cause of death in North America, and the third leading cause of death and disability in the Northern Hemisphere (Report of the Atherosclerosis Task Force, 1982; Mackay and Nias, 1979).

Therapeutic interventions applied early in the course of cerebrovascular ischaemia currently under scrutiny may be classified into two groups: (1) agents which are neuronal protective; and (2) agents known to mediate arterial revascularization. In this chapter, the molecular processes subserving the use of pharmacological revascularization by agents such as tissue plasminogen activator (t-PA) will be considered, and the preliminary results of ongoing trials with these agents in arterial thrombosis of the cerebrovascular system will be presented.

14.2 VASCULAR BASIS OF CEREBRAL ISCHAEMIA

The use of thrombolytic agents to achieve vascular perfusion in patients with symptoms of acute thrombotic stroke derives from the observation that 80–90% of acute strokes result from atherothrombotic or thrombo-embolic processes (Solis *et al.*, 1977). The location and extent of the arterial territory compromised by the acute cerebrovascular occlusive event is determined (at the anatomical level) by the presence, type, and

*This research was supported in part by grant 1-R01-NS26945 from the National Institutes of Health. This is publication No. 6193-MEM from the Research Institute of Scripps Clinic, La Jolla, California

extent of reversible arterial collateral flow. Studies employing acute interventional angiography have confirmed the presence of cerebral arterial occlusions appropriate to the presenting symptoms in the majority of patients studied (Zeumer *et al.*, 1983a, 1983b; Del Zoppo *et al.*, 1986a, 1988, 1989; Hacke *et al.*, 1988).

Atherothrombotic occlusions originate by in situ thrombosis most typically at the proximal internal carotid artery and the region of the flow divider of the (extracranial) common carotid artery (Reneman *et al.*, 1988). Other sites of atheroma-based in situ thrombosis include the carotid siphon (C2–C4) and, in some ethnic populations, the most proximal portion of the M1 segment of the middle cerebral artery (MCA) (Nag and Robertson, 1987). Thromboemboli originating from sites of in situ thrombosis, and from more central cardiac sources (e.g. ventricular thrombi, dysplastic valvular processes) are the other common causes of acute cerebral vascular occlusions.

Cerebral blood flow is maintained at 20–25 ml per 100 g cerebral tissue per minute, minimum, below which electrical, cellular and clinical evidence of ischaemia develops (Astrup *et al.*, 1977; Rehncrona *et al.*, 1981). Regional flow disturbances occur secondary to vascular stenoses and obstructions to flow. Because blood is a non-Newtonian fluid, haemorrheological considerations may play a primary or secondary role in cerebral vascular atherogenesis and thrombogenesis. Atherogenesis occurs preferentially at bifurcations and at sites of significant sheer stress (Nag and Robertson, 1987; Reneman *et al.*, 1988), while sites of unusually high sheer stress occurring along the edges of partial obstructions (stenoses) may result in erythrocyte deformation, platelet activation and endothelial damage (Tevritti *et al.*, 1977). Furthermore, at points of flow division (e.g. the carotid bulb) radially oriented streamlines are established ending in flow nodal points which become sites of platelet accumulation (Schmid-Schonbein and Wurzinger, 1988). Therefore, mechanical sheer stress and turbulence may promote in situ thrombosis at sites of vascular irregularity, by indirect promotion of platelet aggregation and thrombosis. A secondary role for haemorrheological factors in the extension of regional cerebral ischaemia may occur when local or systemic increases in whole blood viscosity contribute to perfusion defects in marginally ischaemic regions centred about occluded vascular segments ('ischaemic penumbra'). Whilc evidence for each phase of these scenarios comes from model studies, the response of cerebral blood flow to decreases in whole blood formed elements (e.g. erythrocytes), and in fibrinogen reduction, as may occur with certain fibrinolytic agents, has been documented (Ehrly, 1976).

14.3 VASCULAR ENDOTHELIUM-BLOOD HEAMOSTASIS AXIS

Plasminogen activators (PAs), in therapeutic context, rely for their efficacy on both vascular and haemostatic factors.

14.3.1 Endothelial protective functions

The intact endothelium provides both passive and active barrier functions for transport of materials to the subendothelium and subjacent tissues (Gimbrone, 1986). Additionally, and importantly for haemostasis, the endothelial cell plays a predominantly antithrombotic role, preventing spontaneous thrombosis in flowing blood, and limiting thrombus growth in regions of endothelial injury (Gimbrone, 1986). Active protective mechanisms include: prostacyclin (PGI_2) synthesis and release, secretion of PA, active uptake and degradation of proaggregatory ADP and vasoactive amines, inactivation and clearance of thrombin, thrombomodulin-dependent activation of protein C and other processes.

14.3.2 Thrombus formation

Arterial thrombosis involves the systems of platelet and coagulation activation, superimposed upon endothelial cell reactivity in a complex multilayered manner. Thrombin generation and its attendant conversion of fibrinogen to fibrin is central to the formation of the fibrin lattice, the stabilizing framework for the growing thrombus. Endogenous PAs (t-PA, single-chain urokinase PA (scu-PA) and to a lesser degree urokinase PA (u-PA)) act directly on the formed and forming fibrin network. The process of thrombin-mediated fibrinogen cleavage to form fibrin is accelerated by factor Xa, bound to its high-affinity platelet receptor, factor Va. Factor Xa is generated from circulating factor X by the 'Tenase' complex, consisting of platelet phospholipid-bound factors V and VIII. In addition to its dependence upon the intrinsic coagulation pathways, factor Xa may also be formed when tissue thromboplastins are released (Zur and Nemerson, 1987).

Platelet activation, with subsequent α and δ granule release, and platelet aggregation may precede or coincide with coagulation system activation in interdependent processes. Exposure of subendothelial structures following endothelial injury stimulate platelet adherence, release and aggregation. Platelet activation is accompanied by secretion of dense (δ)-granules (containing pro-aggregatory ADP) and α-granule contents, as well as stimulation of a platelet membrane phospholipase complex generator of the pro-aggregatory thromboxane A_2 (Bloom, 1987). Coincident with activation, as noted above, platelets may promote the early stages of intrinsic coagulation by accelerating factor X to Xa conversion at the phospholipid-bound factor V/VIII complex. Once initiated, platelet aggregration (in association with thrombosis) is an essentially 'committed' process. Platelet-related factor X conversion is responsible for an explosive (10^5-fold) increase in thrombin formation, thereby accelerating fibrin formation. The fibrin lattice is stabilized by factor XIIIa (Rosenberg, 1987).

In time, via cross-linking of fibrin-associated ε-amino groups, organization of the thrombus occurs (Hermans and McDonagh, 1982). Erythrocyte trapping, incursion of polymorphonuclear (PMN) leucocytes and monocytes/macrophages and participating platelets contribute to individual thrombus characteristics. Generally, thrombi formed in high-sheer-rate environments (e.g. extracranial carotid arteries) are relatively platelet rich/erythrocyte poor, whereas the converse obtains from thrombi formed in low-flow vascular beds (e.g. deep femoral veins). The contribution of the perivascular tissue to thrombus composition and formation rates has not been studied, but could conceivably be of considerable influence. Little is known of the composition of cerebral arterial thromboemboli.

14.3.3 Thrombus lysis (intrinsic)

Agents with fibrin(ogen)olytic activity are endogenous or exogenous in origin (Table 14.1). Endogenous thrombolysis contributes to thrombus growth limitation and thrombus dissolution in vivo. Thrombus dissolution results from the local generation of plasmin from fibrin-bound plasminogen by fibrin-bound t-PA and, perhaps, the local effect of scu-PA and u-PA (Verstraete and Callan, 1986). Plasmin, a serine protease, mediates fibrin(ogen) degradation. The resultant oligopeptide fragments are determined by the specific substrate – fibrinogen (YY, DXD, YD/DY) or fibrin (DD/E, YY, DXD, YD/DY).

Two classes of inhibitors modulate the activity of thrombolytic substances in vivo: specific PA inhibitors (e.g. PAI-1, PAI-2, and PAI-3 from various organ systems); and the circulating plasmin inhibitors, α_2-antiplasmin and α_2-macroglobulin. The former appear to be primarily responsible for inhibition of circulating t-PA; however, they probably play a minor role in the modulation of the thrombus lytic effects of infused t-PAs.

In contrast to the fibrinogenolytic agents, u-PA and streptokinase (SK), an anticoagulant effect is not achieved by t-PA and scu-PA because the latter do not substantially activate circulating factors V and VIII.

14.3.4 Limitation of thrombus growth and mechanisms of thromboembolism

Although there is no firm proof, endogenous thrombolysis may underlie

Table 14.1 Comparison of t-PA with other plasminogen activators

Activator	*Mol. wt. (kDa)*	*Chains*	*Concentration (mg/dl)*	$t_{1/2}$ *(min)*	*Substrates*
Engogenous					
t-PA	68 (59)	1→2	5×10^{-4}	5–8	Fibrin + plasminogen
scu-PA	54 (31)	1→2	$2\text{–}20 \times 10^{-4}$	8	Fibrin + plasmin(ogen)
u-PA	54 (31)	2	8×10^{-4}	9–12 (27–61)	Plasminogen
Exogenous					
Streptokinase	47	-	-	23	Plasminogen, fibrin(ogen)
BRL26921	131	-	-	40	Plasminogen, fibrin(ogen)
BRL33575		-	-	408	Plasminogen, fibrin(ogen)

BRL compounds refer to APSAC, *p*-anisoylated derivatives of lys-plasminogen/streptokinase activator complex.

the apparent spontaneous dissolution of occlusions observed in the cerebrovasculature (Dalal *et al.*, 1965). In in vitro culture, endothelial cells (human umbilical vein) secrete t-PA and its principal inhibitor, PAI-1, in response to the presence of thrombin (Levin, 1986). Some differences have been noted between human endothelial cells of cerebrovascular origin and those from other sources (Spatz *et al.*, 1989); however, little is known concerning the regulation of the fibrinolytic system in the cerebrovasculature, hence the presumption that fibrinolytic responses would be similar to those noted elsewhere. Fortuitous serial angiographic investigations have documented fragmentation and embolization of thrombotic cerebrovascular occlusions (Bruckmann and Ferbert, 1989). It is presumed that thrombus-associated fibrinolysis and microenvironmental flow and haemorrheological forces produce thrombus fragmentation and downstream embolization. Once a thromboembolism has lodged, the process of thrombus growth and dissolution continue at the new site. Augmentation of the dissolution process by regional delivery of a thrombus lytic agent is the primary rationale for the use of agents such as t-PA in clinical trials.

14.4 CONSEQUENCES OF REVASCULARIZATION

Implicit, although not covered, in this discussion are consideration of (1) the significance of collateral flow to cerebral tissue protection, and (2) the contribution of reperfusion injury to ischaemia. Superoxide free-radical-mediated disturbances of cellular membrane function following reperfusion and the 'no-reflow' phenomenon are under rigorous study (Hirshberg and Hofferberth, 1987; Cross *et al.*, 1987).

The contribution of collateral flow to the limitation of the focal ischaemic zone, and the potential for salvage of presumed perifocal ischaemic tissue ('ischaemic penumbra') following cerebral arterial thrombotic occlusion are of great relevance to the use of thrombolytic agents as a therapeutic modality in acute thrombotic stroke. Model studies (Astrup *et al.*, 1977; Buchweitz and Weiss, 1986; Schlockley and LaManna, 1988) suggest that a nascent, but unknown, volume of potential collaterals may be partly responsible for the variation in rapidity of neurological symptom onset and degree of recovery noted in some patients.

Detection of functional and potential collaterals in the acute stroke patient is currently not possible, although angiography may define large vascular collaterals.

Of more relevance to the present discussion are the presumed effects of acute arterial 'revascularization' in the presence of pharmacological quantities of thrombolytic substances such as recombinant t-PA (rt-PA): arterial reflow and cerebral haemorrhage.

14.4.1 Re-establishment of arterial flow

The use of thrombus lysis in acute thrombotic stroke is based upon the unproven premise that re-establishment of arterial flow is essential for recovery of ischaemic neuronal tissue. Suggestive evidence from model-derived and clinical experience with thrombolysis-mediated coronary artery reperfusion supports this hypothesis (DeWood *et al.*, 1980; Passamani *et al.*, 1987; Verstraete, 1986; ISIS-2 Collaborative Group, 1988). However, significant differences between myocardial and cerebral tissues exist. Because of time constraints, cerebral arterial reperfusion would be expected to salvage only tissues in the periphery of a reversibly ischaemic zone ('ischaemic penumbra'). Reversibility of ischaemic deficits in non-human primate and other models of focal ischaemia (Spetzler *et al.*, 1980; Del Zoppo *et al.*, 1986b) and patients suffering transient ischaemic attacks (Barnett, 1979; Ross, 1977) supports this observation. However, at this time, no firm direct evidence exists confirming salvage of cerebral tissue by reperfusion.

14.4.2 Haemorrhagic transformation

The risk of haemorrhage is common to all fibrinolytic agents (Bewermeyer *et al.*, 1984; Aldrich *et al.*, 1985; Del Zoppo *et al.*, 1986a). While a dose-dependent increase in intracerebral haemorrhage following the use of rt-PA in myocardial infarction patients has been documented (Braunwald *et al.*, 1987), the risk of haemorrhage associated with fibrinolytic agents in acute cerebral ischaemia or infarction is not known. It has been suggested that central nervous system haemorrhage following fibrinolytic agents may occur because of lysis of fibrin-stabilized haemostatic plugs (Marder, 1986). Animal model experience also suggests that haemorrhagic transformation in an ischaemic vascular bed occurs most readily from exposure of ischaemia-damaged vessels to increased perfusion pressure upon re-opening of the stem artery (Garcia and Kamijyo, 1974; Garcia *et al.*, 1983; Garcia *et al.*, 1971). Additional (unproven) factors which may augment the risk and severity of haemorrhage in this setting include concomitant use of antiplatelet agents or anticoagulants, and systemic fibrinogen depletion and the anticoagulant effect characteristic of exogenous agents.

Haemorrhagic transformation may be categorized as (1) haemorrhagic infarction, consisting of petechial haemorrhage without clinical sequelae, and (2) parenchymatous haemorrhage, characterized by extravasation of blood into the brain substance with mass effect, and associated clinical deterioration in the time frame of the haemorrhage. Silent haemorrhagic transformation, typified as haemorrhagic infarction, is a common accompaniment to acute cerebral ischaemia (Fisher and Adams, 1951). The mechanisms underlying haemorrhagic transformation of either type remain speculative. Spontaneous haemorrhagic conversion rates in cerebral infarction have been reported between 0 and 43% (Del Zoppo *et al.*, 1986a). While the risk of intracerebral haemorrhage accompanying the use of fibrinolytic agents in the setting of acute cerebral ischaemia remains unknown, limited studies have indicated the incidence to fall within the spontaneous haemorrhagic conversion rate noted above (Del Zoppo *et al.*, 1988; Hacke *et al.*, 1988; Mori *et al.*, 1988). With exogenous fibrinolytic agents, both haemorrhagic infarction and parenchymatous haemorrhages have been observed, without preference for either type. The reported incidence of intracerebral haemorrhage in patients without known cerebrovascular disease is consistently between 0 and 1.5%. On theoretical grounds, rt-PA would not be expected to produce a higher risk of haemorrhage than that of other exogenous fibrinolytic agents at the same effective dose rate.

14.5 TISSUE PLASMINOGEN ACTIVATOR

t-PA is a single-chain 527 amino acid polypeptide glycoprotein, with an unglycosylated molecular weight of approximately 59 kDA (Rijken *et al.*, 1982b; Collen *et al.*, 1982; Bachman and Kruithof, 1984). Proteolytic cleavage of the arginine^{275}–isoleucine^{276} linkage by plasmin converts single-chain t-PA to the two-chain form. Both forms are enzymatically active, although in vitro thrombus lysis experiments indicate the two-chain form to have a slightly higher specific activity in the absence of fibrin to sythetic tripeptide substrates (Ranby *et al.*, 1982). Significant structure–function relationships have been elucidated: an N-terminal finger domain and a glycosylated kringle (kringle-2), the latter with characteristic triple disulphide internal linkages, mediate selective fibrin binding. Mutants devoid of these two structures have markedly reduced fibrin-binding constants (Ehrlich *et al.*, 1987).

The catalytic activity of t-PA derives from its trypsin-like serine protease properties. Its ability to hydrolyse plasminogen to plasmin is enhanced by the association of both t-PA and plasminogen with fibrin. Therefore, both the fibrin-binding and the activity stimulation by association with fibrin are essential features of the thrombus selectivity of this molecule, a property not observed with u-PA. While two-chain t-PA is somewhat more active than the single-chain form in the absence of fibrin in amidolytic assays in vitro, the activity of the single-chain form approaches that of the two-chain form in the presence of fibrin (Ranby, 1982). The relative thrombus selectivity of the two forms to fibrin-rich thrombi is quite comparable (Rijken *et al.*, 1982a). To date, both single-chain and two chain t-PA have been employed in clinical studies of efficacy and safety in the treatment of acute myocardial infarction, deep venous thrombosis, acute pulmonary embolism and, more recently, in acute cerebral ischaemia.

14.6 A BASIS FOR THE USE OF rt-PA IN ACUTE CEREBRAL ISCHAEMIA

14.6.1 Pros and Cons

At present the use of thrombus lytic agents in acute thrombotic stroke as a therapeutic modality is not yet established. Potential advantages of rt-PA, for example, are (1) its relative thrombus selectivity, (2) systemic effect, (3) site-directed activation of the molecule, (4) rapid onset of action and short half-life, (5) the availability of stable purified preparations with minimal antigenicity, (6) limited systemic activation of coagulation factors, and (7) minimal effective perturbation of the vascular system. Several limitations are apparent in work to date with this agent, which include (1) the as yet unknown risk of associated intracerebral haemorrhage in the setting of acute cerebral ischaemia, (2) the absence, at this time, of an effective antidote for an ongoing central nervous system haemorrhagic complication (although PAI-1 is now available in purified form), and (3) the absence of a known specific dose rate(s) of rt-PA appropriate to lyse thrombotic occlusions in acute thrombotic stroke. Further considerations derive from early studies of the use of fibrinolytic agents in stroke patients. Fundamental to all considerations is the still unproven concept that recanalization and clinical neurological improvements in this setting are interdependent.

14.6.2 Clinical experience with exogenous fibrinolytic agents

The outcome of eight clinical trials evaluating the efficacy of intravenous infusion of a number of thrombolytic agents in patients with completed stroke have confirmed the haemorrhagic risk characteristic of such agents (Del Zoppo *et al.*, 1986a). The absence of any consistent beneficial clinical effect in those studies may be explained by the diversity of dose rates and the inconsistency of control of the lytic agents used; the common treatment of patients with *completed* stroke, rather than those early after symptom onset; and methodological limitations related to the absence of more sophisticated diagnostic techniques currently available.

In addition to anecdotal reports, four recent limited prospective studies have employed intra-arterial infusion techniques to deliver urokinase/streptokinase at the site of angiographically defined arterial occlusions in acute stroke patients (Table 14.2) Del Zoppo *et al.*, 1988;

Table 14.2 Intra-arterial thrombolytic therapy

	Patients	*Recanalization*	*Recovery*[a]	*Haemorrhage*[b]	*Deaths*
Carotid territory					
Miyakawa (1984)	2	2	2	0	0
Del Zoppo *et al.* (1988)	20	18	12	4	3
Mori *et al.* (1988)	22	10	10	3 (1)	3
Theron *et al.* (1989)	12	8	6	3	1
Vertebrobasilar territory					
Nenci *et al.* (1983)	4	4	3	0	1
Hacke *et al.* (1988)	43	19	8 (6)	2 (2)	29
Zeumer *et al.* (1989)	7		4	1	2

[a], partial recovery.
[b], contributed to patient demise.

Hacke *et al.*, 1988; Mori *et al.*, 1988; Miyakawa, 1984; Nenci *et al.*, 1983; Zeumer *et al.*, 1989; Theron *et al.*, 1989). The feasibility of fibrinolytic agent delivery by interventional neuroradiological techniques has been demonstrated in both the carotid and vertebrobasilar territories. In the carotid territory, complete recanalization has been achieved in 45–75% of patients treated (Del Zoppo *et al.*, 1988; Mori *et al.*, 1988). Recanalization has been associated with a significant reduction in infarct size noted on serial computed tomography (CT) scans (Mori *et at.*, 1988), while two studies suggest an improvement in clinical outcome in patients demonstrating partial or complete recanalization following fibrinolytic therapy (Del Zoppo *et al.*, 1988; Mori *et al.*, 1988). A significant improvement in clinical outcome and survival has been suggested in two studies of patients with vertebrobasilar territory occlusions following complete recanalization (Hacke *et al.*, 1988; Zeumer *et al.*, 1989).

Haemorrhagic transformation reported in the above studies fell within the broad ranges published for acute stroke patients receiving no ancillary therapy, or anticoagulants with or without haemodilutional agents. Because of the small numbers in those studies, a more precise estimation of haemorrhagic risk could not be made.

14.7 FIBRIN-SELECTIVE AGENTS IN ACUTE THROMBOTIC STROKE

The above experience demonstrated the efficacy of thrombus lysis with the use of fibrin non-selective agents by local delivery, and has provided the basis for the development of recent ongoing studies to examine the effect of fibrin selective agents in acute thrombotic stroke patients.

A study of acute thrombotic stoke patients treated with melanoma-derived t-PA by local intra-arterial perfusion techniques was terminated following a negative outcome in two of three patients. Significant haemorrhagic transformation occurred in another, raising questions regarding interpretation of outcome in patients undergoing an interventional procedure with an agent of as yet unknown efficacy (Del Zoppo, 1987). As the dose rate delivered to ischaemic cerebral tissue in this setting could not be gauged, the design of systemic infusion studies was undertaken.

At least five preliminary studies of various design have been undertaken to evaluate rt-PA (either single-chain or two-chain) in patients presenting early after symptom onset of acute stroke. A multicentre CT scan/angiography-based dose-range study of safety, and recanalization efficacy, in patients receiving a 60 min intravenous infusion of rt-PA infusion within 8 h of the onset of acute symptoms has been initiated. Documentation of unequivocal arterial occlusions in the territory appropriate to the focal neurological symptoms, in the absence of haemorrhage, have been essential requirements for study entry. Anticoagulants have not been used in the initial 24 h. Preliminary results in five incremental dose-rate groups receiving rt-PA (two-chain, Wellcome Biotech) have demonstrated that haemorrhagic events have not been dose dependent, and recanalization has occurred in all dose-rate groups (rt-PA Acute Stroke Study Group, 1988; Del Zoppo *et al.*, 1989). As data analysis is not complete, attributions of vascular anatomical characteristics to the aetiology of haemorrhagic transformation cannot be made at this time. Thus far, it is apparent that (1) angiographic studies in acute stroke patients may be performed safely, (2) haemorrhagic infarction per se is not associated with clinical deterioration, and (3) continuation of this study has been warranted because haemorrhagic events fall within the prospectively established safety criteria. An additional consequence of this study will be the accumulation of new insights into the vascular consequences of cerebral ischaemia in the early hours following a thrombotic event, data hitherto unavailable.

A study of similar design has been undertaken at a single dose rate of rt-PA (single- chain, Boehringer Ingelheim), which has demonstrated recanalization in a limited number of patients (von Kummer *et al.*, 1989). E. Mori and colleagues have undertaken a placebo comparison trial of two dose rates of intravenous rt-PA (two- chain, Sumitomo). This study is a single-centre, angiography-based study of safety and recanalization efficacy.

Two separate non-angiography-based studies are also in progress. A small multicentre trial examines the incidence of haemorrhage following rt-PA administered at several dose levels (single-chain, Genentech) within 90 min of symptom onset (Brott *et al.*, 1988). The time to treatment from symptom onset obviates the use of vascular imaging diagnostic methods; however, all patients are screened by CT scan to rule out intracerebral haemorrhage as a cause of the neurological deficits. To date, this study has demonstrated the feasibility of entering and treating patients within 90 min of the onset of acute stroke symptoms. A second study, of different design, examines the effect of rt-PA (two-chain, Sumitomo) on late consequences of acute stoke (Yamaguchi, 1989). Data from this study are yet forthcoming. Additional studies in Scandinavia have been undertaken, the results of which are not yet available.

The above studies demonstrate the dynamic nature of this recently developed field of investigation. It will owe its success to a paradigm of stroke treatment: the pathophysiology and consequences of ischaemic vascular stroke dictate that patients be evaluated and treated early after the onset of symptoms.

14.8 PRACTICAL ISSUES REGARDING ACUTE STROKE MANAGEMENT

Experience in the above studies has demonstrated a number of issues which must be addressed for appropriate patient evaluation, and formulation of study hypotheses. While early acute stroke patient evaluation is feasible at some centres (Barsan *et al.*, 1989), patient and family ignorance regarding the meaning of acute symptoms, transportation difficulties and delays in hospital emergency receiving areas often confound attempts to treat patients early. Because clinical symptoms may not reveal the aetiology of a given territorial neurological event, the minimum of a screening CT cerebral scan to eliminate a haemorrhagic event as the cause for the symptoms is necessary. This and further testing procedures (e.g. angiography, non-invasive Doppler ultrasonography, somatonsensory evoked potentials) will increase the time interval necessary to patient treatment. However, such diagnostic procedures may be warranted by the nature of the hypothesis to be tested, and the agent to be evaluated (Del Zoppo *et al.* 1986a). Finally, discrete measures of outcome are necessary to provide a common evaluation of treatment-associated events. Because of the paucity of prospective outcome analyses, neurological scoring instruments at this time are probably not sufficient alone to provide an indication of efficacy, particularly recanalization efficacy. It should be remembered that the primary hypothesis, that reflow in a previously occluded cerebral artery may promote neuronal functional recovery from ishaemic injury, is yet unproven.

The impetus for early intervention in acute stroke patients comes from the availability of agents with potential efficacy in this disorder. Logistical difficulties which may limit patient accrual must be solved by community-based education programmes, patient responsibility, establishment of minimum distance transportation networks, and preparation of emergency rooms to receive such patients. In hospital-or practice-based referral systems, emergency personnel must be apprised of the importance of such patients so that minimum time paths to diagnosis and treatment may be followed. In the receiving facility, a team of professional investigators including neurologists, neuroradiologists, haematologist–internists and appropriate nursing personnel are essential to the success of early treatment. The role of neurological intensive care units is yet to be evaluated in the current climate of acute stroke treatment. With increasing interest and impetus to evaluate and treat patients early after the onset of acute focal neurological symptoms, the above issues will be addressed. It will undoubtedly be true that local solutions to each of these problems will require differences in approach and tactic.

14.9 COMMENTARY

The potential benefits of fibrin-selective agents in the treatment of acute thrombotic stroke must be weighed against the unknowns. Present data from ongoing trials have not altered the view that careful dose-rate studies of rt-PA are necessary to determine both the safety and the recanalization efficacy of this agent. Because of the labour-intensiveness of such studies, only studies with a clear-cut hypothesis and clear outcome assessment methodologies should be encouraged. A travesty will be averted when outcome assessments are of sufficient relevance and accuracy to prevent dispensing with a study as irrelevant on the basis of weak methodologies. Discussions of the benefits and concerns regarding thrombolysis and acute stroke have been presented elsewhere (Del Zoppo, 1987). These concerns suggest caution in future studies with fibrinolytic agents in acute stroke. Therefore, carefully designed and conducted prospective studies in several stages are envisioned: (1) dose-rate escalation studies to determine the specific dose rate capable of achieving recanalization without haemorrhage, as preparation for (2) larger, controlled, blind (placebo-based) studies of clinical outcome and safety of a given agent at a pre-selected dose rate. Until the results of these studies become available, the use of fibrin-selective agents in a non-study setting should not be encouraged.

REFERENCES

Aldrich, M.S., Sherman, S.A., Greenberg, H.S. *et al.* (1985) *JAMA*, **253**, 1777–9.

Astrup, J., Symon, L., Branston, N.M. and Lassen, N. (1977) *Stroke*, **8**, 51–7.

Bachman, F. and Kruithof, I.E.K.O. (1984) *Semin. Thromb. Hemost.*, **10**, 6–17.

Barnett, H.J.M. (1979) *Med. Clin. North Am.*, **63**, 649–79.

Barsan, W.G., Brott, T.G., Clinger, C.P. and Marler, J.R. (1989) *Ann. Intern. Med.*, **111**, 449-512.

Bewermeyer, H. Schumacher, A., Neveling, M. and Heirslo, D. (1984) *Dtsch. Med. Wochenschr.*, **109**, 1653–9.
Bloom, A.L. (1987) *Haemostasis and Thrombosis*, Churchill Livingstone, Edinburgh.
Braunwald, E., Knatterud, G.L., Passamani, E. (1987) *J.Am. Coll. Cardiol.*, **10**, 970.
Brott, T., Haley, E.L., Lew, D.E. (1988) *Ann. Emerg. Med.,* **17**, 1202–5.
Bruckmann, H. and Ferbert, A. (1989), **31**, 95–7.
Buchweitz, E. and Weiss, H.R. (1986) *Brain Res.*, **377**, 105–11.
Collen, D., Rijken, D.C., Van Damme, J. and Billiau, A. (1982) *Thromb. Haemost.*, **48**, 294–6.
Cross, C.E., Halliwell, B., Borish, E.T. (1987) *Ann. Intern. Med.*, **107**, 526–45.
Dalal, P.M., Shah, P.M., Sheth, S.C. and Deshpande, C.K. (1965) *Lancet*, **i**, 61–4.
Del Zoppo, G.J. (1987) in *New Trends in Diagnosis and Management of Stroke*, (eds K. Poeck, E.B. Ringlstein *et al.*), Springer-Verlag, Berlin, p. 115–27.
Del Zoppo, G.J., Zeumer, H. and Harker, L.A. (1986a) *Stroke*, **17**, 595–607.
Del Zoppo, G.J., Copeland, B.R., Harker, L.A. *et al.* (1986b) *Stroke*, **17**, 638–43.
Del Zoppo, G.J., Ferbert, A., Otis, S.L. *et al.* (1988) *Stroke*, **19**, 307–13.
Del Zoppo, G.J., Poeck, L., Pessin, M.S. *et al.* (1989) *Thromb. Haemost.* **62**, 27.
DeWood, M.A., Spores, J., Notske, R. *et al.* (1980) *N. Engl. J. Med.*, **303**, 897–902.
Ehrlich, H.J., Bang, N.U., Little, S.P. *et al.* (1987) *J. Fibrinolysis*, **2**, 75–81.
Ehrly, A. (1976) *Angiology*, **27**, 188–96.
Fisher, C.M. and Adams, R.D. (1951) *J. Neuropathol. Exp. Neurol.*, **10**, 92–4.
Garcia, J.H. and Kamijyo, Y. (1974) *J. Neuropathol. Exp. Neurol.*, **33**, 408–21.
Garcia, J.H., Cox, J.V., Hudgins, W.R. *et al.* (1971) *Arch. Neuropathol,* **18**, 273–85.
Garcia, J.H., Lowig, S.L., Briggs, L. *et al.* (1983) *Cerebrovascular Diseases*, (eds M. Reivich and H.I. Hartig) Raven Press, New York, pp. 169–79.
Gimbrone, M.A. Jr. (1986) in *Vascular Endothelium in Hemostasis and Thrombosis* (ed M.A. Gimbrone Jr), Churchill Livingstone, Edinburgh, pp. 1–13.
Hacke, W., Zeumer, H., Ferbert, A. *et al.* (1988) *Stroke*, **19**, 1216–22.
Hermans, J. and McDonagh, J. (1982) *Semin. Thromb. Hemost.,,* **8**, 11–24.
Hirschberg, M. and Hofferberth, B. (1987) *Neurology*, **37**, 133.
ISIS-2 Collaborative Group (1988) *Lancet*, **ii**, 349–60.
Levin, E.G. (1986) *Blood*, **67**, 1309–13.
Mackay, A. and Nias, B.C. (1979) *J. R. Coll. Physicians Lond.*, **13**, 106–12.
Marder, V.J. (1986) in *Cerebrovascular Diseases: Fourteenth Research Conference* (eds F. Plum and W.A. Pulsinelli), Raven Press, New York, p. 241–7.
Miyakawa, T. (1984) *Rinsho Ketsueki*, **25**, 1018–26.
Mori, E., Tabuchi, M., Yoshida, T. and Yamadori, A. (1988) *Stroke*, **19**, 802–12.
Nag, S. and Robertson, D.M. (1987) in *Cerebral Blood Flow* (ed. J.H. Wood). McGraw-Hill, New York, pp. 59–74.
Nenci, G.G., Gresele, P., Taramelli, M. *et al.* (1983) *Angiology*, **34**, 561–71.
Passamani, E., Hodges, M., Herman, M. *et al.* (1987) *J. Am. Coll. Cardiol.*, **10**, 51B–64B.
Ranby, M. (1982), *Biochem. Biophys. Acta*, **704**, 461–9.

Ranby, M., Bergsdorf, N. and Nilsson, T. (1982) *Thromb. Res.*, **27**, 175–83.
Rehncrona, S., Rosen, I. and Siessö, B . (1981) *J. Cereb. Blood Flow Metab.*, **1**, 297–311.
Reneman, R.S., van Merode, T., Smeets, F.A.M. and Hoeks, A.P.G. (1988) in *Carotid Artery Plaques* (eds M. Hennerici G. Sitzer and Weger, H.D.), Karger, Basel, pp.143–62..
Report of the Atherosclerosis Task Force (1982) NIH Publication.
Rijken, D.C., Hoylaerts, M. and Collen, D. (1982a) *J. Biol. Chem.*, **257**, 2920–5.
Rosenberg, R.D. (1987) in *The Molecular Basis of Blood* (eds B. Stamatoyannopoulos, A.W. Neinhuis, P. Leder *et al.*), Saunders, Philadelphia, pp. 534–74.
Ross, R.T. (1977) *Can. J. Neurol. Sci.*, **4**, 143–50.
rt-PA Acute Stroke Study Group (1988) *Stroke*, **19**, 134.
Schockley, R.P. and LaManna, J.C. (1988) *Brain Res.*, **454**, 170–8.
Schmid-Schonbein, H. and Wurzinger, L.J. (1988) in *Carotid Artery Plaques* (eds M. Hennerici, G. Sitzer and Weger, H.D.), Karger, Basel, pp. 64–91.
Solis, O.J., Robertson, G.R., Taderas, J.M. *et al.* (1977) *Revist. Interam. Radial.*, **2**, 19–25.
Spatz, M., Bauc, F., McCarron, R.M. *et al*, (1989) *J. Cereb. Blood Flow Metab.*, **9**, S393.
Spetzler, R.F., Selman, W.R., Weinstein, P. *et al.* (1980) *J. Neurosurg.*, **7**, 257–61.
Tevritti, V.T., Miggli, R. and Baumgartner, N.R. (1977) *NY Acad. Sci.*, **2**, 84–102.
Theron, J., Coutheoux, P., Casaseo, A. *et al.* (1989) *Am. J. Neuroradiol.*, **10**, 753–65.
Verstraete, M. (1986) *Lancet*, **i**, 397–401.
Verstraete, M. and Collen, D. (1986) *Blood*, **67**, 1529–41.
von Kummer, R., Hutschenreuter, M., Wildemann, B. and Hacke, W. (1989) *J. Cereb. Blood Flow Metab.*, **9** (suppl.) 5721.
Yamaguchi, T. (1989) (Abstract) Tokyo Satellite Symposium of XIIth Congress of the ISTH.
Zeumer, H., Ringelstein, E.B. and Hacke, W. (1983a) *Fortschr. Roentgenstr.*, **139**, 467–75.
Zeumer, H., Ringelstein, E.B., Hassell, M. *et al.* (1983b) *Dtsch. Med. Wochenschr.*, **108**, 1103–5.
Zeumer, H., Freilag, H.J., Grzyka, U. and Neunzig, H.P. (1989) *Neuroradiology*, **31**, 336–40.
Zur, M. and Nemerson, Y. (1987) in *Haemostasis and Thrombosis* (eds A.L. Bloom and D.P. Thomas), Churchill Livingstone, Edinburgh, pp. 148–64.

15 Lysosomal storage of a mitochondrial protein in Batten's disease (ceroid lipofuscinosis)

S.M. MEDD,[1] J.E. WALKER,[1] I.M. FEARNLEY,[1] R.D. JOLLY[2] AND D.N. PALMER[2]
[1]Medical Research Council, Laboratory of Molecular Biology, Hills Road, Cambridge CB2 2QH, UK; [2]Department of Veterinary Pathology and Public Health, Massey University, Palmerston North, New Zealand

15.1 INTRODUCTION

The neuronal ceroid lipofuscinoses are a group of recessively inherited lysosomal storage disease of children and animals. Clinical signs include blindness, seizures and dementia, culminating in premature death. There are three main forms of human ceroid lipofuscinosis (Batten's disease). These are the infantile, late infantile and juvenile forms, distinguished from each other by the age of onset and the clinical course of the disease. Other variant types and an adult form (Kuf's disease) have also been reported (Rider and Rider, 1988; Boustany *et al.*, 1988; Wisniewski *et al.*, 1988; Dyken, 1988; Eto *et al.*, 1988; Lake, 1984). Collectively they are probably the most common lysosomal storage diseases. The incidence has been estimated as high as 1 in 12 500 (Rider and Rider, 1988). Ceroid lipofuscinoses also occur in a number of animals. In particular, a flock of sheep with the disease has been maintained and studied as a model of the human diseases (Jolly *et al.*, 1980, 1982; Graydon and Jolly, 1984; Mayhew *et al.*, 1985). The clinical course of ovine ceroid lipofuscinosis most closely resembles the juvenile form of the human disease.

Characteristically the neuronal ceroid lipofuscinoses are associated with retinal and brain atrophy, and the accumulation of fluorescent storage bodies of lysosomal origin in neurones and a wide variety of other cells. It is the similarity of these storage bodies to the lipopigments ceroid and lipofuscin that gave rise to the name (Zeman and Dyken, 1969). To date, the underlying biochemical defect causing the ceroid lipofuscinoses remains elusive. Various hypotheses have been advanced that suggest that the disease arises because of abnormalities in peroxidation of lipids (Zeman, 1974; Siakotos *et al.*, 1988), or because of a deficiency of a peroxidase (Armstrong *et al.*, 1974), disturbance of fatty acid metabolism (Svennerholm *et al.*, 1975; Pullarkat *et al.*, 1982), defects in retinoic acid and dolichol metabolism (Wolfe *et al.*, 1977, Ng Ying Kin *et al.*, 1983) or by a defect in iron metabolism (Gutteridge *et al.*, 1982). More recently proposals have been made that the disease is related to reduced activities of proteases such as cathepsin D (Pullarkat *et al.*, 1988), cathepsin B (Dawson and Glaser, 1987), a thiol endo-protease (Wolfe *et al.*, 1987), and a protease involved in the processing of amyloid precursor protein (Wisniewski and Maslinska, 1989). It has also been suggested that the infantile form of the human disease could be considered as a separate type of disease and the name 'poly-unsaturated fatty acid lipidosis' has been proposed for it (Svennerholm, 1976).

Significant progress towards our understanding the cause of the ceroid lipofuscinoses has come from studying the biochemical nature of the storage bodies in humans and sheep affected with the disease. The studies are reviewed briefly, and are followed by a description of experiments to investigate the genetic cause of these diseases.

15.2 COMPOSITION OF STORAGE BODIES

In order to determine the nature of the stored component, storage bodies have been isolated from fresh tissues of affected sheep (Palmer *et al.*, 1986a, b, 1988). They were shown to be relatively free from contaminating material by electron microscopic examination. As summarized in Fig. 15.1, they contain normal lysomal phospholipids including bis(monoacylglycero)phosphate, which is unique to lysosomes. Also the levels of neutral lipids, notably dolichol, ubiquinone and dolichyl esters were typical of those in lysosomal membranes. These two lipid classes are present in similar amounts and make up less than one-third of the mass of the storage bodies (Palmer, 1986a, 1986b). Minor components include metal ions which are also indicative of a lysosomal origin for the storage bodies (Palmer *et al.*, 1988), and dolichol pyrophosphate-linked oligosaccharides (Hall *et al.*, 1989). The remaining two-thirds of the mass of the ovine bodies is protein.

Storage bodies have also been isolated from the brains and pancreas of children affected with juvenile, late infantile and infantile forms of the disease. As in sheep, the human bodies retain an ultrastructural appearance that is indistinguishable from that observed in electron micrographs of tissues. Consistent with the similarity of the human and ovine diseases, the human storage bodies appear to be similar in both lipid and protein contents to those of the sheep (Wolfe *et al.*, 1987; Palmer *et al.*, 1988). Dolichyl pyrophosphate-linked oligosaccharides are a minor component (Hall *et al.*, 1989).

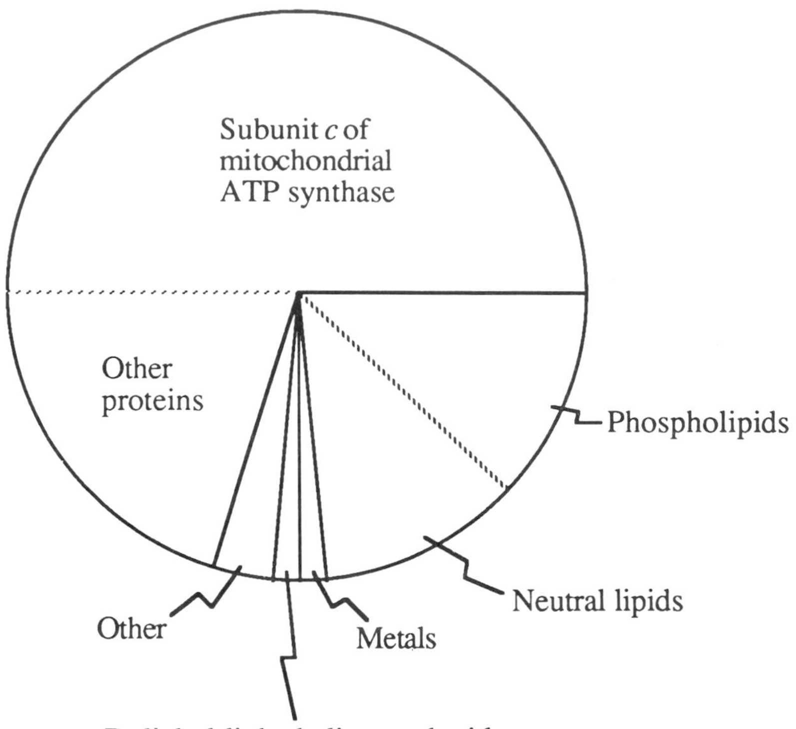

Fig. 15.1 Composition of the storage bodies in ovine ceroid lipofuscinosis.

15.3 ANALYSIS OF THE STORAGE BODY PROTEINS

Ovine storage body proteins contain a major component with an apparent molecular weight of 3.5 kDa as estimated by polyacrylamide gel electrophoresis (see Fig. 15.2; Palmer *et al.*, 1986b, 1988; Fearnley *et al.*, 1990). Similarly, storage bodies from victims of the juvenile and late infantile diseases also contain a component of the same size, but this component was not detected in storage bodies isolated from the brain of a child with the infantile form of the disease. Two other minor components of 14.8 kDa and 24 kDa were detected frequently in samples from ovine pancreas in variable amounts. Both of these higher-molecular-weight components have been shown by protein sequencing to be oligomers of the 3.5 kDa component (Fearnley *et al.*, 1990).

Sequence analysis of total ovine storage body proteins revealed that the major component had an N-terminal sequence that is identical with that of subunit *c* (the DCCD reactive proteolipid) of human and bovine mitochondrial ATP synthase, at least over the first 40 amino acids (Sebald

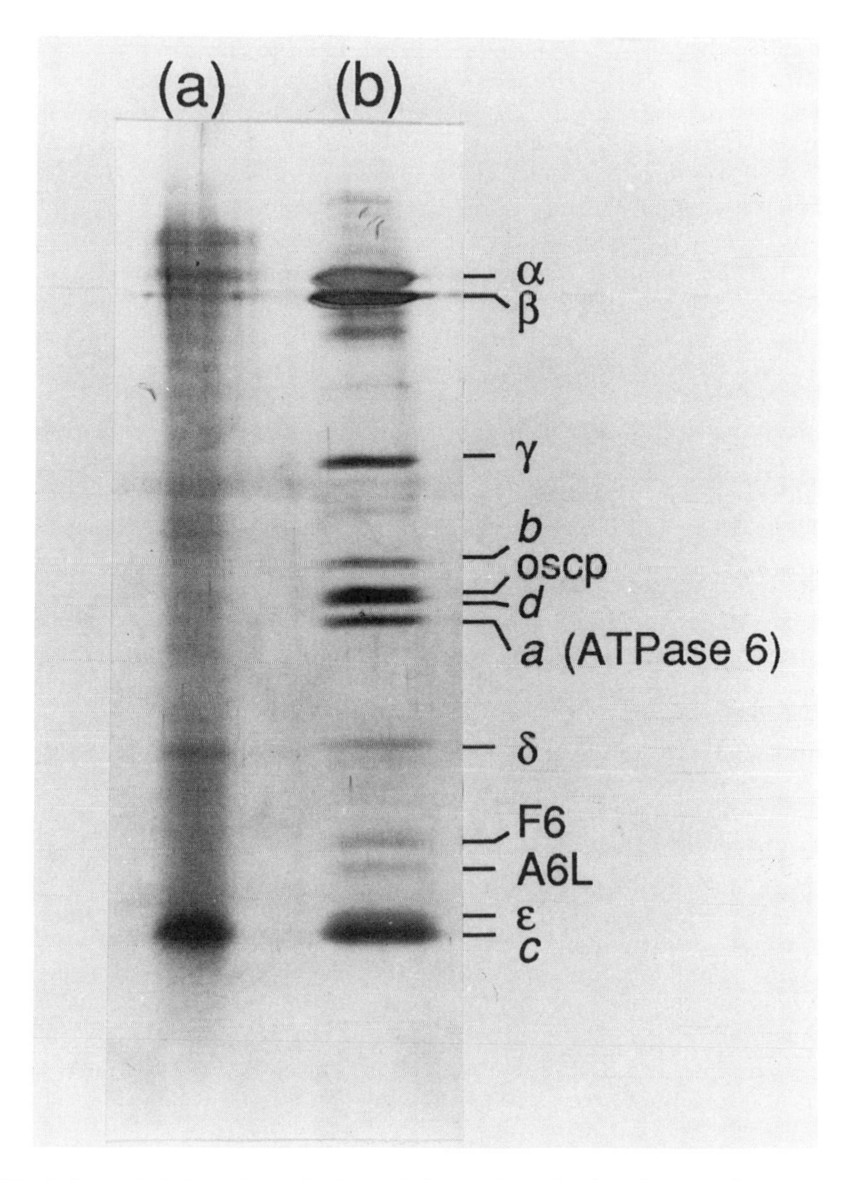

Fig. 15.2 Analysis by polyacrylamide gel electrophoresis of total protein from storage bodies isolated from the liver of sheep with ceroid lipofuscinosis. Lane (a) contains total storage body protein solubilized in lithium dodecyl sulphate, and lane (b) contains purified ATP synthase complex from bovine heart mitochondria. Molecular weights of the subunits of F^1-ATPase are as follows: α, 55 164; ß, 51 595; γ, 30 141; δ, 15 065; and ϵ, 5652.

et al., 1979; Gay and Walker, 1985; Palmer *et al.*, 1989). A minor sequence was detected in ovine bodies isolated from pancreas, but not from any other tissue. This sequence is related to that of a proteolipid isolated from bovine chromaffin granule H^+-ATPase (Mandel *et al.*, 1988), and to the N-terminal sequence of a 17 kDa protein isolated from mouse gap junction preparations (Walker *et al.*, 1986). It is not clear whether it represents an intrinsic component of the storage bodies or whether it originates from a minor cellular contaminant.

Furthermore, storage bodies associated with the juvenile and late infantile forms of the human disease also contained a predominant sequence identical with the N-terminal sequence of mitochondrial ATP synthase subunit *c* (Fearnley *et al.*, 1990; Palmer *et al.*, unpublished results). These biochemical data strengthen the ovine disease as a valid model for the juvenile and late infantile forms of the human disease. It has also been found that storage bodies isolated from the brains of an English setter and tissues of Devon cattle affected with ceroid lipofuscinosis also contain large amounts of subunit *c* (Martinus *et al.*, unpublished work). However, it was not detected in storage bodies from the brain of a victim of the infantile disease.

More extensive protein sequence studies showed that the major protein stored in the ovine bodies is identical with mitochondrial subunit *c* that is assembled into the ATP synthase complex, and appears to be unmodified as far as can be ascertained (Fearnley *et al.*, 1990). Similar experiments conducted on human storage bodies have shown that the complete subunit *c* is stored in the juvenile and late infantile diseases. The N-terminus of the stored protein is the same as that of the mature mitochondrial protein and so the mitochondrial pre-sequences have been precisely removed before storage.

Subunit *c* is an intrinsic membrane protein in the mitochondrial inner membrane, and each ATP synthase complex contains 9–12 copies. It is an essential component of the trans-membrane proton channel. The term proteolipid has often been applied to this protein because it is highly hydrophobic and has lipid-like properties, but it contains no covalently attached lipid. In common with the protein present in the storage bodies, mitochondrial subunit *c* migrates on polyacrylamide gels with an apparent molecular weight of about 3500 and higher-molecular-weight aggregates can also be detected (see Fig.15.2; Walker *et al.*, 1991).

From protein-sequencing experiments and amino acid composition, it has been calculated that subunit *c* constitutes at least 73% of the protein content of the ovine storage bodies (Fearnley *et al.*, 1989; Palmer *et al.*, 1989). Since the bodies are at least two-thirds protein, this means that subunit *c* accounts for greater than 50% of the total storage body mass. Similar calculations made for the storage bodies from the juvenile and late infantile ceroid lipofuscinoses gave amounts ranging from 20 to 85% of storage body protein. It has been reported that the storage bodies from sheep with ceroid lipofuscinosis and Batten's patients have a characteristically similar fluorescence distinct from that of the age pigment lipofuscin (Dowson *et al.*, 1982). Several mechanisms by which proteins can be modified to form fluorophores have been suggested, including peroxidation reactions forming malonaldehyde, which can react with the amino groups of proteins to yield a fluorescent Schiff base product (Chio *et al.*, 1969; Koobs *et al.*, 1978), and the glycation of lysine residues by glucose to produce fluorescent heterocyclic compounds (Pongor *et al.*, 1984). Both of these mechanisms involve the modification of amino acids and could also cross-link polypeptide chains. There is no evidence that such a mechanism is the basis of the fluorescence of the storage bodies and no modified amino acids were detected during protein sequence analysis of the subunit *c* isolated from the storage bodies. Nevertheless, it should be borne in mind that minor levels of modification would not have been detected in the sequencing experiments, and that the aggregates of subunit *c* observed on polyacrylamide gels could conceivably be the cross-linked products of such reactions.

15.4 STUDIES OF GENES FOR SUBUNIT *c* OF MITOCHONDRIAL ATP SYNTHASE

In humans and cattle, subunit *c* has two nuclear genes, called P1 and

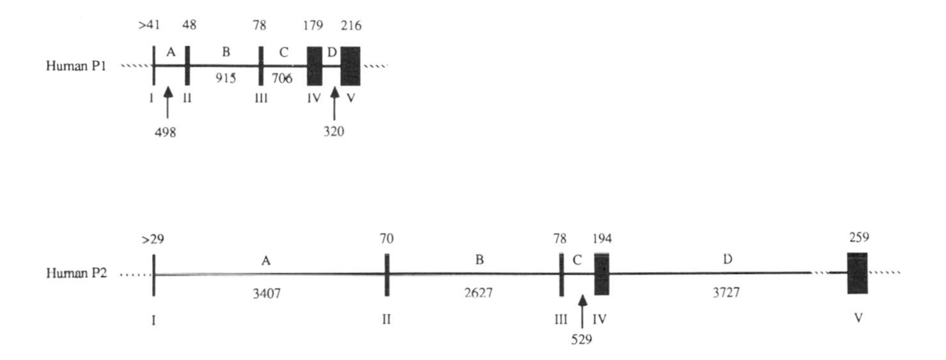

Fig. 15.3 Structures of the human P1 and P2 genes for subunit c of mitochondrial ATP synthase. The exons and introns of these genes are represented by filled boxes and adjoining lines, respectively. The exons are numbered with roman numerals and their sizes in bases are shown above the box. Introns are labelled with capital letters and their sizes are given below the lines. From Dyer and Walker (1990).

P2, that encode precursor proteins containing the same mature protein, but with different N-terminal presequences (Gay and Walker, 1985; Dyer *et al.*, 1989; Dyer and Walker, 1990). These pre-sequences are sufficiently conserved and characteristically similar to the import pre-sequences of other mitochondrial proteins, to suggest that they both direct the protein to mitochondria. Import sequences are removed by one or more proteases located in the mitochondrial matrix during import into the organelle. The human genes for P1 and P2 have been cloned and sequenced and they have closely related structures (Dyer and Walker, 1990; see Fig. 15.3). In addition, numerous spliced and partly spliced pseudogenes related to P1 or P2 have been discovered in man and cattle, and many are also present in sheep (Dyer *et al.*, 1989; Dyer and Walker, 1990). In cattle, both the P1 and P2 genes are expressed in all tissues, but the ratio of their expression differs according to the embryonic origin of the tissue. The ratio of P1 : P2 is 1 : 1 in tissues of mesodermal origin and 1 : 3 in those of endodermal and ectodermal origin (Gay and Walker, 1985; Walker *et al.*, 1988).

A priori, it was possible that the disease could be caused by a defect in the targeting of one of the precursors for subunit *c*. In order to test this hypothesis cDNA libraries were constructed from sheep mRNA, and cDNAs for the P1 and P2 genes were cloned by screening these libraries and by using the polymerase chain reaction on total cDNA from both normal sheep and sheep with ceroid lipofuscinosis. Sequence analysis

```
                     10                  20                30
P1 M Q T T G A L L I S P A L I R S C T R G L I R P V S A S F L S R P E I P S V Q P
   :   :               :   : : :           :   :     : :         : : :
P2 M Y T C A K F V S T P S L I R R T S T L L S R S L S A V V V R R P E T L T D E S
                     10                  20                  30

                     50                                60                  70
P1 S Y S S G P L Q V A R - - - - - - - R E F Q T S V V S R D I D T A A K F I G A G
       :                               :   : : : :     : : : : : : : : : : : : : :
P2 H S S L A V V P R P L T T S L T P S R S F Q T S A I S R D I D T A A K F I G A G
                     50                  60                  70

               80                  90                  100                 110
P1 A A T V G V A G S G A G I G T V F G S L I I G Y A R N P S L K Q Q L F S Y A I L
   : : : : : : : : : : : : : : : : : : : : : : : : : : : : : : : : : : : : : : : :
P2 A A T V G V A G S G A G I G T V F G S L I I G Y A R N P S L K Q Q L F S Y A I L
                     90                  100                 110

               120                 130
P1 G F A L S E A M G L F C L M V A F L I L F A M
   : : : : : : : : : : : : : : : : : : : : : : :
P2 G F A L S E A M G L F C L M V A F L I L F A M
                     130                 140
```

Fig. 15.4 Alignment of ovine P1 and P2 subunit *c* precursors. The colons indicate amino acid differences and the dashes denote insertions in the sequence of the P1 precursor. The mitochondrial import sequences are shown in the open boxes and the mature polypeptides are in shaded boxes.

of these clones showed that the coding region sequences of the P1 and P2 cDNAs cloned from the diseased sheep were identical with those of the normal sheep (see Fig. 15.4). Also, the amino acid sequence of mature subunit *c* predicted from the cDNA sequence agrees precisely with that determined by protein sequencing. Therefore, both the P1 and P2 precursors are normal in diseased animals, and the disease is not caused by mutation of the mitochondrial import sequences.

A second possible cause of the disease was that one or both of the genes for subunit *c* was over-expressed, leading to excess synthesis of the protein and consequent accumulation in lysosomes. In order to compare the expression of the genes for subunit *c* in normal sheep and sheep with ceroid lipofuscinosis, total RNA from normal and diseased sheep liver was fractioned by agarose gel electrophoresis and then

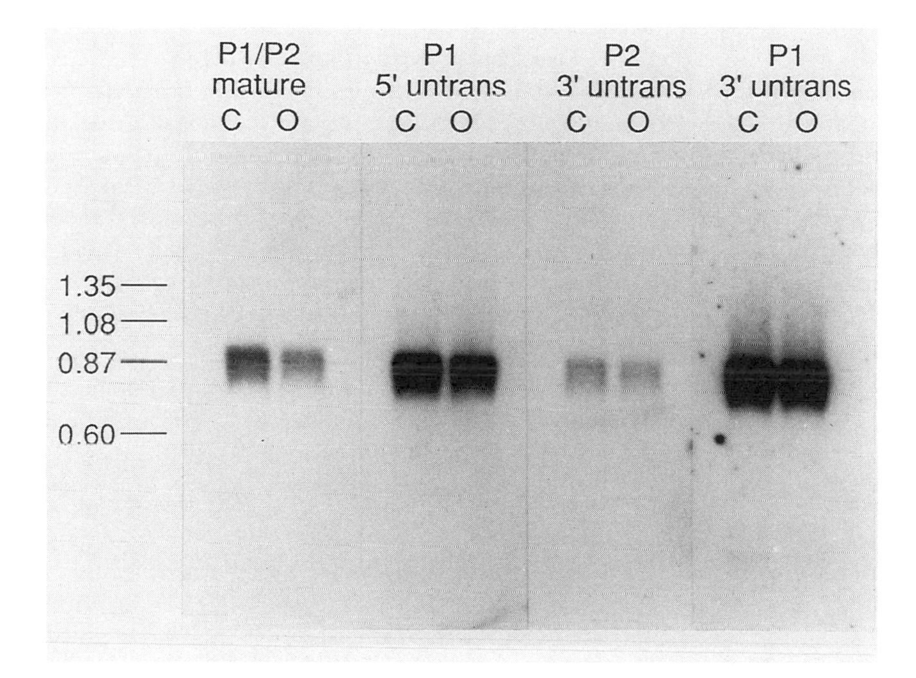

Fig. 15.5 Hybridization of total liver RNA from normal sheep and sheep with ceroid lipofuscinosis with P1 and P2 specific probes. Size markers on the left of the gel are in kilobases. Four identical pairs of total RNA samples probed with different DNA probes are aligned. In each blot 'C' and 'O' refer to total RNA from normal (control) sheep and diseased sheep (0). The P1/P2 mature probe is bases 196 to 556 of the bovine P1 cDNA and hybridizes to both P1 and P2. The P1 5′ untranslated region probe is bases 1 to 195 of the bovine P1 cDNA. The P2 3′ untranslated region probe is bases 406 to 615 of the bovine P2 cDNA and the P1 3′ untranslated region probe is bases 404 to 558 of the bovine P1 cDNA (see Gay and Walker, 1985).

hybridized with DNA probes that were specific for either P1 or P2 (see Fig. 15.5). This showed that there is no gross difference in the amounts or sizes of the mRNAs for P1 and P2 in normal and diseased sheep. Therefore, it appears to be unlikely that subunit *c* is being synthesized in the diseased animals in greater amounts than is normally required for assembly of the multi-subunit enzyme ATP synthase in mitochondria.

During the course of the cDNA cloning experiments, a mutated cDNA related to P2 was isolated from the cDNA library made from a sheep with ceroid lipofuscinosis. This cDNA arose by expression of a spliced P2 pseudogene; a normal sheep also expressed a closely related pseudogene. However, the pseudogene from the normal sheep encodes a 31 amino acid peptide identical with the first 31 amino acids of the P2 precursor, but the peptide encoded by the pseudogene from the sheep with ceroid lipofuscinosis is one amino acid different, arginine-23 being changed to glutamine (see Fig. 15.6). Spliced pseudogenes are thought to be retroposons that have been copied from mRNA by reverse transcription during evolution, and that the resulting cDNA-like molecule has been introduced into the genomic DNA (Weiner *et al.*, 1986). Transcription of a pseudogene, as in the present case, requires that potential promoter elements are present to the 5′ side of the site of insertion.

Further experiments have shown that the normal and diseased sheep investigated so far are homozygous for their respective pseudogenes. This raises two possibilities: the expressed pseudogene in the diseased sheep could be the cause of ovine ceroid lipofuscinosis, or the pseudogenes may be closely linked to the gene causing the disease.

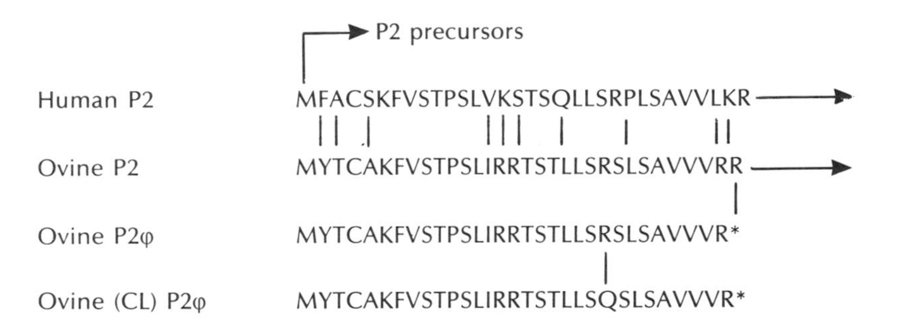

Fig. 15.6 Alignment of the amino acid sequences P2 precursors encoded by the human P2 gene, the ovine P2 gene, an expressed ovine P2 pseudogene from normal sheep and an expressed ovine P2 pseudogene from sheep with ceroid lipofuscinosis (CL). The sequence of the human P2 precursor is taken from Dyer and Walker (1990). Vertical lines denote differences in amino acid difference between adjacent sequences, an asterisk denotes a stop codon, and the arrows indicate that the amino acid sequence continues further.

It is difficult to assess the possibility that the mutation of the pseudogene in the diseased sheep is the cause of ovine ceroid lipofuscinosis. Hybridization experiments on total RNA have established that transcripts from the pseudogenes of normal and diseased animals exist in very low amounts. Should these transcripts be translated, the products could conceivably interfere with mitochondrial import by blocking receptors or inhibiting the processing of the precursors. However, if the pseudogene in diseased sheep is the cause of ovine ceroid lipofuscinosis, the the peptide it encodes would have to be different and capable of causing the accumulation of subunit *c* in diseased cells. This could be investigated by transfecting the sequences of both pseudogenes in suitable expression vectors into appropriate cell lines and observing what effects, if any, ensue.

Experiments are in progress to establish whether the pseudogenes are closely linked to the gene causing ceroid lipofuscinosis. These involve amplifying the pseudogenes from the DNA of normal sheep, sheep which are heterozygous carriers of the disease, and diseased sheep using the polymerase chain reaction, and then digesting these products with restriction enzymes that distinguish between the sequences of the two pseudogenes. Should the pseudogene previously cloned from a diseased sheep always segregate with the disease, then it would be reasonable to assume that this pseudogene is close to or involved in the genetic cause of ovine ceroid lipofuscinosis.

Candidate genes for the genetic lesion in the ceroid lipofuscinosis might also include genes for proteins involved in the degradative pathway of subunit *c* of ATP synthase. The processes of mitochondrial protein turnover and degradation are poorly understood. It is known that lysosomes possess many proteases with a wide range of specificities, and that they can digest whole mitochondria by a process that involves autophagocytosis (Pfeifer, 1987). There is also evidence that the primary lysosomes of myocytes can fuse with mitochondria to initiate mitochondrial turnover (Skepper and Navaratnam, 1987). However, the results of turnover studies on a variety of mitochondrial proteins have shown that their turnover rates can vary widely and that even mitochondrial inner membrane proteins, of which subunit *c* is an example, can differ severalfold in their turnover rates (Lipsky and Pedersen, 1981; Hare and Hodges, 1982a, 1982b). This suggests that mitochondria contain at least some proteases capable of digesting their own proteins. One such protease is the ATP-dependent protease of the mitochondrial matrix which appears to be capable of extensive proteolysis (Desautels and Goldberg, 1982). Whether this protease or any other mitochondrial protease is involved in the degradation of subunit *c* is unknown.

ACKNOWLEDGEMENTS

S.M.M. was supported by an MRC Research Studentship and a Beit Memorial Fellowship for Medical Research. D.N.P. and R.D.J. were supported in part by grant no. NS 11238 of the United States National Institute of Neurological and Communicative Disorders, and a New Zealand MRC–Wellcome Trust Travel Fellowship (D.N.P.).

REFERENCES

Armstrong, D., Dimmett, S. and Van Wormer, D.E. (1974) *Arch. Neurol.*, **30**, 144–52.

Boustany, R.N., Alroy, J. and Kolodny, E.H. (1988) *Am. J. Med. Genet. Suppl.*, **5**, 47–58.

Chio, K.S., Reiss, U., Fletcher, B. and Tappel, A.L. (1969) *Science*, **166**, 1535–6.

Dawson, G. and Glaser, P. (1987) *Biochem. Biophys. Res. Commun.*, **147**, 267–74.

Desautels, M. and Goldberg, A.L. (1982) *J. Biol. Chem.* **257**, 11673–9.

Dowson, J.H., Armstrong, D., Koppang, N. *et al.* (1982) *Acta Neuropathol. (Berl.)*, **58**, 152–6.

Dyer, M.R. and Walker, J.E. (1990) *J. Mol. Biol.*, in press.

Dyer, M.R., Gay, J.N. and Walker, J.E. (1989) *Biochem. J.* **260**, 249–58.

Dyken. P.R. (1988) *Am. J. Med. Genet. Suppl.*, **5**, 69–84.

Eto, Y., Tsuda, T., Ohhashi, T. *et al.* (1988) *Am. J. Med. Genet. Suppl.*, **5**, 59–68.

Fearnley, I.M., Walker, J.E., Jolly, R.D. *et al.* (1990) *Biochem J.*, **268**, 751–8.

Gay, N.J. and Walker, J.E. (1985), *EMBO J.*, **4**, 3519–24.

Graydon, R.J. and Jolly, R.D. (1984) *Invest. Ophthalmol. Visual Sci.*, **25**, 294–301.

Gutteridge, J.M.C., Rowley, D.A., Halliwell, B. and Westermarck, T. (1982) *Lancet*, **ii**, 459–60.

Hall, N.A. and Patrick, A.D. (1985) *J. Inherit. Metab. Dis.*, **8**, 178–83.

Hall, N.A., Lake, B.D., Palmer, D.N. *et al.* (1990) *Adv. in Exp Med and Biol.*, **266**, 225–42.

Hare, J.F. and Hodges, R. (1982a) *J. Biol. Chem.*, **257**, 12950–3.

Hare, J.F. and Hodges, R. (1982b) *J. Biol. Chem.*, **257**, 3575–80.

Jolly, R.D., Janmaat, A., West, D.M. and Morrison, I. (1980) *Neuropathol. Appl. Neurobiol.*, **6**, 219–28.

Jolly, R.D., Janmaat, A., Graydon, R.J. and Clemett. R.S. (1982) in *Ceroid-lipofuscinosis (Batten's Disease)* (eds D. Armstrong, N. Koppang and J.A. Rider), Elsevier, Amsterdam, pp. 219–28.

Jolly, R.D., Shimada, A., Dopfmer, I. *et al.* (1989) *Neuropathol. Appl. Neurobiol.*, **15**, 371–83.

Koobs, D.H., Schultz, R.L. and Jutzy, R.V. (1978) *Arch. Pathol. Lab. Med.*, **102**, 66–8.

Lake, B.D. (1984) in *Greenfield's Neuropathology*, (eds J. Hulme-Adams, J.A.N. Corsellis and L.W. Duchen) Edward Arnold, London, pp. 491–572.

Lipsky, N.G. and Pedersen, P.L. (1981) *J. Biol. Chem.*, **256**, 8652–7.

Mandel. M., Moriyama, Y., Hulmes, J.D. *et al.* (1988) *Proc. Natl Acad. Sci. USA*, **85**, 5521–4.

Mayhew, I.G., Jolly, R.D., Pickett, B.T. and Slack, P.M. (1985) *Neuropathol. Appl. Neurobiol.*, **11**, 273–390.
Ng Ying Kin, N.M.K., Palo, J., Haltia, M. and Wolfe, L.S. (1983) *J. Neurochem.*, **40**, 1465–73.
Palmer, D.N., Husbands, D.R., Winter, P.J. *et al.* (1986a) *J. Biol. Chem.*, **261**, 1766–72.
Palmer, D.N., Barns, G., Husbands, D.R. and Jolly, R.D. (1986b) *J. Biol. Chem.*, **261**, 1773–7.
Palmer, D.N., Martinus, R.D., Barns, G. and Jolly, R.D. (1988) *Am. J. Med. Genet. Suppl.*, **5**, 141–58.
Palmer, D.N., Martinus, R.D., Cooper, S.M. *et al.* (1989) *J. Biol. Chem.*, **264**, 5736–40.
Pfeifer, U. (1987) in *Lysosomes: Their Role in Protein Breakdown*, Academic Press, London, pp. 3–59.
Pongor, S., Ulrich, P.C., Bencsath, F.A. and Cerami, A. (1984) *Proc. Natl Acad. Sci. USA*, **81**, 2684–8.
Pullarkat, R.K., Reha, H., Patel, V.K. and Goebel, H.H. (1982) in *Ceroid-Lipofuscinosis (Batten's Disease)* (eds D. Armstrong, N. Koppang and J.A. Rider), Elsevier, Amsterdam, pp. 335–43.
Pullarkat, R.K., Kim, K.S., Sklower, S.L. and Patel, V.K. (1988) *Am. J. Med. Genet. Suppl.*, **5**, 243–52.
Rider, J.A. and Rider, D.L. (1988) *Am. J. Med. Genet. Suppl.*, **5**, 21–6.
Sebald, W., Wachter, E. and Tzagoloff, A. (1979) *Eur. J. Biochem.*, **100** 599–607.
Siakotos, A.N., Bray, R., Dratz, E. *et al.* (1988) *Am. J. Med. Genet. Suppl.*, **5**, 171–82.
Skepper, J.N. and Navaratnam, V. (1987) *J. Anat.*, **150**, 155–67.
Svennerholm, L. (1976) in *Current Trends in Sphingolipidoses and Allied Disorders*, (eds B.W. Volk and L. Schnek), Plenum Press, New York, pp. 389–402.
Svennerholm, L., Hagberg, B., Haltia, M. *et al.* (1975) *Acta Pediatr. Scand.*, **64**, 489–96.
Walker, J.E., Fearnley, I.M. and Blows, R.A. (1986) *Biochem. J.*, **237**, 73–84.
Walker, J.E., Cozens, A.L., Dyer, M.R. *et al.* (1988) in *Molecular Basis of Biomembrane Transport*, Elsevier, Amsterdam, pp. 209–16.
Walker, J.E., Lutter, R., Dupuis, A. and Runswick, M.J. (1989) *Biochemistry*, submitted for publication.
Weiner, A.M., Deininger, P.L. and Estratiadis, A. (1986) *Annu. Rev. Biochem.*, **55**, 631–61.
Wisniewski, K.E. and Malinska, D. (1989) *N. Engl. J. Med.*, **20**, 256–7.
Wisniewski, K.E., Rapin, I. and Heaney-Kieras, J. (1988) *Am. J. Med. Genet. Suppl.*, **5**, 27–46.
Wolfe, L.S., Ng Ying Kin, N.M.K., Baker, R.R. *et al.* (1977) *Science*, **195**, 1360–2.
Wolfe, L.S., Ivy, G.O. and Witkop, C.J. (1987) *Chem. Scripta*, **27**, 79–84.
Zeman, W. (1974) *J. Neuropathol. Exp. Neurol.*, **33**, 1–12.
Zeman, W. (1976) in *Progress in Neuropathology* (ed H.M. Zimmerman), Grune and Stratton, New York, pp. 207–23.
Zeman, W. and Dyken, P. (1969) *Paediatrics*, **44**, 570–83.

16 Leber's hereditary optic neuropathy: from the clinical to the neurobiochemical and molecular findings for understanding the pathogenesis of the disorder

A. FEDERICO
Istituto di Scienze Neurologiche e Centro Per lo Studio delle Encefalo-Neuro-Miopatie Genetiche, Università degli Studi di Siena, Italy

16.1 INTRODUCTION

Leber's hereditary optic neuropathy (LHON) is a well-defined disorder clinically characterized by acute or subacute bilateral visual loss, usually in young, otherwise healthy men. Women can also be affected. The disease is clearly maternally transmitted and, in contrast to X-inheritance, the descendants of men are never affected. This particular type of genetic transmission directed the attention to the mitochondrial pathogenesis of the disease, which has recently been demonstrated (Wallace *et al.*, 1988).

Here we report data on the clinical aspects of LHON and on the biochemical pathogenesis of the disease.

16.2 PURE LHON AND COMPLICATED LHON SYNDROMES

LHON is clinically characterized by a subacute bilateral visual loss at a young age. The initial reduction of visual acuity is frequently accompanied by a violent bout of headache. Furthermore, during the period prior to the acute onset, patients may have a noticeable sensation of decreased vision each time they undergo physical exertion (Uhthoff's symptom).

The evolution of the ophthalmoscopic picture has been described by Nikoskelainen *et al.* (1983): in the prodromic period the optic disc appears hyperaemic with the presence of abnormal peripapillary arterioles and capillaries. With the appearance of symptoms the optic disc is swollen with indistinct margins, while hyperaemia and peripapillary telangiectatic angiopathy are increased. The peripapillary vessels became turgid and tortuous. Small pre-retinal haemorrhages may appear and the surrounding retina seems glistening and sometimes with visible striations.

The terminal atrophic stage is characterized by degrees of atrophy which vary from subject to subject, with extremely attenuated arterial vessels and possible persistence of tortuousness in the venous vessels. This vasculopathy seems to be the primary lesion and optic atrophy is the secondary change. The peripapillary microangiopathy reported by Nikoskelainen *et al.* (1982) occurs also in asymptomatic members in the female line of families with the disease.

The possibility of multisystem involvement in the disease has been extensively investigated. In the original description, Leber (1871) reported that cardiac palpitations occurred in some patients with the disease. Palpitations and episodic loss of consciousness were later reported by Wilson (1965) in one patient and a cardiomyopathy was reported in another patient at autopsy. Rose *et al.* (1970) found electrocardiographic abnormalities in two patients with LHON. More recently Nikoskelainen *et al.* (1985) reported a high incidence of a pre-excitation syndrome in affected as well as in symptom-free subjects. These data have been also confirmed by our group (Federico *et al.*, 1987).

Other evidence of multisystem involvement in LHON cases without other central nervous system (CNS) signs is the presence of abnormalities of brain stem auditory evoked responses that have been reported by our group in 62% of the studied subjects (seven affected, one at risk of being affected, two asymptomatic carriers), all without hearing defect (Mondelli *et al.*, 1990).

Abnormalities in other systems, including muscle (Nikoskelainen *et al.*, 1984; Federico *et al.*, 1988) and liver (Pallini *et al.*, 1988), in cases of pure LHON will be discussed in section 16.3.

Despite the presence of signs suggesting a minor involvement of other organs, the ocular aspect is predominant and the extraocular signs are minor. However, in the literature cases are reported in which LHON, with typical clinical fluorangiographic aspects and genetic transmission, is combined with other severe neurological signs.

McLeod *et al.* (1978) reported a family in which LHON was associated with Charcot–Marie–Thoot disease: in his article he concluded that the two conditions were independently inherited. Pages and Pages (1983) reported a case of LHON with spastic paraparesis and peripheral neuropathy. Novotny *et al.* (1986) studied a family in which eight members were affected by LHON, 14 by a progressive generalized dystonia attributed to striatal degeneration seen by computed tomography (CT) scan or nuclear magnetic resonance imaging and one had both the disorders. Wallace (1970) reported a large family that has been more recently biochemically investigated (Parker *et al.*, 1989) in which LHON is associated in many subjects with a variety of neurological disorders, including ataxia, spasticity, skeletal deformities, posterior column signs, encephalopathy, psychiatric symptoms and death in childhood for neurological diseases. Leber (1871) in his original article noted the regular presence of psychiatric stigmata and minor neurological symptoms in patients and in some of their unaffected relatives. De Weerdt and Went (1971) found that in a total of 35 pedigrees (from 23 articles) in which 343 patients with LHON were reported, 230 patients were examined and in 84 of them neurological and psychiatric symptoms have been reported, including dystonia, headache and migraine, tremors, nystagmus, epilepsy, pyramidal signs and mental retardation, Wilson (1965) described ten patients with LHON (six belonging to one large pedigree), all with marked neurological abnormalities, which developed after the onset of optic atrophy. In another 30 patients with 'uncomplicated' LHON, minor psychiatric symptoms were evident, which may be aspecific.

Bruyn and Went (1964) and Went (1972) reported in a large pedigree of seven generations 16 patients with LHON, of whom eight showed a severe but not progressive neurological disorder intermediate between spastic paraparesis and Hallerworden–Spatz disease. In this family seven other subjects presented the same neurological symptoms except optic atrophy. One of the two female patients in this family died from a disease that was diagnosed as multiple sclerosis. A similar case has also been reported by Lees *et al.* (1964)

On the basis of these clinical data, we can conclude that two types of conditions exist:

1. The more common LHON, in which an isolated affection of the optic nerve is present, and in which minor subclinical neurological or psychiatric changes may occur, as already described by Leber (1871). Multisystem involvement is only characterized by heart changes (Nikoskelainen *et al.*, 1984; Federico *et al.*, 1987) or brainstem acoustic evoked potentials (BAEP) alterations (Mondelli *et al.*, 1990).
2. A form in which the ophthamological abnormalities of isolated LHON is one of the signs of a more diffuse disorder of the CNS. In such families cases can be present in which optic atrophy alone or progressive neurological disorders involving different systems without optic atrophy or the combined conditions are observed.

Whether these two types of disorders are due to the same pathogenetic mechanism is not known at present.

16.3 BIOCHEMICAL PATHOGENESIS OF LHON

The first evidence that a disorder of cell energy metabolism at the mitochondrial level could be important for the pathogenesis of LHON was reported by Went (1964), who found changes in serum pyruvate and lactate after glucose and glucagon tests. In 1965 Wilson observed that in most cases of LHON the severity of the disease could be related to tobacco smoking. Because tobacco is the major exogenous source of cyanide in Europe, he suggested cyanide as a possible precipitant factor of visual loss in LHON. This hypothesis has been reinforced by the observation that optic neuritis occurs as a result of cyanide intoxication from ingestion of bitter cassava in man (Cliff *et al.*, 1985). Wilson (1965) observed significantly increased urinary thiocyanate excretion in healthy smokers as compared to that in non-smokers. In contrast, no difference between smokers and non-smokers was found in LHON patients, suggesting that they were unable to respond to an increased intake of cyanide by increased detoxication. A relatively low level of thiocyanate was observed in LHON patients also by Rogers (1977).

The key enzyme for cyanide detoxication is rhodanese (thiosulphate sulphur transferase, EC 2.8.1.1), a mitochondrial matrix enzyme. When this enzyme is inactive, CN^- ion forms a stable complex with iron, decreasing its electron carrier properties. There is inhibition of cytochrome oxidase activity with consequent decrease of aerobic metabolism and increase of lactate and pyruvate. Cyanide also forms complexes with other haem compounds such as methaemoglobin, catalase and peroxidase, and non-haem compounds such as hydroxycobalamine, succinic and pyruvic acid dehydrogenases, tyrosinase and ascorbic oxidase (Rieders, 1977; Pagani and Galante, 1983).

A significant decrease of rhodanese enzyme has been found by Cagianut *et al.*, (1984) in the liver and by Poole and Kind (1986) in the rectal mucosa of patients with LHON, while normal levels have been found in the liver by Wilson (1965), in the muscle by Nikoskelainen (1984), in leucocyte, fibroblasts and liver by our group (Pallini *et al.*, 1987, 1988, 1989, in red cells by Syme *et al.* (1983) and in leucocytes by Berninger *et al.* (1989).

Immunoblotting using rabbit polyclonal antibodies against bovine liver rhodanese, able to recognize the human enzyme in various tissues, showed no differences in electrophoretic mobility and immunoreactivity of rhodanese between patients' liver and controls (Pallini *et al.*, 1988).

To confirm the possible pathogenetic role of a dysfunction in cyanide metabolism, Berninger *et al.* (1989) reported an increased level of blood cyanide in an LHON patient in the acute stage. Cyanide level returned and remained normal 5 months after the acute onset of the disease. Normal cyanide blood levels were found in LHON subjects in the chronic stage.

Histochemical and biochemical studies on the mitochondrial pathogenesis of LHON have been performed in muscle, where Nikoskelainen *et al.* (1984) first reported ultrastructural abnormalities of mitochondria in 'pure' LHON cases.

We studied respiratory chain enzyme activity in isolated mitochondria from three LHON cases, in which histological and ultrastructural abnormalities had been found (increase of subsarcolemmal mitochondria): an increased activity of all the examined enzymes and particularly of succinate cytochrome c reductase was found (Federico *et al.*, 1988).

Parker *et al.* (1989) reported that in four members of the very complex family described by Wallace (1970) a decreased activity of reduced nicotinamide adenine dinucleotide–coenzyme Q (NADH–CoQ) reductase in platelets occurred. In this family, as previously mentioned, the clinical symptoms reported in many patients (including stroke-like episodes and encephalopathy) suggest a similarity to MELAS syndrome, a disorder in which a defect of complex I has been reported (Petty *et al.*, 1986).

The mutation of mitochondrial DNA associated with LHON has been identified by Wallace *et al.* (1988) and converts an evolutionarily highly conserved arginine to a histidine in ND4, one of the seven subunits of NADH dehydrogenase encoded by mitochondrial DNA. The mutation (G to A at base pair 11 778) eliminates as SfaNI site and thus provides a method for detection of the mutation by restriction fragment length polymorphism analysis. Wallace *et al.* (1988) found the mutation in 9 of 11 LHON families and in none of 45 controls. More recently, in a more extensive study, the mutation has been found in 10 of 19 Finnish families (Vilkki *et al.*, 1989), providing evidence of a genetic heterogeneity.

The mutation has been reported in 'pure' LHON families from different ethnic origin (Wallace *et al.*, 1988; A. Harding, personal communication, 1989; Hotta *et al.*, 1989; Oldfors *et al.*, 1989; Yoneda *et al.*, 1989; Vilkki *et al.*, 1989). The mutation was not found in the family in which LHON was associated with dystonia (Novotny *et al.*, 1986; Wallace *et al.*, 1988).

In the families showing the SfaNI site mutation, the mutation is homoplasmic in all individuals irrespective of their disease status, suggesting that the intrafamilial variations in the clinical expression, well documented in many reports, is not due to the different ratios of the mutant versus normal mitochondrial DNA.

This evidence implies that additional factors are involved in the expression of the mutant phenotype.

16.4 CONCLUSIONS

Although a mutation has been found in several families and in a very particular family a defect of NADH-CoQ reductase has been found that well agrees with the molecular change reported in other families, many questions are still open that can be summarized as follows:

1. What is the alternative mutation in families with 'pure' LHON that are negative for the SfaNI mutation?
2. What is the explanation of the clinical heterogeneity in the presence of a homoplasmic DNA mutation in all the affected or unaffected family members?
3. Is the SfaNI mutation present in the family of Wallace (1970) in which Parker *et al.* (1989) have found a deficiency of NADH-CoQ reductase, and in many other families with 'complicated' LHON syndromes?
4. Why and how is the optic nerve so severely vulnerable?
5. How can a widespread mitochondrial DNA mutation giving rise to a respiratory chain defect cause the subacute development of blindness in adult life?

All the recent data are not inconsistent with the previous hypothesis emphasizing cyanide intoxication in the pathogenesis of the disease. Cyanide is a known inhibitor of complex I activity and it may be that patients with a potentially symptomatic mutation involving the activity of complex I (as in the case of the SfaNI mutation) are more likely to become symptomatic when the defective enzyme complex is further inhibited by cyanide.

If this is the case, metabolic therapies that are able to decrease cyanide level or to increase respiratory metabolism might reduce the risk of development of symptoms in, as yet, unaffected family LHON members, showing the DNA mutation.

ACKNOWLEDGEMENTS

Research was in part funded by a grant from the Regione Toscana, Florence.

REFERENCES

Berninger, T.A., Meyer, L.V., Siess, E. *et al.* (1989) *Br. J. Ophthalmol.*, **73**, 314–16.
Bruyn, G.W. and Went, L.N. (1964) *J. Neurol. Sci.*, **1**, 59–80.
Cagianut, B., Schnebli, H.P., Rhyner, K. and Furrer, J. (1984) *Klin. Wochenschr.*, **62**, 850–4.
Cliff, J., Lundquist, P., Martensson, J. *et al.* (1985) *Lancet*, **i**, 1211–13.
De Weerdt, C.J. and Wend, L.N. (1971) *Acta Neurol. Scand.*, **47**, 541–54.
Federico, A., Aitiani, P., Lo Monaco, B. *et al.* (1987) *J. Inherited Metab. Dis.* **10**, (Suppl. 2), 256–9.

Federico, A., Manneschi, L., Meloni, M. *et al.* (1988) *J. Inherited Metab. Dis.*, **11** (Suppl.2), 193–7.
Hotta, Y, Hayaakawa, M., Saito, K. *et al.* (1989) *Am. J. Ophthalmol.*, **108**, 601–2.
Leber, Th. (1871) *Albrecht v. Graefes Arch. Ophthalmol.*, **17**, 249–91.
Lees, F., MacDonald, A.M.E. and Turner, J.W.A. (1964) *J. Neurol. Neurosurg. Pschiat.*, **27**, 415–21.
McLeod, J.G., Low, P.A. and Morgan, J.A. (1978) *Neurology*, **28**, 179–84.
Mondelli, M., Rossi, A., Scarpini, C. *et al.* (1990) *Acta Neurol. Scand.*, **81**, 349–53.
Nikoskelainen, E. (1984) *Neurology*, **34**, 1482–8.
Nikoskelainen, E., Hoyt, W.F. and Nummelin, K. (1982) *Arch. Ophthalmol.*, **100**, 1597–602.
Nikoskelainen, E., Hoyt, W.F. and Nummelin, K. (1983) *Arch. Ophthalmol.*, **101**, 1059–68.
Nikoskelainen, E., Hassinen, I.E., Raljarri, L. *et al.* (1984) *Lancet*, **ii**, 1474.
Nikoskelainen, E., Savontaus, M.L., Wanne, O. *et al.* (1987) *Arch. Ophthalmol.*, **105**, 665–71.
Nikoskelainen, E., Wanne, O. and Dahl, M. (1985) *Lancet*, **i**, 696.
Novotny, E.J., Singh, G., Wallace, D.C. *et al.* (1986) *Neurology*, **36**, 1053–60.
Oldfors, A., Holme, E., Kristiansson, B. *et al.* (1989) Abstract, Meeting Molecular Basis of Neurological Disorders and their Treatment, Selva di Fasano, Sept. 11–14.
Pagani, G. and Galante, Y.M. (1983) *Biochem. Biophys. Acta*, **742**, 278–84.
Pages, M. and Pages, A.M. (1983) *Eur. Neurol.*, **22**, 181–5.
Pallini, R., Martelli, P., Bardelli, A.M. *et al.* (1987) *Neurology*, **37**, 1878–80.
Pallini, R., Dotti, M.T., Di Natale, P. *et al.* (1988) *Brain Dysfunct.*, **1**, 331–4.
Pallini, R., Cannella, C., Bardelli, A.M. *et al.* (1989) *Bull. Mol. Biol. Med.*, **14**, 1–10.
Parker, W.D., Jr, Oley, C.A. and Parks, J.K. (1989) *N. Engl. J. Med.*, **320**, 1331–1333.
Petty, R.K.H., Harding, A.E. and Morgan Hughes, J.A. (1986) *Brain*, **109**, 915–38.
Poole, C.J.M. and Kind, P.R.N. (1986) *Br. Med. J.*, **292**, 1229–30.
Rieders, F. (1977) *Trattato di farmacologia Medica di Drill*, Piccin, Padova, pp. 1154–60.
Rogers, A.R. (1977) *Aust. J. Ophthalmol.*, **5**, 111–19.
Rose, F.C., Bowden, A.N. and Bowden, P.M.A. (1970) *Br. J. Ophthalmol.*, **54**, 388–93.
Syme, I.G., Bronte-Stewart, J., Foulds, W.S. *et al.* (1983) *Trans. Ophthalmol. Soc. UK*, **103**, 556–9.
Vilkki, J., Savantaus, M.L. and Nikoskelainen, E.K. (1989) *Am. J. Hum. Genet.*, **45**, 206–11.
Wallace, D.C. (1970) *Brain*, **93**, 121–32.
Wallace, D.C., Sing, G., Lott, M.T. *et al.* (1988) *Science*, **242**, 1427–30.
Went, L.N. (1972) in *Handbook of Clinical Neurology*, Vol. 13 (eds P.J. Vinken and G.W. Bruyn), North-Holland, Amsterdam, pp. 94–110.
Wilson, J. (1965) *J. Clin. Sci.*, **29**, 505–15.
Yoneda, M., Tsuji, S., Yamauchi, T. *et al.* (1989) *Lancet*, **ii**, 1076–7.

17 Types and mechanism of mitochondrial DNA mutations in mitochondrial myopathy and related diseases

T. OZAWA, M. TANAKA, W. SATO, K. OHNO, M. YONEDA AND T. YAMAMOTO
Department of Biomedical Chemistry, Faculty of Medicine, University of Nagoya, Japan

17.1 INTRODUCTION

17.1.1 Physiological role of mitochondrial DNA

Human mitochondrial genome is a closed circular DNA of 16 569 bp (Anderson *et al.*, 1981). Mitochondrial DNA (mtDNA) codes for 13 hydrophobic polypeptides, which play a central role in energy-transducing

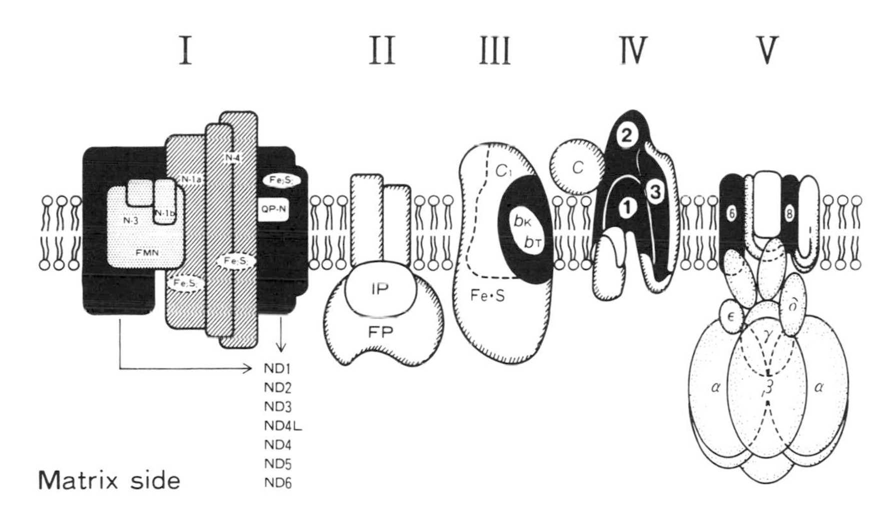

Fig. 17.1 Model of five complexes in the mitochondrial oxidative phosphorylation system. Dark areas represent subunits encoded by the mtDNA: ND1, ND2, ND3, ND4L, ND4, ND5 and ND6 subunits in Complex I; cytochrome b subunit in Complex III: CO1, CO2 and CO3 subunits in Complex IV; and subunits 6 and 8 in Complex V.

function of four multi-subunit enzyme complexes (Complexes I, III, IV and V) of the oxidative phosphorylation system (Fig. 17.1) and for two mitochondrial rRNAs and 22 organelle-specific tRNAs, which are essential for the assembly of a functional mitochondrial protein-synthesizing system (Attardi, 1981; Chomyn *et al.*, 1985).

17.1.2 Pathological role of mtDNA mutations

One of the several unusual features of mtDNA is its capacity for rapid mutation. We proposed that the accumulation of mitochondrial genome mutation during the whole life of an individual is an important contributor to ageing and degenerative diseases (Linnane *et al.*, 1989). This hypothesis is based on the following: the high frequency of gene mutation in mtDNA; the small and economically packed size of the mitochondrial genome and its high information content; the lack of a repair mechanism for mtDNA, unlike nuclear DNA; the accumulation of oxidative mtDNA damage by reactive oxygen species produced in the mitochondria; the established features of the somatic segregation of individual mtDNA genomes during eukaryotic cell division; and findings on molecular lesions underlying several human mitochondrial disorders. Recent reports demonstrate that a number of neuromuscular diseases have structural abnormalities of mitochondria as their common feature (Tanaka *et al.*, 1986; Tanaka *et al.*, 1988; Yoneda *et al.*, 1989a) and these appear to be caused by mutations in mtDNA. In some cases, the diseases, like mtDNA itself, are maternally inherited. Maternal inheritance of mutated mtDNA was reported by our group in a family of chronic progressive external ophthalmoplegia (Ozawa *et al.*, 1988) and in a family of Leber's hereditary optic neuropathy (Yoneda *et al.*, 1989b). In others, mutations appear to rise spontaneously during the whole life of an individual.

17.1.3 Three types of mtDNA mutations

Recent studies have clarified at least three types of mtDNA mutations in human diseases (Fig. 17.2). Type A is homoplasmy of mutant mtDNA with point mutation within a single gene. A single base transition, G to A, converting a highly conserved arginine residue to a histidine in the ND4 gene for complex I in the electron transfer chain, was found in patients with Leber's hereditary optic neuropathy (Yoneda *et al.*, 1989b; Wallace *et al.*, 1989). Type B is heteroplasmy of mutant and normal mtDNAs. Southern blot analysis reveals coexistence of the normal-sized mtDNA and mutant mtDNA with a large deletion in patients with Kearns–Sayre syndrome (Ozawa *et al.*, 1988). Type C is pleioplasmy.

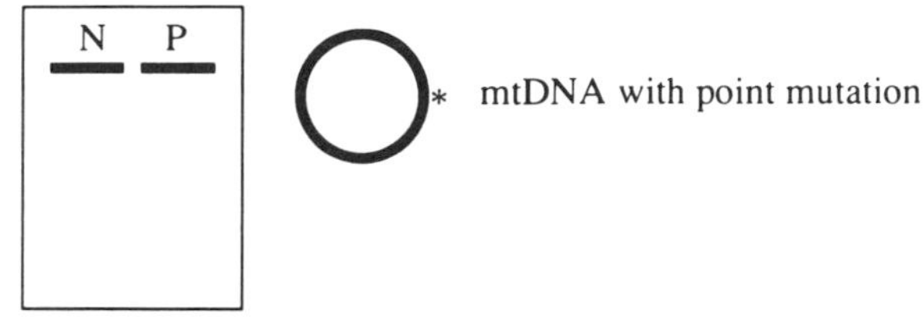

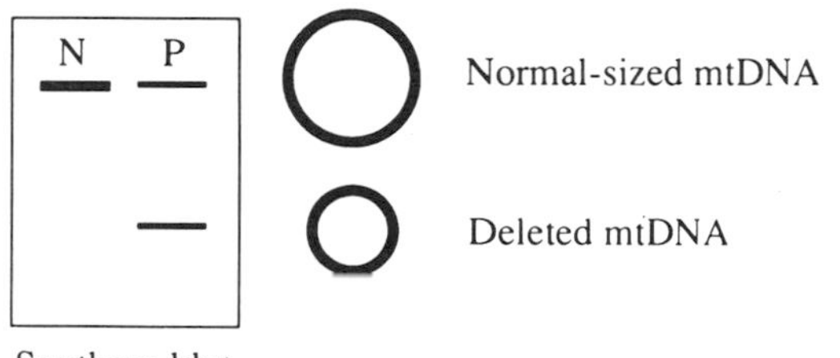

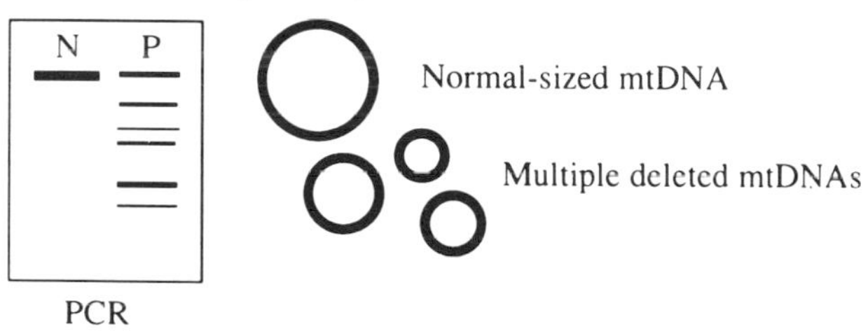

Fig. 17.2 Three types of mutations of mtDNA.

In this type, multiple mutant mtDNAs with various deletions coexist with normal-sized mtDNA. Multiple deletions of mtDNA can be detected by the gene amplification method using polymerase chain reaction (Sato *et al.*, 1989).

17.2 TYPE A MUTATION (HOMOPLASMY)

Fig. 17.3 shows the Southern blot analysis of mtDNA from a Japanese

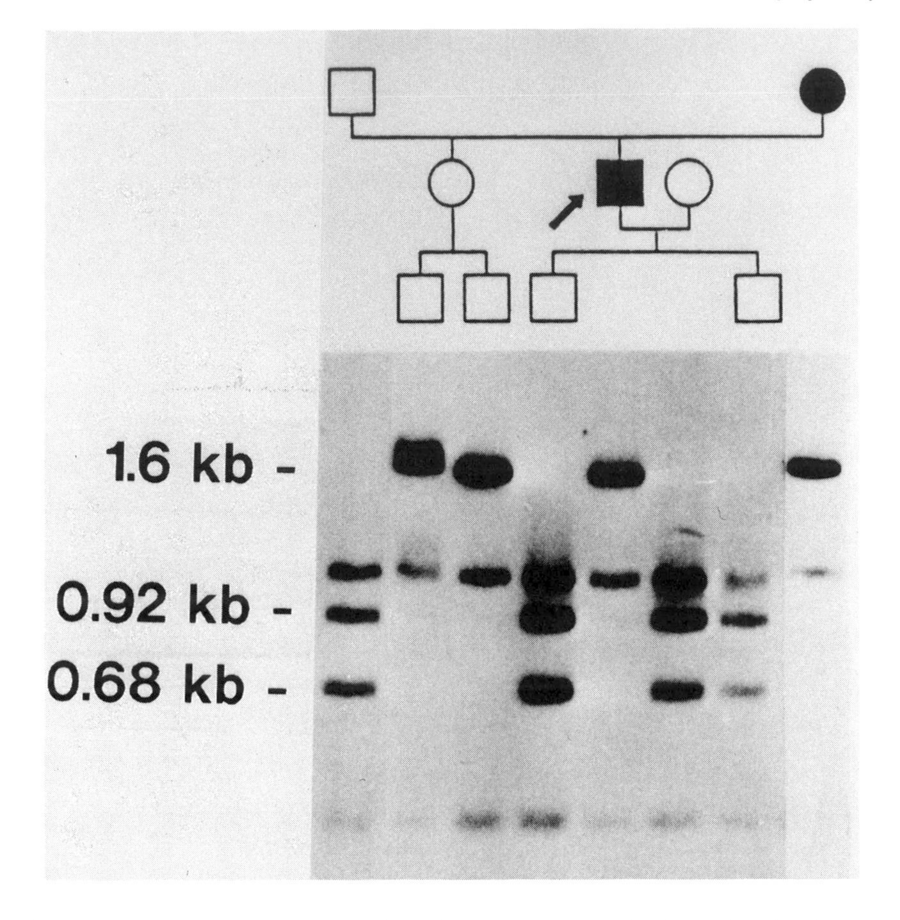

Fig. 17.3 Southern blot analysis of mtDNA in a Japanese family with Leber's optic neuropathy. From Yoneda *et al.* (1989b).

family with Leber's hereditary optic neuropathy (Yoneda *et al.*, 1989b). Wallace *et al.* (1989) have pointed out that a base transition from G at nucleotide number 11 778 to A converts the highly conserved 340th arginine to histidine in the ND4 subunit of complex I in patients with Leber's disease. The base transition can be detected as a site loss of a restriction enzyme, *Sfa*NI. In the proband (indicated with an arrow), a 1.6 kb DNA fragment resulting from the site loss was detected. But the mutation was not present in his children, in whom two fragments of 0.92 kb and 0.68 kb were transmitted from their normal mother. The analysis shows that the two nephews of the proband have the mutation specific for Leber's disease. Careful follow-up of these boys is recommended.

Because the maternally inherited mutation was found both in the North American and Finnish pedigrees and in the Japanese pedigrees, we could confirm that this mutation is responsible for the disorder.

17.3 TYPE B MUTATION (HETEROPLASMY)

The second type of mutation is heteroplasmic mtDNA deletion (Holt *et al.*, 1988; Lestienne and Ponsot, 1988; Ozawa *et al.*, 1988). Heteroplasmic existence of mutant mtDNA with a multi-gene deletion with the normal-sized mtDNA was found in one-third of patients with Kearns-Sayre syndrome and chronic progressive external ophthalmoplegia (Holt *et al.*, 1988; Zeviani *et al.*, 1988; Moraes *et al.*, 1989).

17.3.1 Strategy for the analysis of deleted mtDNA

To investigate the mechanism of mtDNA deletion, first we localized the deleted region using the combination of polymerase chain reaction (PCR) and S_1 nuclease digestion, PCR plus S_1 method (Tanaka-Tamamoto *et al.*, 1989). Then, we directly sequenced the cross-over regions of the deleted mtDNA without cloning (Tanaka *et al.*, 1989).

The principle of PCR plus S_1 analysis

The principle of PCR plus S_1 method for analysing the deleted region is schematically shown in Fig. 17.4 (Tanaka-Yamamoto *et al.*, 1989). Fragment A–B, which includes the starting point of the deletion, is amplified from the normal-sized mtDNA, and fragment A–D, which includes both ends of the deletion, is amplified from the mutant mtDNA. These fragments are mixed and subjected to heteroduplex formation. The complementary region of the heteroduplex formed between fragments A–D and A–B as well as the homoduplexes formed from the self-reannealing of each fragment are protected against S_1 nuclease. The size of the complementary region of the heteroduplex indicates the distance between point A and the starting point of the deletion (asterisks in the figure) in the mutant mtDNA. Similarly, the end-point of the deletion can be determined by S_1 nuclease analysis of the heteroduplex formed from fragments C–D and A–D.

(b) PCR plus S_1 analysis of deleted mtDNA

Fig. 17.5 shows the electrophoretic patterns of fragments before and after the S_1 nuclease digestion (Tanaka-Yamamoto *et al.*, 1989). For determination of the starting point of the deletion, a fragment of 3.2 kb (position 8201–11 380) which was amplified from the normal-sized

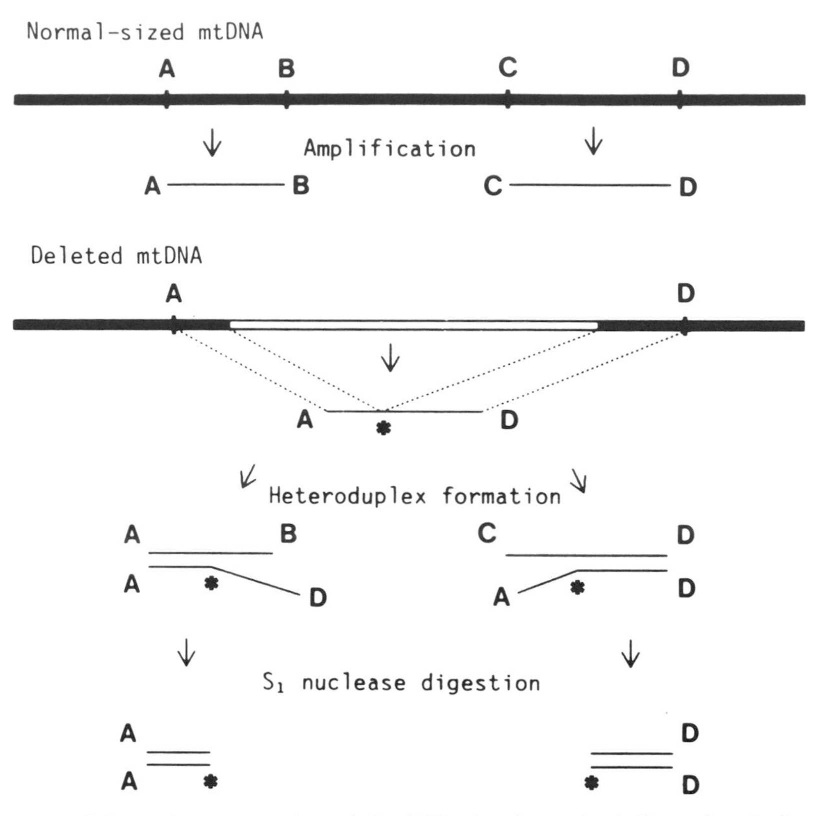

Fig. 17.4 Schematic presentation of the PCR plus S_1 method. Open box indicates the region of the deletion in mutant DNA. From Tanaka-Yamamoto *et al.* (1989).

mtDNA using the primers L820 and H1136 was mixed with a fragment of 1.9 kb which was amplified from the deleted mtDNA using the primers L820 and H60 (Fig. 17.5, lane 1). These fragments were then subjected to heteroduplex formation. It is essential to cool down the mixture quickly in order to obtain a sufficient amount of heteroduplexes. After the S_1 nuclease digestion, a fragment of 0.45 ± 0.05 kb appeared (lane 2), assuming the error in estimation of the fragment size to be 10%. Since the primer L820 starts at position 8201, the start-point of the deletion should be located at position 8650 ± 50 bp within the ATPase subunit 6 gene, To determine the end-point of the deletion, a 1.7 kb fragment (position 14 511–16 209) which was amplified from the normal-sized mtDNA using

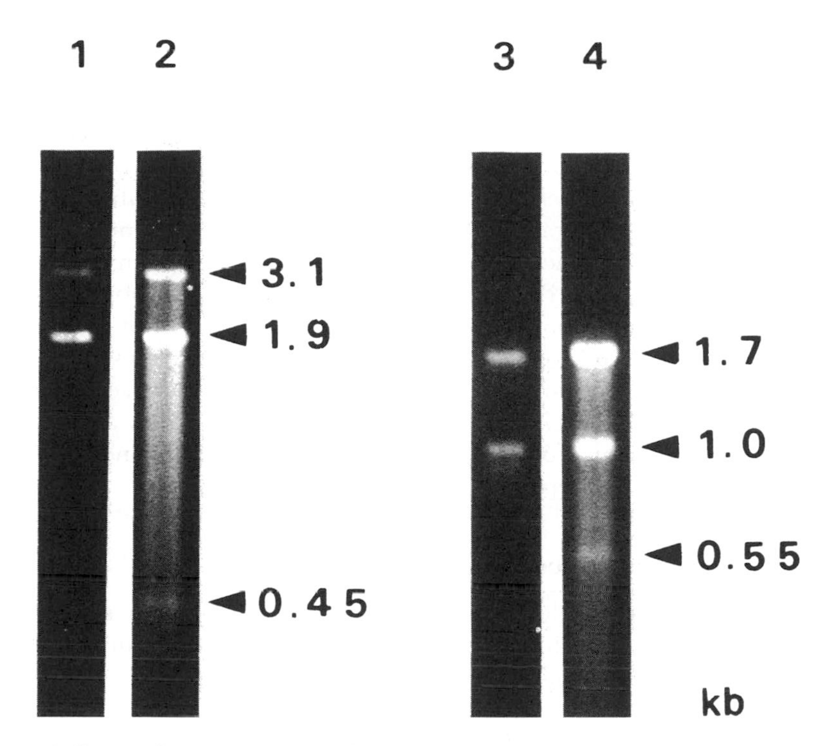

Fig. 17.5 Electrophoretic patterns of PCR-amplified mtDNA fragments after heteroduplex formation and S_1 nuclease digestion. From Tanaka-Yamamoto *et al.* (1989).

the primers of L1451 and H1619 was mixed with a 1.0 kb fragment which was amplified from the deleted mtDNA using the primers of L820 and H1619 (lane 3). A fragment of 0.55 ± 0.06 kb appeared after the S_1 nuclease digestion (lane 4). Since the primer H1619 starts at position 16 209, the end-point of the deletion should be located at position $15\,660 \pm 60$ within the cytochrome b gene.

(c) Advantages of PCR plus S_1 method

The PCR plus S_1 method has the following advantages over the Southern blot method.

1. This method requires no radioisotopes, and can be performed in ordinary clinical laboratories.
2. The new method is fast. Using the Southern blot method, it takes at least a couple of days to determine the deleted region. Using the

new method, we can complete the analysis of the deleted region within 5 h: 2 h for amplification, 1 h for heteroduplex formation, 20 min for S_1 nuclease digestion, and 1 h for electrophoresis and detection.
3. This method is so sensitive that total DNA extracted from only 5 mg of muscle tissue is sufficient for the determination of the deleted region of mtDNA. Therefore, this method is of value especially when only a small amount of clinically biopsied sample is available.
4. This method is accurate. The deleted region can be determined within ±60 bp. Because the experimental error in estimation of electrophoresed fragments depends on the sizes of fragments obtained after S_1 nuclease digestion, the accuracy in determination of the deleted regions can be improved by choosing appropriate pairs of primers so that smaller fragments are obtained after S_1 nuclease digestion.
5. The present method is simple, because we can selectively and directly amplify fragments from the deleted mtDNA or from the normal-sized mtDNA without separating the two populations.

17.3.2 PCR direct sequencing of deleted mtDNA

According to the result of the PCR plus S_1 analysis, a sequencing primer was chosen for the direct sequencing of the deleted mtDNA. The cross-over sequence was found to be a directly repeated sequence containing four cytosines (5′-ATCCCCA-3′), flanked by AT-rich regions (Tanaka *et al.*, 1989). The directly repeated sequence was located in the boundaries of the deletion between the ATPase6 gene and the cytochrome b gene. When the wild-type sequences of the ATPase6 and the cytochrome b genes were compared to each other without adding a gap, almost no sequence homologies were obtained except in the cross-over sequence. But when they were compared with addition of a gap, additional homologies were found in 11 of 12 bases adjacent to the 3′-side of the directly repeated sequence. Therefore the deletion spanned 7039 bp probably starting from position 8624 on the 3′-side of the directly repeated sequence within the ATPase6 gene, and ending at position 15 662 on the 3′-side of the directly repeated sequence within the cytochrome b gene. In patient 2, a 3 bp sequence of 5′-CCT-3 was found in the boundaries of deleted segment spanning 3717 bp between the ATPase6 and the ND5 genes (Fig. 17.6). Because additional homologies were found adjacent to the directly repeated sequence, the deletion might be promoted not only by the short directly repeated sequences but also by the homologous sequences surrounding them (Fig. 17.7). In patient 3, a 13 bp sequence of 5′-ACCTCCCTCACCA-3′ was found in the boundaries of deleted segment spanning 4977 bp between the ATPase8 and the ND5 genes. The same deletion was

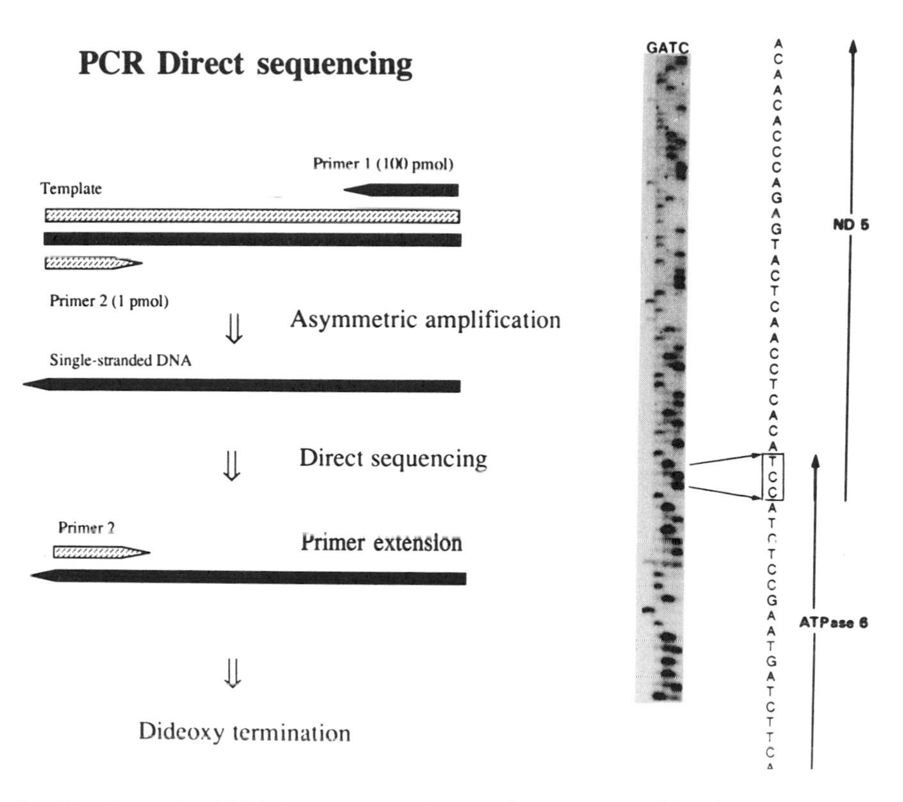

Fig. 17.6 Principle of PCR direct sequencing and determination of the directly repeated sequence involved in the deletion of mtDNA in patient 2 with chronic progressive external ophthalmoplegia. From Tanaka *et al.* (1989).

reported in a patient with Pearson's marrow/pancreas syndrome by Rotig *et al.* (1988, 1989) and in 19 out of 29 patients with Kearns–Sayre syndrome and progressive external ophthalmoplegia by Schon *et al.* (1989).

17.3.3 mtDNA deletions with or without frame shift

We can distinguish two types of mtDNA deletions (Tanaka *et al.*, 1989). Type 1 deletion results in frame shift at the junction of the two genes. As shown in Fig. 17.7, the sequence of the mutant mtDNA of patient 1 with type 1 deletion forms an open reading frame which predicts a 12 kDa hybrid protein composed of 32 amino acid residues from the N-terminal side of the ATPase subunit 6 and of 75 amino acid residues

Type 1 deletion

Patient 1: 7039 bp deletion

8624

Wild-type sequence
ATPase 6 gene 5′-CCCCCTCTATTG**ATCCCCA**-CCTCCAAATATCTCA-3′
** * ******* ****** *** * *
Cytochrome b gene 5′-ATCCTAGCAATA**ATCCCCA**TCCTCCATATACCAAA-3′

15 662

Deleted mtDNA sequence
hybrid gene 5′-CCC CCT CTA TTG **ATC CCC A**TC CTC CAT ATA TCC AAA-3′
hybrid protein -Pro Pro Leu Leu Ile Pro Ile Leu His Met Ser Lys-

Patient 2: 3717 bp deletion

9192

Wild-type sequence
ATPase 6 gene 5′-CTTCTAGTAAGCCTCTA**CCT**GCACGACAACACATAA-3′
*** * * * *** *** **** *** *
ND5 gene 5′-GCCTTAGCATGATTTAT**CCT**A CA CTCCAACTCATGA-3′

12 908

Deleted mtDNA sequence
hybrid gene 5′-CTT CTA GTA AGC CTC TA**C CT**A CAC TCC AAC TCA TGA-3′
hybrid protein -Leu Leu Val Ser Leu Tyr Leu His Ser Asn Ser Trp-

Type 2 deletion

Patient 3: 4977 bp deletion

8483

Wild-type sequence
ATPase 8 gene 5′-AACTACCACCTA**CCTCCCTCACCA**AAAGCCCATAAA-3′
** * * ************* ** *
ND5 gene 5′-CTCTCACTTCA**ACCTCCCTCACCA**TTGGCAGCCTAG-3′

13 459

Deleted mtDNA sequence
hybrid gene 5′-AAC TAC CAC CT**A CCT CCC TCA CCA** TTG GCA GCC TAG-3′
hybrid protein -Asn Tyr His Leu Pro Pro Ser Pro Leu Ala Ala **stop**

Fig. 17.7 Comparison of the nucleotide sequences of the wild-type genes with those of the hybrid genes in the deleted mtDNAs from patients 1–3 (from Tanaka *et al.*, 1989). The sequence of the wild-type genes is shown above the sequence of the hybrid gene resulting from deletion of mtDNA in each patient. Bold letters indicate the directly repeated sequences. Underlined is the excised sequence in mtDNA deletion in each patient. Numerals indicate the nucleotide numbers (Anderson *et al.*, 1981).

from the C-terminal side of the cytochrome b protein. The sequence of the mutant mtDNA of patient 2 with type 1 deletion forms an open reading frame which predicts a 69 kDa hybrid protein composed of 222 amino acid residues from the N-terminal side of the ATPase subunit 6 and of 412 amino acid residues from the C-terminal side of the ND5 protein. Type 2 deletion of patient 3 introduces a stop codon 12 nucleotides aside from the boundary, resulting in a premature termination of the ND5 protein. Therefore, the sequence of the mutant mtDNA predicts a 5 kDa abnormal protein composed of 42 amino acid residues from the N-terminal side of the ATPase subunit 8 and of 3 amino acid residues resulting from a frame shift in the ND5 gene. The ATPase subunits 6 and 8 in Complex V, the ND5 protein in Complex I, and the cytochrome b protein in Complex III play central roles in the energy-transducing function of each complex. If they are translated, the effect of hybrid proteins in the molecular assembly of the oxidative phosphorylation complexes would be greater than that of the abnormally short proteins, because the hybrid proteins would affect two complexes at the same time. It should be determined whether the message for these abnormally short proteins and hybrid proteins could be translated into proteins and whether they could disturb the molecular assembly of these energy-transducing complexes or not. We hypothesize that the hybrid protein of two genes is perhaps capable of providing an abnormal mitogenic stimulus to mitochondria resulting in deregulated proliferation observed in chronic progressive external ophthalmoplegia.

17.4 TYPE C MUTATION (PLEIOPLASMY)

In this type, multiple mutant mtDNAs with various deletions are present in the patient tissues (Sato *et al.*, 1989). The population of each mutated mtDNA was so small that the deletions were often not detected by Southern blot analysis but by PCR amplification.

17.4.1 Primer shift PCR method

For the analysis of pleioplasmic mtDNA deletions, we have developed a novel method – the primer shift PCR method (Sato *et al.*, 1989). The principle of the method is presented on the left-hand side of Fig. 17.8. This method can detect small populations of deleted mtDNA, which are undetectable by the conventional Southern blot method. The presence of deleted mtDNAs can be confirmed, if we compare the shift in the sizes of the amplified fragments with the shift in the positions of the primers used for the amplification. We applied the primer shift PCR method for the analysis of skeletal muscle from a patient with familial

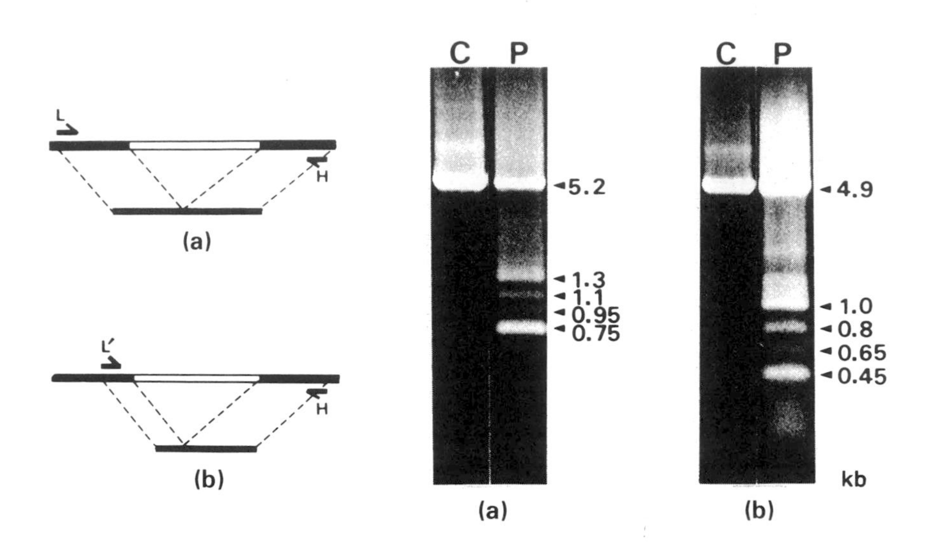

Fig. 17.8 Principle of the primer shift PCR method and its application to the analysis of multiple deletions of mtDNA in the skeletal muscle of a patient with chronic progressive external ophthalmoplegia (from Sato *et al.*, 1989). (a) PCR amplification using primers L820 (L) and H1338 (H). (b) PCR amplification using primers L853 (L′) and H1338 (H). Sizes of amplified fragments are indicated in kilobases. A single band derived from the normal-sized mtDNA is seen in the control (C). Four additional bands derived from the deleted mtDNAs are seen in the patient (P).

ocular myopathy. When we used a pair of primers, the distance between which was 5.2 kb on the mitochondrial genome, multiple bands, representing at least four populations of differently deleted mtDNAs, were detected in the patient muscle. When we used another pair of primers, the distance between which was 4.9 kb, the size of each band shifted by 0.3 kb as expected. This patient, age 55, had external ophthalmoplegia from age 49. Multiple mtDNA deletions might be somatically accumulated as an accelerated ageing process.

17.4.2 Mechanism of multiple mtDNA deletions

In order to elucidate the mechanism of pleioplasmic mtDNA deletions, we examined one of the mutant mtDNAs of two patients with multiple mtDNA deletions. Direct sequencing of the mutant mtDNAs revealed the same directly repeated sequence of 13 bp found in patients with heteroplasmic mtDNA deletion. Therefore, it is likely that directly repeated sequences are involved not only in the heteroplasmic mtDNA deletion but also in the pleioplasmic mtDNA deletions. Zeviani *et al.* (1989) described multiple mtDNA deletions starting exclusively at the D-loop region in a family with autosomal dominant mitochondrial myopathy. In contrast, we found multiple mtDNA deletions outside the D-loop region both in the present patients and in the previous patient with familial ocular myopathy (Sato *et al.*, 1989). Therefore, the mechanism responsible for the multiple mtDNA deletions described here apparently differs from the mechanism proposed by Zeviani *et al.* (1989) for the multiple mtDNA deletions in their cases. Since directly repeated sequences of relatively short base pairs are abundant in mtDNA, these sequences could provide numerous chances for mtDNA deletions.

17.4.3 Difference between heteroplasmic and pleioplasmic mtDNA deletions

We analysed mitochondrial DNA from biopsied skeletal muscles of 15 patients with chronic progressive external ophthalmoplegia, as listed in Table 17.1. Kearns–Sayre syndrome and myopathy. A single population of deleted mtDNA was detected by the Southern blot method in patients 1–7, and two populations of deleted mtDNA in patient 8. Although no deleted mtDNAs were detected by the Southern blot method in patients 9–15, small extra fragments indicating that they had multiple deletions were detected by the PCR method. These results demonstrate the presence of two distinct modes of mtDNA deletions in patients with

Table 17.1 Characteristics of patients

Patient		*Age (years)/ sex*	*Age of onset*	*Family history*	*Clinical diagnosis*[a]	*Deletion*
1	SK	29/M	11	–	KSS	Single
2	HA	40/M	12	–	KSS	Single
3	SO	42/F	27	–	CPEO	Single
4	YO	15/F	12	–	CPEO	Single
5	TH	25/M	21	–	CPEO	Single
6	KK	17/M	14	–	CPEO	Single
7	KO	32/F	14	+	CPEO	Multiple
8	KW	74/M	45	+	KSS	Multiple
9	MS	55/F	49	+	CPEO	Multiple
10	JW	44/M	34	+	CPEO	Multiple
11	YO	46/M	39	–	CPEO	Multiple
12	NT	63(M	62	–	Myopathy	Multiple
13	MY	67/F	66	–	CPEO	Multiple
14	TS	54/F	44	+	CPEO	Multiple
15	MY	39/F	32	–	CPEO	Multiple

[a]KSS, Kearns-Sayre syndrome; CPEO, chronic progressive external ophthalmoplegia.

mitochondrial myopathy. In the first type, the muscle tissue contains a single population of deleted mtDNA (heteroplasmy). In the second type, the tissue contains multiple populations of deleted mtDNA (pleioplasmy). We confirmed that all patients with chronic progressive external ophthalmoplegia and Kearns–Sayre syndrome had either types of mtDNA deletions. The patients with the heteroplasmic mtDNA deletion were of mean age 16.2 years (range 11–27) at the onset of clinical symptoms. In contrast, the patients with pleioplasmic mtDNA deletions were of mean age 42.3 years (range 14–66), almost over 30 years, at the onset of clinical symptoms. In nine patients with multiple mtDNA deletions, five patients had family history, in contrast to no family histories in patients with single mtDNA deletion.

17.4.4 Multiple mtDNA deletions and maternal inheritance

We observed mtDNA deletions in both a mother and daughter with chronic progressive external ophthalmoplegia (Fig. 17.9); although the

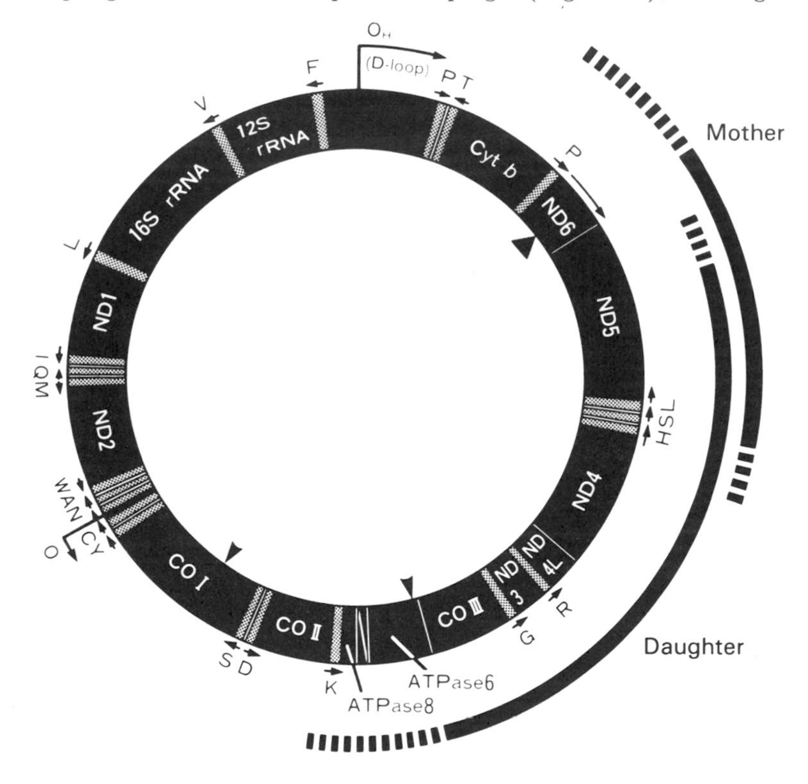

Fig. 17.9 Differently deleted mtDNA in a mother and her daughter with chronic progressive external ophthalmoplegia (from Ozawa *et al.*, 1988). The physical map of mtDNA and the deleted regions of mtDNA in the mother and the daughter are shown. Solid line, defined deleted region of mtDNA; dotted line, undefined deleted region.

deletions overlapped substantially, they were not identical (Ozawa *et al.*, 1988). This observation is consistent with the possibility of somatic extensions of a smaller lesion, which is perhaps genetically transmitted from the mother to the daughter. Alternatively, we can speculate that pleioplasmic mtDNA deletions might be responsible for the difference in the location of the deletions between the mother and daughter. We examined the hypothesis using PCR. When primers were selected so that the deleted mtDNA of the daughter was amplified as a 0.5 kb fragment, several additional fragments were detected in the mother, suggesting that she had multiple mtDNA deletions. When primers were selected so that the deleted mtDNA of the mother was amplified as a 1.5 kb band, no corresponding band was detected in the daughter. These results suggest that the mother had pleioplasmic mtDNA deletions and that one of the multiple deletions in the mother was maternally transmitted and increased its population in the daughter.

17.4.5 Multiple mtDNA deletions in exertional myoglobinuria

We have found multiple deletions of muscle mtDNA in patients with

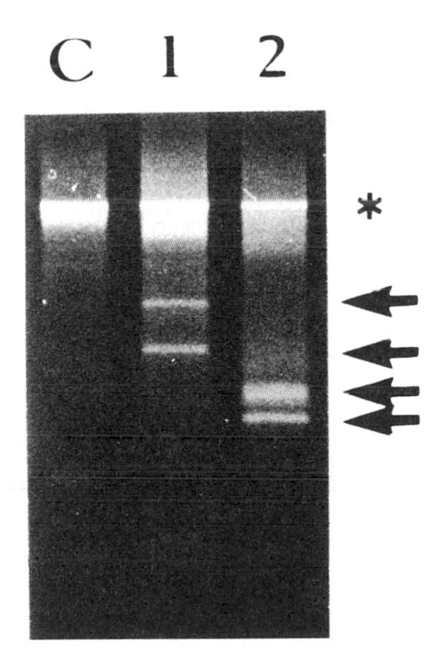

Fig. 17.10 PCR analysis of the skeletal muscle DNA from brothers with exertional myoglobinuria due to multiple deletions of mtDNA (from Ohno *et al.*, 1991). PCR detected several bands (arrows) amplified from deleted mtDNA in both patients along with the normal-sized mtDNA fragment (4.9 kb, indicated by asterisk). It is demonstrated that the two brothers had different populations of deleted mtDNAs. C, control; 1 and 2, patients.

exertional myoglobinuria but without external ophthalmoplegia (Ohno *et al.*, 1991). Patients were brothers. They experienced generalized muscular pain with myoglobinuria, which was provoked by alcohol intake or strenuous exercise. Both Southern blot and PCR analyses identified multiple populations of deleted mtDNA (Fig. 17.10). In contrast to the multiple deletions starting at the D-loop region in autosomal dominant cases described by Zeviani *et al.* (1989), the deletions detected in our cases were present on the outside of the D-loop region. Exertional myoglobinuria is a fairly common disorder, and frequently induces acute renal failure. Defects in the enzymes involved in glycolysis, glycogenolysis and lipid metabolism have been reported to cause exertional myoglobinuria (Rowland, 1984). Our study demonstrates that multiple mtDNA deletions, which lead to mitochondrial respiratory enzyme defects, provide a novel genetic entity in recurrent myoglobinuria.

17.4.6 Multiple mtDNA deletions in cardiomyopathy

The cardiomyopathies are often of unknown aetiology, and are recognized as a significant cause of morbidity and mortality. To examine the possibility that some cases of cardiomyopathies are caused by multiple mtDNA deletions, we have studied the autopsied cardiac tissues from five patients with cardiomyopathies. In the Southern blot analysis, no deletions of mtDNA could be detected in any of the patients. In the PCR analysis, multiple deletions of mtDNA were detected in patient 1 with hypertrophic cardiomyopathy, patient 2 with cardiomyopathy associated with familial conduction block, and patient 3 with hypertrophic cardiomyopathy, but were detected neither in patient 4 with dilated cardiomyopathy in puerperium, patient 5 with cardiac infarction and diabetes, nor in the controls. We further confirmed the multiple mtDNA deletions by using the primer shift PCR method (Sato *et al.*, 1989), in which the shift in sizes of amplified fragments was compared with the shift in positions of primers used for amplification, and the PCR–Southern method, in which the PCR products were hybridized to various mtDNA fragments. These results suggest that multiple mtDNA deletions are an important genetic defect, not only in mitochondrial myopathies but also in cardiomyopathies. We propose 'mitochondrial cardiomyopathy due to multiple mtDNA deletions' as a novel disease entity among cardiomyopathies. Müller-Höcker (1989) demonstrated that the number of cytochrome-c-oxidase-deficient cardiomyocytes increases with age. Because cytochrome-c-oxidase-negative fibres are a common finding in the skeletal muscle of patients with mtDNA deletions (Ozawa *et al.* 1988), the increase in the number of cytochrome-c-oxidase-deficient cardiomyocytes with age may be related to the increase in the population of deleted mtDNA.

17.5 ACCUMULATION OF mtDNA MUTATIONS IN AGEING AND DEGENERATIVE DISEASES

We have proposed a hypothesis that accumulation of somatic mitochondrial gene mutations will eventually lead to accumulation of partially or grossly bioenergetic defective cells with time as a key facet of the processes of ageing and degenerative diseases (Linnane *et al.*, 1989).

17.5.1 Fundamentals of mtDNA mutations

Four fundamental facts support the hypothesis. First, the small and essential genome mtDNA is situated inside the mitochondria where oxygen metabolism is actively carried out. Second, because the human mtDNA has no introns, mutations of mtDNA directly affect the function of the oxidative phosphorylation system. Third, the genome is continuously exposed to oxygen radicals, and prone to mutation. Fourth, there are multiple copies of mtDNA in each cell. Accumulation of mtDNA mutations results in a state of coexistence of normal mtDNA and various mutated mtDNAs, which is clled pleioplasmy (Sato *et al.*, 1989). Segregation of mutated mtDNA among cells results in mosaicism of bioenergetically normal cells and bioenergetically defective cells.

17.5.2 Possible involvement of multiple mtDNA deletions in ageing and degenerative diseases

Among the various mutations that are accumulated in the mitochondrial genome, we propose that deletions of mtDNA play the most important role in degenerative disorders and in decreased function of the brain and the skeletal and cardiac muscles in senescence. Exposure of mtDNA to oxygen radicals results in formation of oxidized products, such as 8-hydroxydeoxyguanosine (Richter *et al.*, 1988). Deletion of mtDNA may occur as the consequence of such oxidative damage of the mitochondrial genome. The small populations of multiple point mutations resulting from oxidative damage are hard to analyse quantitatively. In contrast, small populations of deleted mtDNAs can be detected with a high sensitivity and can be analysed semiquantitatively using the PCR method. In the present review, we have presented several pieces of evidence supporting the hypothesis that pleioplasmic mtDNA deletions are involved in several degenerative diseases: late-onset ophthalmoplegia, exertional myoglobinuria and cardiomyopathy. Further study must be focused on the analysis of small populations of deleted mtDNA in other degenerative disorders and the ageing process.

ACKNOWLEDGEMENTS

This work was supported in part by the Grants-in-Aid for General Scientific Research (62570128) to M.T. and for Scientific Research on Priority Areas (Bioenergetics, 63617002) to T.O. from the Ministry of Education, Science and Culture, Japan, and by Grant 88-02-39 from the National Center for Nervous, Mental and Muscular Disorders of the Ministry of Health and Welfare, Japan, to T.O.

REFERENCES

Anderson, S., Bankier, A.T., Barrell, B.G. *et al.* (1981) *Nature*, **290**, 457–65.
Attardi, G. (1981) *Trends Biochem. Sci.*, **89**, 100–3.
Chomyn, A., Mariottini, P., Cleeter, M.W.J. *et al.* (1985) *Nature*, **314**, 592–7.
Egger, J. and Wilson, J. (1983) *N. Engl. J. Med.*, **309**, 142–6.
Holt, I.J., Harding, A.E. and Morgan-Hughes, J.A. (1988) *Nature*, **331**, 717-19.
Lestienne, P. and Ponsot, G. (1988) *Lancet*, **i**, 885.
Linnane, A.W., Marzuki, S., Ozawa, T and Tanaka, M. (1989) *Lancet*, **i**, 642-5.
Moraes, C.T., DiMauro, S., Zeviani, M. *et al.* (1989) *N. Engl. J. Med.*, **320**, 1293–9.
Müller-Höcker, J. (1989) *Am. J. Pathol.*, **134**, 1167–73.
Ohno, K., Tanaka, M., Sahashi, K. *et al.* (1991) *Ann. Neurol.*, in press.
Ozawa, T., Yoneda, M., Tanaka, M. *et al.* (1988) *Biochem. Biophys. Res. Commun.*, **154**, 1240–7.
Richter, C., Park, J.-W. and Ames, B.N. (1988) *Proc. Natl Acad. Sci. USA*, **85**, 6405–7.
Blanche, S. *et al.* (1988) *Lancet*, **ii**, 567–8.
Rotig, A., Colonna, M., Blanche, S. *et al.* (1988) *Lancet*, **ii**, 567–8.
Rotig, A., Colonna, M., Bonnefont, J.P. *et al.* (1989) *Lancet*, **i**, 902–3.
Rowland, L.P. (1984) in *Merritt's Textbook of Neurology*, 7th edn (ed. L.P. Rowland), Lea and Febiger, Philadelphia, pp. 585–90.
Sato, W., Tanaka, M., Ohno, K. *et al.* (1989) *Biochem. Biophys. Res. Commun.*, **162**, 664–72.
Schon, E.A., Rizzuto, R., Moraes, C.T. *et al.* (1989) *Science*, **244**, 346–9.
Tanaka, M., Nishikimi, M., Suzuki, H. *et al.* (1986) *Biochem. Biophys. Res. Commun.*, **140**, 88–93.
Tanaka, M., Miyabayashi, S., Nishikimi, M. *et al.* (1988) *Pediatr. Res.*, **24**, 447–57.
Tanaka, M., Sato, W., Ohno, K. *et al.* (1989) *Biochem. Biophys. Res. Commun.*, **164**, 156–63.
Tanaka-Yamamoto, T., Tanaka, M., Ohno, K. *et al.* (1989) *Biochem. Biophys. Acta*, **1009**, 151–5.
Wallace, D.C., Singh, G., Lott, M.T. *et al.* (1989) *Science*, **242**, 1427–30.
Yoneda, M., Tanaka, M., Nishikimi, M. *et al.* (1989a) *J. Neurol. Sci.*, **92**, 143–58.
Yoneda, M., Tsuji, S., Yamauchi, T. *et al.* (1989b) *Lancet*, **i**, 1076–7.
Zeviani, M., Moraes, C.T., DiMauro, S. *et al.* (1988) *Neurology*, **38**, 1339–46.
Zeviani, M., Servidei, S., Gellera, C. *et al.* (1989) *Nature*, **339**, 309–11.

18 Direct tandem duplications of mitochondrial DNA in mitochondrial myopathy

J. POULTON, M.E. DEADMAN, M. SOLYMAR, S. RAMCHARAN AND R.M. GARDINER
University of Oxford Department of Paediatrics, John Radcliffe Hospital, Headington, Oxford OX3 9DU, UK

18.1 INTRODUCTION

Some of the mitochondrial myopathies show a maternal pattern of inheritance and/or a biochemical defect of respiratory chain components encoded at least in part by mitochondrial (mt)DNA, and could thus arise from a mutation in the mitochondrial genome. Holt *et al.* (1988) described heteroplasmy (more than one population of mtDNAs within the same individual) with large deletions (2–7 kb) in muscle mtDNA in about one-third of patients with mitochondrial myopathy.

18.2 METHODS AND RESULTS

We have studied two sporadic cases of Kearns–Sayre syndrome in whom Southern hybridization showed heteroplasmy in both blood and muscle, with an abnormal population of mtDNAs which were 24 kb in length. Restriction enzyme analysis using 14 regional human mtDNA probes in M13 (courtesy of Professor Attardi) (Poulton *et al.*, 1989) showed that these were direct tandem duplications (Fig. 18.1).

In both cases the duplicated regions included the D-loop (which contains both promoters and the origin of replication of the heavy strand), the reading frames for subunits 1 and 2 of reduced nicotinamide adenine dinucleotide (NADH) dehydrogenase (MTND 1 and 2), the genes for several transfer RNAs and both ribosomal RNAs. Mapping data were confirmed by sequence analysis using the polymerase chain reaction (PCR) to amplify the junction between the duplicated segments. In each case the gene for cytochrome oxidase subunit 1 (MTCOX1) was interrupted, with the gene for apocytochrome b out of frame in patient 1, and with $tRNA^{thr}$ gene in patient 2. In the case of patient 1, the reading frame for MTCOX1 was interrupted after base no. 6130 with base no. 15 056 of apocytochrome b (MTCYB) sequence out of frame without any intervening sequence (Fig. 18.2).

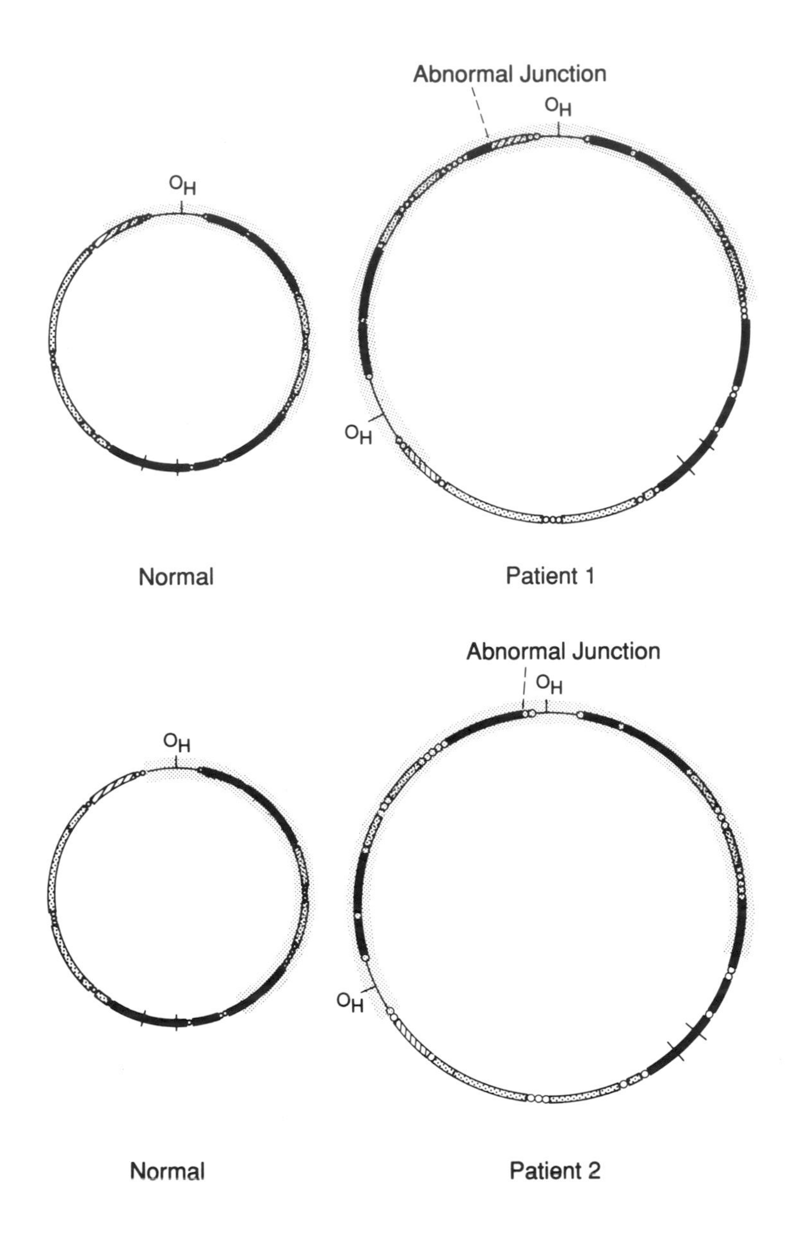

Fig. 18.1 Diagram to show the location of mtDNA duplications in patients 1 and 2. In each case the shaded region represents the duplicated portion. 0_H, origin of replication of the heavy strand. For genes involved in the duplicated region see text.

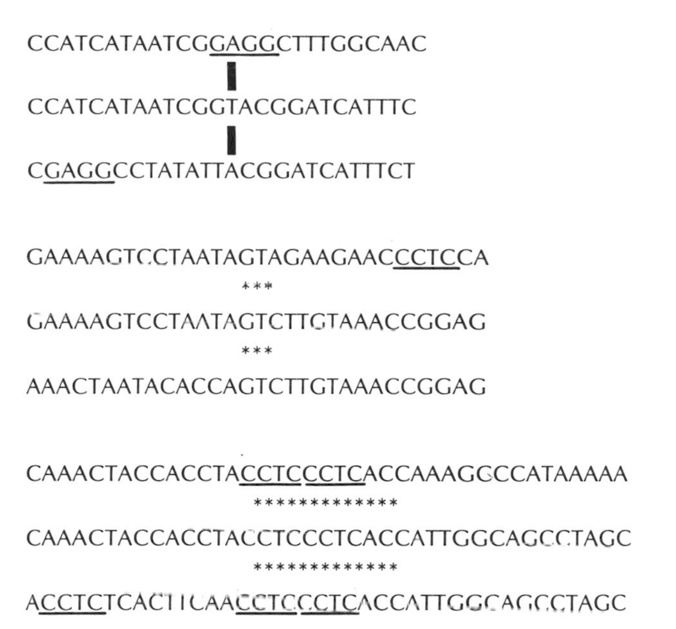

Fig. 18.2 Comparison of junction sequences in duplications and the common deletion. Top line: in patient 1, cytochrome oxidase subunit 1 sequence at bp 6130 adjoins bp 15 056 (apocytochrome B). Middle line: in patient 2, at bp 7354 cytochrome oxidase subunit 1 sequence ajoins bp 15 913. Bottom line: common deletion. Bold vertical lines represent breakpoint (patient 1); asterisk indicates homologous region: (patient 2 and common deletion); underlines indicate Picolli sequence (CCTC or GAGG) (Picolli *et al.*, 1984).

If transcribed and translated these arrangements would result in truncated subunits of cytochrome oxidase subunit 1. In patient 1 this would consist of the first 76 residues of MTCOX1 sequence followed by 20 abnormal residues unrelated to the normal MTCOX1 or MTCYB amino acid sequence before the first stop codon is generated. In patient 2, MTCOX1 sequence is interrupted after base no. 7354 with base no. 15 913 of the $tRNA^{thr}$ sequence without any intervening sequence (Fig. 18.2). There was a homologous portion three base pairs in length in both strands. If transcribed and translated this would result in a peptide corresponding to the first 497 amino acid residues of MTCOX1 followed by one abnormal residue before the first stop codon is generated.

The relative proportions of normal and abnormal genomes in various tissues was measured by laser densitometry (Fig. 18.3(a)). Clonality within a tissue was examined using PCR to detect duplicated genomes on granulocyte clones (Potter and Capellini, 1983) and hairs (Higuchi *et al.*, 1988) (X inactivation data suggest that individual hairs arise from

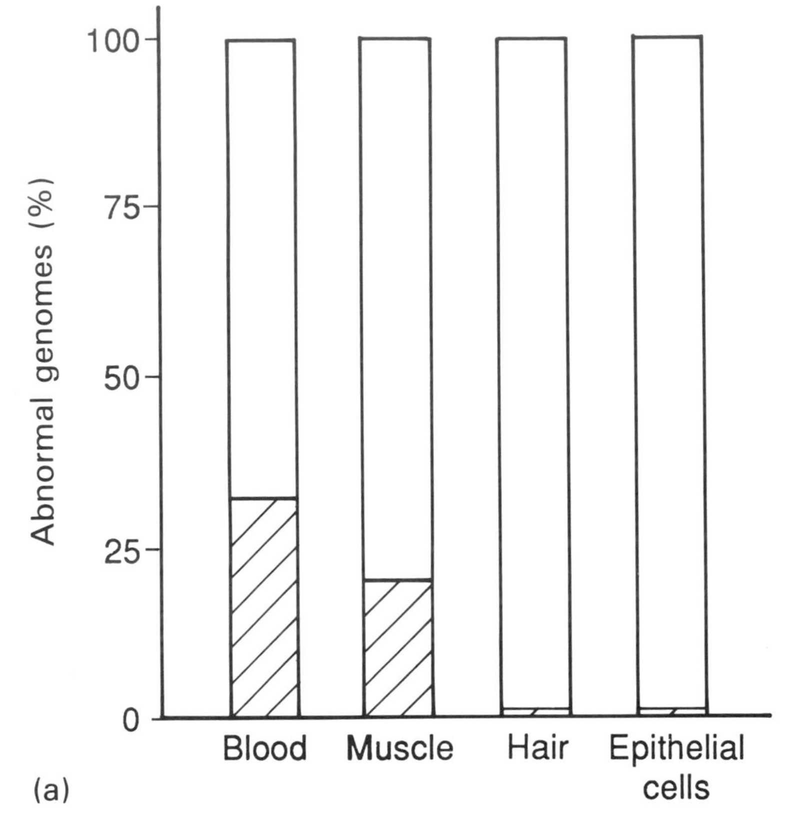

Fig. 18.3 Bar charts showing varying proportions of duplicated mtDNAs in patient 1 in (a) different tissues (vertical axis: percentage abnormal genomes measured by laser densitometry); (b) different clones within a tissue (vertical axis: percentage of single hairs or white cell clones in which duplication sequence was detectable). Hatched regions, duplicated genomes; unhatched regions, normal genomes.

single cells; Gartler and Riggs, 1983) in patient 1 (Fig, 18.3(b)). This suggested that this patient is a mosaic, i.e. different tissues contain different ratios of normal to abnormal genomes.

Northern analysis of total cellular RNA from lymphocyte lines reveals an abnormal transcript in one case, probably corresponding to an RNA from the region of the abnormal junction. cDNA was synthesized from RNA which had been treated with DNAse, and the product amplified with primers either side of the abnormal junction (Chelly *et al.*, 1988). In both cases a product of appropriate size and sequence was obtained, confirming that the abnormal fusion genes at the junction between duplicated segments is indeed transcribed.

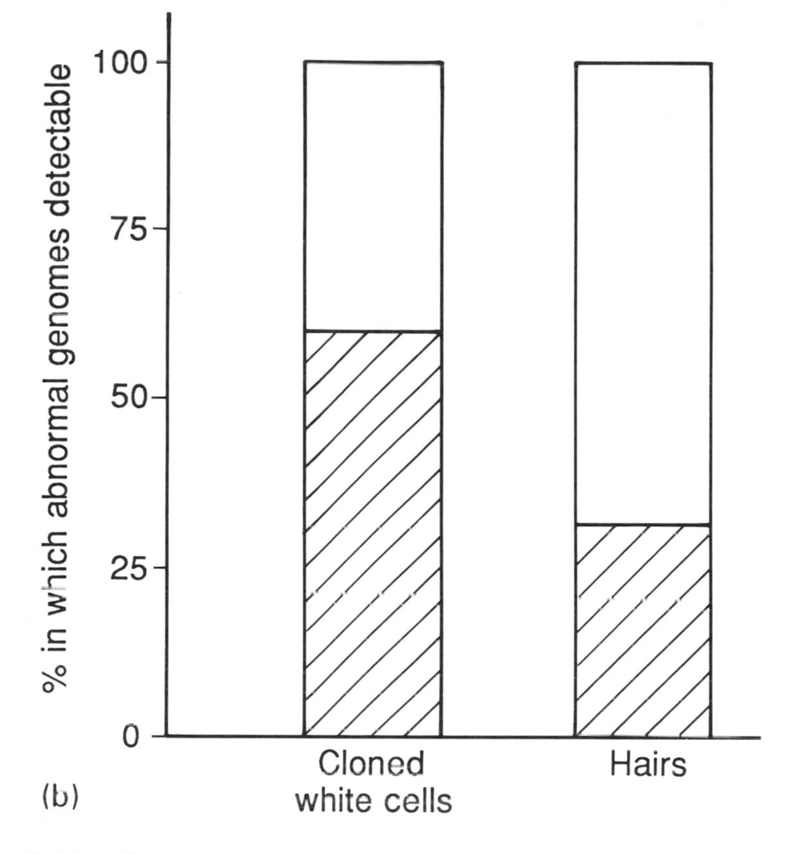

Fig. 18.3 (contd)

18.3 DISCUSSION

As these abnormalities are specific to the two patients, and have not been observed in unaffected family members or in normals despite extensive population surveys (Brown *et al.*, 1979; Johnson *et al.*, 1983; Horai *et al.*, 1984; Brega *et al.*, 1986; Santachiara-Benerecetti *et al.*, 1988), we believe that they are causative. We have shown that the fusion genes are transcribed, and if translated these would result in abnormal truncated subunits of CO1. It is possible that either abnormal transcripts or peptides could impair mitochondrial function by competing with the normal gene products. Alternatively an excess of normal products from the duplicated regions could be harmful. The fact that the phenotype of plants and lizards (Moritz and Brown, 1987) does not appear to be affected by some duplications makes the former slightly more likely; analysis of RNA and proteins may clarify this further.

Of the tissues investigated, duplicated mtDNAs were only detectable using Southern hybridization in blood and muscle. PCR revealed that they were present in every other tissue investigated, and although not strictly quantitative it is reasonable to suppose that the variation in quantity of product between different hairs and granulocyte clones may represent differences in the proportion of abnormal genomes in different clonal lines. Surprisingly, the proportion of abnormal genomes was higher in blood than muscle in patient 1. However, it must be remembered that the muscle sample was obtained 9 years before the blood sample, during which time her muscle power deteriorated. It may be that the proportion of duplicated genomes increases over time because of factors such as a replicative advantage over the normal. Moreover, as only one muscle was biopsied it might not be representative of the overall distribution.

Determinants of the tissue distribution of abnormal genomes probably include the exact stage of development when the duplication occurred, and factors influencing their propagation. The latter may act at the level of the individual mtDNA, the mitochondrion, the cell or the tissue. Studies of yeast 'petit' mutants suggest that genomes which are shorter and have more origins of replication are more 'suppressive', or replicate faster (De Zamaroczy *et al.*, 1981). Deletions and duplications might only reach significant levels if they have a replicative advantage over the normal genomes. Thus deleted genomes may be more likely to become clinically apparent if neither origin of replication is included in the deleted segment. Similarly, duplications including both origins of replication might be at a replicative advantage over the normal. The effect that duplicated mtDNAs have on the ability of a mitochondrion or cell to divide or tissue to develop is unknown. The effect of high levels of abnormal mtDNAs on a given tissue might depend on its energy requirements (Wallace *et al.*, 1988), and could be incompatible with life. In these two patients it would be particularly interesting to obtain pancreatic and central nervous system tissue.

The mechanism by which the duplications have arisen is not clear. Tandem direct duplications are equivalent to deleted dimers, in this case deleted by 8–9 kb. Dimers may rarely occur during normal replication (Clayton and Vinegrad, 1987), and an 8–9 kb deletion in one of these could have given rise to these duplications by whatever mechanism causes the deletions described by Holt *et al.* Alternatively, duplication could result from recombination with a transducing deleted mtDNA (Anderson and Roth, 1977). However, mtDNAs with the reciprocal deletion which would be predicted in either case were not detectable using 30 cycles of PCR, and if present must be a low proportion. As such mtDNAs would not contain any origins of replication and might not replicate, their apparent absence does not exclude either mechanism.

The location of the duplications is of interest because the non-duplicated or 'deleted' region was similar to, but larger than the deletions described in Kearns-Sayre syndrome or chronic progressive external ophthalmoplegia. Furthermore, duplications in a similar region have been described in lizards (Moritz and Brown, 1987). This suggests that there are recombination 'hotspots' in this region, but that perhaps different constraints determine the size of the deleted segment in the two different types of rearrangement. For instance, it is likely that some of the deletions which occur have not yet been described because they are lethal or the defective genomes are unable to replicate.

The sequence data which have now been obtained from patients with deletions of mtDNA suggest a molecular mechanism for recombination. For instance, in the so-called 'common deletion' (Schon *et al.*, 1988) there is an exact 13 bp homologous portion in each strand at the breakpoint (Fig. 18.2(c)). Zeviani *et al.* (1989) described a family with an autosomal dominant tendency towards multiple different deletions with one constant and one variable end. Homologous regions ranged from 3 to 12 bp in length. Neither of the duplications occur exactly at one of these 'hotspots'. In patient 2 there is a 3 bp homology in each strand at the breakpoint, similar to those found in the shorter homologies of Zeviani *et al* (1989). However, as there is no homology at the breakpoint in patient 1, other mechanisms must play a part. Inspection of the sequences found at these recombination breakpoints shows a higher than expected incidence of 'CCTC' or 'GAGG', and this is statistically significant ($p < 0.02$; see Appendix). This sequence is present in both duplications: in both strands in patient 1 and in one strand in patient 2 within 12 bp of the breakpoint (see Fig. 18.2). It has recently been implicated in rearrangements of nuclear DNA by Picolli *et al* (1984). Furthermore, it is present either singly or in repeats within 13 bp of the breakpoint in at least six separate chromosomal translocations (Hurst and Porteous, personal communication). Although the frequency of the Picolli sequence at sites of recombination may just be a chance finding in a region of high GC content (Dieckmann and Gandy, 1987), it could be causally involved. For instance, it might be a low-affinity binding site for a DNA-modifying protein. One candidate for this role in mitochondria is an endonuclease which has recently been described whose function is unknown (Low *et al.*, 1988).

In conclusion, there is now good evidence that mutations in mitochondrial DNA cause disease in man. Many questions remain unanswered. Further work on rearrangements of mtDNA in health and disease will provide new insights into mitochondrial function.

ACKNOWLEDGEMENTS

Financial support was provided by the National Fund for Research into Crippling Diseases, the Muscular Dystrophy Group, British Medical Association (Vera Down Award), the EP Abraham Trust, the Medical Research Council and the Medical Research Fund of Oxford University. We thank the patients, their families, and their physicians (Dr N. Evans, Dr W.J. Turner, Dr B. Lake for kindly supplying muscle biopsy material, Professor G. Attardi and Dr M. King for allowing us to use the mitochondrial probes, Professor A.C. Wilson and associates for providing primers and instruction in the use of PCR, Dr C. Potter for growing granulocyte colonies, T. Kocher for making the alternating-temperature water bath, Drs H. Jacobs and M. Hurst for helpful discussions and Professor E.R. Moxon for his support and encouragement.

REFERENCES

Anderson, R.P. and Roth, J.R. (1977) *Annu. Rev. Microbiol.*, **31**, 473–505.
Brega, A., Gardella, R., Semino, O. *et al.* (1986) *Am. J. Hum. Genet.*, **39**, 502–12.
Brown, M.W., George, M. and Wilson, A,C. (1979) *Proc. Natl. Acad. Sci. USA*, **76**, 1967–71.
Chelly, J., Kaplan, J.-C., Maire, P. *et al.* (1988) *Nature*, **333**, 858–60.
Clayton, D.A. and Vinegrad, J. (1987) *Nature*, **216**, 652–7.
De Zamaroczy, M., Marotta, R., Faugeron-Fonty, G. *et al.* (1981) *Nature*, **292**, 75–8.
Dieckmann, C.L. and Gandy, B. (1987) *EMBO J.*, **6**, 4197–203.
Gartler, S.M. and Riggs, A.D. (1983) *Ann. Rev. Genet.*, **17**, 155–90.
Higuchi, R., von Beroldingen, C.H., Sensabaugh, G.F. and Erlich, H.A. (1988) *Nature*, **332**, 543–6.
Holt, I.J., Harding, A.E. and Morgan-Hughes, J.A. (1988) *Nature*, **331**, 717–19.
Horai, S., Gojobori, T. and Matsunaga, E. (1984) *Hum. Genet.*, **68**, 324–33.
Johnson, M.J., Wallace, D.C., Ferris, S.D. *et al.* (1983) *J. Mol. Evol.*, **14**, 255–71.
Low, R.L., Buzan, J.M. and Couper, C.L. (1988) *Nucl. Acids Res.*, **16**, 6427–45.
Moritz, C. and Brown, W.M. (1987) *Proc. Natl Acad. Sci. USA*, **84**, 7183–7.
Picolli, S.P., Caimi, P.G. and Cole, M. (1984) *Nature*, **310**, 327–30.
Potter, C.G. and Capellini, M.D. (1983) *Br. J. Haematol.*, **54**, 153–61.
Poulton, J., Deadman, M.E. and Gardiner, R.M. (1989) *Lancet*, **i**, 236–40.
Santachiara-Benerecctti, A.S., Scozzari, R. *et al.* (1988) *Ann. Hum. Genet.*, **52**, 39–56.
Schon, E.A., Rizzuto, S., Moraes, C.T. *et al.* (1988) *Science*, **244**, 346–9.
Turnbull, D.M., Johnson, M.A., Dick, D.J. *et al.* (1985) *J. Neurol. Sci.*, **70**, 93–100.
Wallace, D.C., Zheng, X., Lott, M. *et al.* (1988) *Cell*, **55**, 601–10.
Zeviani, M., Servidei, S., Gellera, C. *et al.* (1989) *Nature*, **339**, 309–11.

APPENDIX: STATISTICAL METHODS

The mitochondrial genome contains 199 copies of the sequence CCTC or GAGG, so the probability of this sequence starting at any particular base number is 199/16 569. This sequence was found near several of the rearranged sites up to 13 bp away. The number of occurrences of the sequence in either strand in the length extending from 13 bp before to 13 bp after the start of the rearrangement was counted. Where it occurred in the homologous sequence it was counted twice, once for each strand (thus 5 out of a possible 78 in the case of the common deletion, see Table 18.1). The number of counts obtained was compared to the expected number (for simplicity it was assumed that these were independent events), using a χ^2 test (Table 18.2).

Table 18.1

Rearrangement	*No. of CCTC/GAGG*	*Expected number*
Duplications		
Patient 1	2	52 × 199/16 569
Patient 2	1	58 × 199/16 569
Deletions		
Common	5	72 × 199/16 569
Patient CM[a]	2	78 × 199/16 569
Zeviani *et al.* (1989)	3	353 × 199/16 569
Total	14	7.3

[a]The sequence of patient CM's deletion has not yet been published. He has Kearns-Sayre syndrome (Turnbull *et al.*, 1985).

Table 18.2 χ^2 test

	No. of base positions where CCTC/GAGG present	*No. of base positions where CCTC/GAGG absent*	*Total*
Observed	14	599	613
Expected	7.3	605.7	613

Sum of $(O - E)^2/E = 6.1 + 0.07 = 6.17$ ($p < 0.02$)

19 Mutations of mitochondrial DNA: the molecular basis of mitochondrial encephalomyopathies and ageing?

B. KADENBACH,[1] P. SEIBEL,[1] M.A. JOHNSON[2] AND D. TURNBULL[2]
[1]Fachbereich Chemie, Philipps-Universität, Hans-Meerwein-Strasse, D-3550 Marburg, Germany; [2]Department of Neurology, University of Newcastle upon Tyne, UK

19.1 INTRODUCTION

The energy consumed by mammalian organisms is mainly generated in mitochondria by respiration-dependent (oxidative) phosphorylation. Mitochondria are partly autonomous cell organelles, because they contain their own DNA (mtDNA). A eukaryotic cell contains hundreds of mitochondria with three to five copies of mtDNA per organelle. In contrast to the nuclear DNA, mtDNA is exclusively maternally inherited (Giles *et al.*, 1980) and in general identical in all cells of a mammalian organism (Greenberg *et al.*, 1983). Recently, in various cases of human mitochondrial diseases, muscle weakness could be correlated with the occurrence of individual respiration-defective cells in the muscle (mitochondrial myopathy). In some cases (Kearns–Sayre syndrome (KSS) and chronic progressive external ophthalmoplegia (CPEO) deletions or insertions were found in the mtDNA of affected tissues. In a systematic study by Müller-Höcker (1988, 1989) a similar pattern of respiration-deficient cells were also found in muscle, but in particular in the heart of healthy individuals with increasing age. The similar progressive appearance of a mosaic pattern of respiration-defective cells in heart and muscle tissues of KSS and CPEO patients, as well as with increasing age in healthy human beings, suggest a similar molecular mechanism underlying the mitochondrial myopathies and the process of ageing which results in respiration-defective cells.

19.2 MITOCHONDRIAL DISEASES

19.2.1 Multiple forms of mitochondrial defects

Mitochondrial diseases, formerly termed 'mitochondrial myopathies', include a clinically and biochemically very heterogeneous group of diseases which are based on a defective energy (ATP) synthesis in mitochondria of either the muscle (mitochondrial myopathy) or another tissue. Very often the defect is found in more than one tissue (mitochondrial encephalomyopathy, cardiomyopathy, etc.). The various diseases can be classified either according to the clinical (DiMauro *et al.*, 1985) or biochemical picture (DiMauro *et al.*, 1987). The molecular basis can be a mutation of nuclear (Müller-Höcker *et al.*, 1989) or mitochondrial DNA. The mutation of mtDNA can be inherited as demonstrated for myoclonic epilepsy associated with ragged red fibres (MERRF) (Wallace *et al.*, 1988b) and Leber's hereditary optic neuropathy (LHON) (Wallace *et al.*, 1988a) or acquired (somatic mutation). Recently deletions (Holt *et al.*, 1988; Zeviani *et al.*, 1988, 1989; Ozawa *et al.*, 1988; Lestienne and Ponsot, 1988; Moraes *et al.*, 1989; Nelson *et al.*, 1989) and insertions of mtDNA (Poulton *et al.*, 1989) from muscle and other tissues of patients with KSS, CPEO and other forms of ocular myopathy, but not with MERRF, mitochondrial encephalomyopathy with lactic acidosis and stroke-like episodes (MELAS) or fatal infantile myopathy (Moraes *et al.*, 1989), were described. Since KSS occurs mainly at juvenile and CPEO at adult age, and since both diseases have a progressive character, the mutations of mtDNA of these patients are assumed to be acquired (but see Nelson *et al.*, 1989).

19.2.2 Various mutations in mtDNA of patients

We have analysed the mtDNA from skeletal muscle biopsies of patients with various forms of mitochondrial myopathies (see Table 19.1) as shown in Fig. 19.1. Isolated mtDNA was linearized with the restriction endonucleases Pvu II and Bam HI, which both have only a single cleavage site on human mtDNA (Anderson *et al.*, 1981). The Southern blot of the gel electrophoretic-separated mtDNA indicates a single band with control mtDNA, but an additional band of higher molecular weight (inserted DNA) in the mtDNA of two patients with ptosis and myopathy (lanes 15, 16 and 18) or of lower molecular weight (deleted DNA) in the mtDNA of two patients with CPEO (lanes 1–4). The DNA of one patient with KSS showed a broad band of lower molecular weight, indicating variable sizes of deleted DNA, as has been described before (Zeviani *et al.*, 1988).

The mtDNA of three patients with CPEO showed only a single band of normal mtDNA (lanes 5–10). This result, however, does not exclude the presence of multiple forms of deleted mtDNA with large size variation.

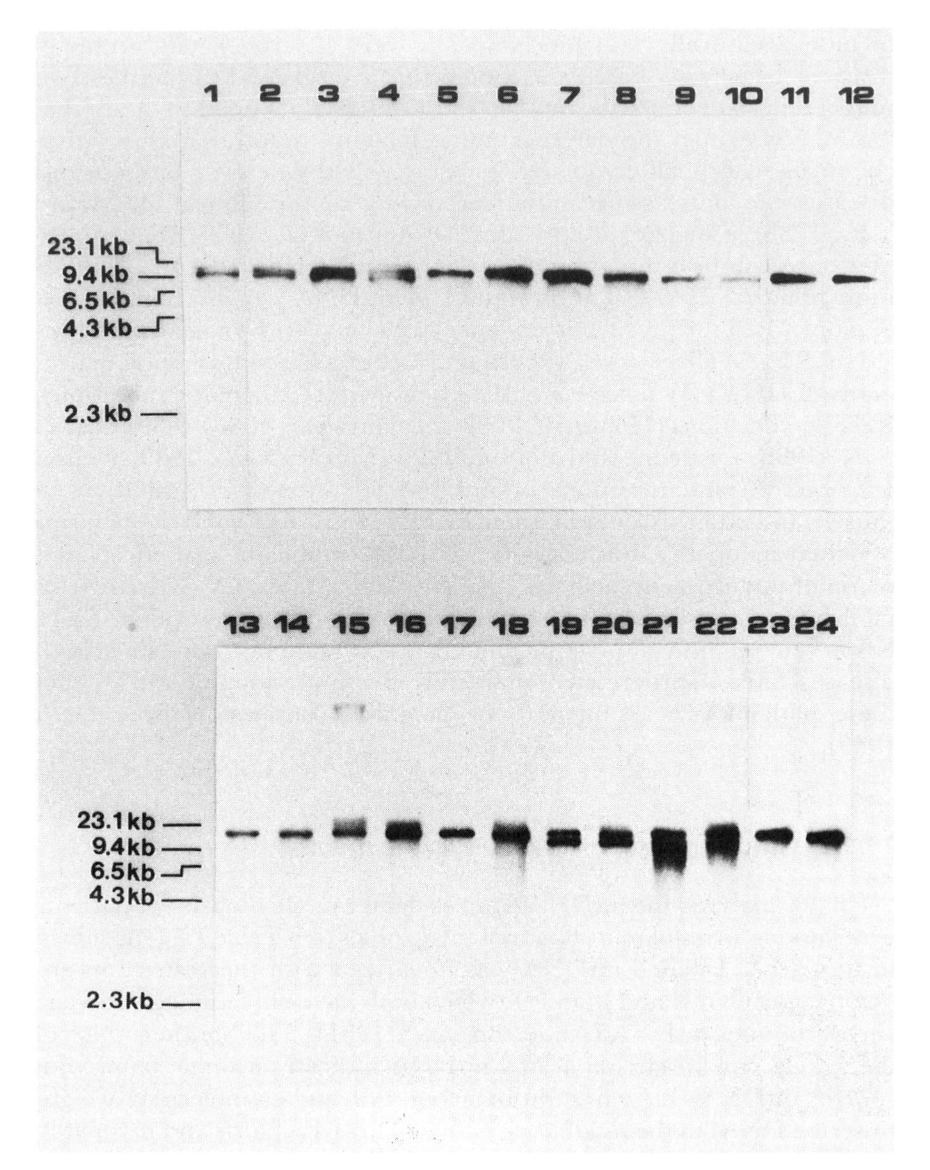

Fig. 19.1 Hybridization pattern of mtDNA from patients with different types of mitochondrial diseases. From about 50 mg muscle biopsy probes mtDNA was isolated (Wallace *et al.*, 1988a), digested with Pvu II (lanes with odd numbers) or Bam HI (lanes with even numbers), electrophoresed through a 1.0% agarose gel, transferred to nylon membranes and hybridized with labelled placental mtDNA. Lanes 1-10 and 13-22, see Table 19.1; lanes 11, 12 23 and 24, control placental mtDNA. Marker sizes (lambda digested with Hind III) are shown at the left.

Table 19.1 Clinical presentation of the ten patients with mitochondrial diseases analysed in Fig. 19.1

Patient	*Age*	*Sex*	*Diagnosis*	*Lanes of Fig. 19.1*
D.B.	35	F	CPEO and myopathy	1, 2
R.L.	33	M	CPEO and myopathy	3, 4
G.M.	32	M	CPEO and myopathy	5, 6
M.R.	64	F	CPEO and myopathy	7, 8
E.G.	67	M	CPEO and myopathy	9, 10
J.W.	42	F	Myopathy	13, 14
D.P.	48	M	Ptosis and myopathy	15, 16
M.M.	50	M	Ptosis and myopathy	17, 18
D.R.	11	F	Ptosis and myopathy	19, 20
K.L.	7	F	KSS	21, 22

19.2.3 Immunohistochemical studies of muscle sections from CPEO patients

In a systematic immunohistochemical study of 17 patients with CPEO, between 6 and 66% of the skeletal muscle fibres showed absence of or decreased activity of cytochrome c oxidase (Johnson *et al.*, 1988). The defective fibres included type I and type II fibres and showed a mosaic-like pattern in cross-section. The immunohistochemical analysis of serial sections with specific antisera to human heart cytochrome c oxidase subunits indicated the partial absence of mitochondria and/or nucleus-encoded subunits. But the pattern of missing subunits in those fibres with absent or decreased cytochrome c oxidase activity varied. Fibres with normal enzyme activity showed a normal immunohistochemical reaction with all antisera (directed against subunits II/III, IV, Vab, VIa, VIbc, VIIabc and VIIbc). The different pattern of missing enzyme subunits in each respiration-defective fibre, and their mosaic-like appearance, suggest in each affected fibre a different mutation of mtDNA, as the molecular basis of the disease. This assumption is supported by the observation that 80% of respiration-defective fibres were lacking the immunoreactivity of the mitochondria-encoded subunit II, but only 20% were lacking that of the nucleus-encoded subunit IV (Johnson *et al.*, 1988).

19.3 RESPIRATION-DEFECTIVE CELLS IN AGEING

As the human heart ages, a growing number of cells lacking cytochrome c oxidase activity was recently found by histochemical analysis (Müller-Höcker, 1988, 1989). The mosaic-like appearance of enzyme-defective cells was independent of underlying heart diseases (coronary arteriosclerosis, heart hypertrophy) and of the treatment of histological tissue sections. Electron microscopic analysis revealed no morphological alterations of the fine structure in cardiomyocytes lacking cytochrome c oxidase activity, including number and shape of mitochondria (Müller-Höcker, 1989). The loss of enzyme activity was always restricted to the area of single cells limited by the plasma membrane. Similar observations were made in the diaphragm and in limb muscle, but respiratory defective fibres appeared later with increasing age in diaphragm, and in skeletal muscle (Müller-Höcker, 1988).

19.4 mtDNA

19.4.1 Structure and information content

Each mammalian cell contains many hundreds of normally identical copies of double-stranded circular mtDNA molecules with three to six copies per mitochondrion (Neubert *et al.*, 1975). Human mtDNA has a size of 16 569 bp (Anderson *et al.*, 1981) and codes for only 13 proteins, two rRNAs and 22 tRNA molecules. The mitochondrial genome contains very few intergenic nucleotides and no detectable introns (Anderson *et al.*, 1981; Ojala *et al.*, 1980; Battey and Clayton, 1980). But for the expression of the 13 genes more than 100 nucleus-encoded proteins are required.

All 13 mitochondria-encoded proteins are components of energy-transducing membrane complexes of the respiratory chain and ATP synthase: seven subunits of reduced nicotinamide adenine dinucleotide (NADH) dehydrogenase, cytochrome b of ubiquinone cytochrome c oxidoreductase, three subunits of cytochrome c oxidase and two subunits of ATP synthase (Chomyn *et al.*, 1985). A mutation in mtDNA accompanied by defective synthesis of one or all of the 13 proteins will result in impaired mitochondrial ATP synthesis, which contributes 95% of total cellular ATP production. Recently it could be demonstrated that mammalian cells can grow in tissue culture also after loss of their mitochondrial genome, if uridine and pyruvate are included in the culture medium (Desjardins *et al.*, 1986; Attardi, personal communication). Under these conditions ATP is generated by glycolysis accompanied by a large production of lactate. In fact patients with defective cytochrome c oxidase in muscle and/or other tissues are characterized by an elevated serum lactate level (e.g. fatal infantile mitochondrial myopathy, Leigh's disease) (DiMauro *et al.*, 1986).

19.4.2 Mutations of mtDNA

The mtDNA is more amenable to chemical attack than nuclear DNA, which is protected by histones. Mitochondria are also lacking an adequate DNA repair system present in the nucleus (Fukanaga and Yielding, 1979; Miyaki *et al.*, 1977). A review on cell ageing by injury of the mitochondrial genome has been given by Fleming *et al.* (1982). On the other hand, extensive treatment of HeLa cells with potent mutagens did not result in frequent mutations of mtDNA, as analysed by cloning and sequencing of mtDNA from treated cells (Mita *et al.*, 1988).

In addition to chemical attack, mutations of mtDNA could arise during replication, due to mistakes of the mitochondrial γ-DNA polymerase. Multiple deletions of similar size within the same region of mtDNA from a patient with KSS have been ascribed to modification of γ-DNA polymerase or another component of the replicating system, which are encoded on nuclear DNA (Zeviani *et al.*, 1989). A mechanism for deletions of the mtDNA has been suggested, based on its displacement-type synthesis (Schon *et al.*, 1989; Holt *et al.*, 1989; Schoffner *et al.*, (1989).

19.4.3 Turnover of mtDNA

Mitochondria grow and divide in mammalian cells continuously, independent of cellular growth and division (Neubert *et al.*, 1975; Gross *et al.*, 1969). Since their total number remains constant, in non-dividing cells of adult organisms, mitochondria are continuously degraded by lysosomal phagocytosis (Swift and Hruban, 1964). While in the brain, heart and skeletal muscle of adult animals and humans nuclear DNA has almost no turnover, the half-life of mtDNA has been determined as 6.7 days in heart, 9.4 days in liver, 10.4 days in kidney and 31 days in brain in adult rats (Gross *et al.*, 1969). Similar data were found for the half-life of cytochrome c (Kadenbach, 1969). The data corresponded well with the half-lives of the corresponding total mitochondrial proteins, indicating that mitochondria turnover in totality (Gross *et al.*, 1969). The turnover of mitochondria was found to be independent of the age of the rats (Menzies and Gold, 1971).

We would like to point out that the general uniformity of mtDNA in mammalian organisms (Greenberg *et al.*, 1983) is mainly due to its continuous turnover. In the population of many hundreds of mtDNA molecules in a cell, a single mutation will normally be eliminated. Only if a mutated mtDNA has a selective advantage, i.e. a higher rate of replication, for example due to a smaller size of deleted mtDNA, could its higher turnover result in the replacement of the normal mtDNA of that cell. In a non-dividing tissue the mutation will be restricted to a single cell or fibre, because mitochondria cannot penetrate the plasma membrane. It should be pointed out that in muscle sections of patients with KSS (Müller-Höcker *et al.*, 1983, 1986) as well as with CPEO (Johnson *et al.*, 1983, 1988) cytochrome c oxidase-negative areas are always limited by the plasma membrane. The complete replacement of the intact mtDNA, however, may last for months or years.

19.5 mtDNA AND AGEING

KSS, CPEO and certain other types of mitochondrial myopathies (DiMauro *et al.*, 1986) as well as respiratory-deficient cardiac cells and muscle fibres in healthy humans have in common the following features: (1) onset of the defect at juvenile or adult age; (2) progressive character of the defect with increasing age; (3) mosaic-like or focal pattern of cytochrome c oxidase deficiency of affected cells (fibres).

We suggest the following events causing respiration-deficient cells (fibres): (1) mutation of a single mtDNA molecule; the type of mutation will differ in healthy human and in patients with mitochondrial disease, where specific mutations (deletions or insertions) could be supported by silent mutations of mtDNA or of nuclear genes for proteins involved in mtDNA replication; (2) replacement of normal by mutated mtDNA, due to its selective advantage, i.e. increased rate of replication, in each affected cell; (3) defective mitochondrial protein synthesis will result in lack of respiration and ATP synthesis in mitochondria.

We hypothesize that mutation of mtDNA in normal healthy humans, resulting in affected cells, contributes to ageing and could finally limit the life-span of human beings. Individual respiratory-deficient, but glycolytically active cardiomyocytes, will reduce the maximal energy output of the heart, because lactic acid production may deteriorate the function of neighbouring intact cells. In fact acidic pH was found to decrease the Ca^{2+} sensitivity of cardiac myofilaments (Fabiato and Fabiato, 1978), and a fourfold higher Ca^{2+} sensitivity was required in skinned golden hamster cardiac muscle for generating the same tension at pH 6.5 as compared to pH 7.0 (Gulati and Babu, 1989). In skeletal muscle this difference in Ca^{2+} concentration dependence at the two pHs was much less pronounced. We suggest that after accumulation of a certain percentage of non-respiring cardiomyocytes, under stress situations, when maximal energy output is required, heart insufficiency could limit the individual human life.

Recently Linnane *et al.* (1989) published a hypothesis on mtDNA mutations as an important contributor to ageing and degenerative diseases. These authors, however, did not consider the continuous turnover of mtDNA in non-dividing cells as the essential factor for either elimination or propagation of mtDNA mutations in cells or fibres of adult mammals.

ACKNOWLEDGEMENTS

This work was supported by the Deutsche Forschungsgemeinschaft (Ka 192/17-4) H, Fonds der Chemischen Industrie and Thyssen-Stiftung.

REFERENCES

Anderson, S., Bankier, A.T., Barrell, B.G. *et al.* (1981) *Nature*, **290**, 457–65.
Battey, J. and Clayton, D.A. (1980) *J. Biol. Chem.*, **255**, 11599–606.
Chomyn, A., Mariottini, P., Cleeter, M.W.J. *et al.* (1985) in *Achievements and Perspectives of Mitochondrial Research, Vol. II, Biogenesis* (eds E. Quagliariello, E.C. Slater, F. Palmieri *et al.*), Elsevier, Amsterdam, pp. 259–75.
Desjardins, P. de Muys, J.-M. and Morais, R. (1986) *Somat. Cell Molec. Genet.*, **12**, 133–9.
DiMauro, S., Bonilla, E., Zeviani, M., *et al.* (1985) *Neurology*, **17**, 521–38.
DiMauro, S., Zeviani, M., Servidei, S. *et al.* (1986) *Ann. NY Acad. Sci.*, **488**, 19–32.
DiMauro, S., Bonilla, E., Zeviani, M. (1987) *J. Inherited Metab. Dis.*, **10** (Suppl. 1), 113–28.
Fabiato, A. and Fabiato, F. (1978) *J. Physiol.*, **276**, 233–55.
Fleming, J.E., Miquel, J., Cotterell, S.F. *et al.* (1982) *Gerontology*, **28**, 44–53.
Fukanaga, M. and Yielding, K.L. (1979) *Biochem. Biophys. Res. Commun.*, **90**, 582–6.
Giles, R.E., Blanc, H., Cann, H.M. and Wallace, D.C. (1980) *Proc. Natl Acad. Sci. USA*, **77**, 6715–19.
Greenberg, B.D., Newbold, J.E. and Sugino, A. (1983) *Gene*, **21**, 33–49.
Gross, N.J., Getz, G.S. and Rabinowitz, M. (1969) *J. Biol. Chem.*, **244**, 1552–62.
Gulati, J. and Babu, A. (1989) *FEBS Lett.*, **245**, 279–82.
Holt, I.J., Harding, A.E. and Morgan-Hughes, J.A. (1988) *Nature*, **331**, 717–19.
Holt, I.J., Harding, A.E. and Morgan-Hughes, J.A. (1989) *Nucl. Acids Res.* **17**, 4465–9.
Johnson, M.A., Turnbull, D.M., Dick, D.J. and Sherratt, H.S.A. (1983) *J. Neurol. Sci.*, **60**, 31–53.
Johnson, M.A., Kadenbach, B., Droste, M. *et al.* (1988) *J. Biol. Sci.*, **87**, 75–90.
Kadenbach, B. (1969) *Biochem. Biophys. Acta*, **186**, 399–401.
Lestienne, P. and Ponsot, G. (1988) *Lancet*, **i**, 885.
Linnane, A.W., Marzuki, S., Ozawa, T. and Tanaka, M. (1989) *Lancet*, **i**, 642–5.
Menzies, R.A. and Gold, P.H. (1971) *J. Biol. Chem.*, **246**, 2425–9.
Mita, S., Monnat, R.J. and Loeb, L.A. (1988) *Cancer Res.*, **48**, 4578–83.
Miyaki, M., Yatagai, K. and Ono, T. (1977) *Chem. Biol. Interactions*, **17**, 321–9.
Moraes, C.T., DiMauro, S., Zeviani, M. *et al.* (1989) *N. Engl. J. Med.*, **320**, 1293–9.
Müller-Höcker, J., (1988) *Verh. Dtsch. Ges. Pathol.*, **72**, 552–65.
Müller-Höcker, J. (1989) *Am. J. Pathol.*, **134**, 1167–73.
Müller-Höcker, J., Pongratz, D. and Hübner, G. (1983) *Virchows Arch. Pathol. Anat.*, **402**, 61–71.

Müller-Höcker, J., Johannes, A., Droste, M. (1986) *Virchows Arch. Cell Pathol.*, **52**, 353–67.
Müller-Höcker, J., Droste, M., Kadenbach, B. *et al.* (1989) *Human Pathol.*, **20**, 666–72.
Nelson, I., Degoul, F., Obermaier, B. *et al.* (1989) *Nucl. Acids Res.*, **17**, 8117–24.
Neubert, D., Gregg, C.T., Bass, R. and Merker, H.-J. (1975) in *The Biochemistry of Animal Development*, Vol. 3 (ed. R. Weber), Academic Press, New York, pp. 387–464.
Ojala, D., Merkel, C., Gelfand, R. and Attardi, G. (1980) *Cell*, **22**, 393–403.
Ozawa, T., Yoneda, M., Tanaka, M. *et al.* (1988) *Biochem. Biophys. Res. Commun.*, **154**, 1240–7.
Poulton, J., Deadman, M.E. and Gardiner, M. (1989) *Lancet*, **i**, 236–40.
Schon, E.A., Rizzuto, R., Moraes, C.T. *et al.* (1989) *Science*, **244**, 346–9.
Shoffner, J.M., Lott, M.T., Voljavec, A.S. *et al.* (1989) *Proc. Natl Acad. Sci. USA*, **86**, 7952–6.
Swift, H. and Hruban, Z. (1964) *Fed. Proc.*, **23**, 1026.
Wallace, D.C., Zheng, X., Lott, M.T. *et al.* (1988a) *Cell*, **55**, 601–10.
Wallace, D.C., Singh, G., Lott, M.T. *et al.* (1988b) *Science*, **242**, 1427–30.
Zeviani, M., Moraes, C.T., DiMauro, S. *et al.* (1988) *Neurology*, **38**, 1339–46.
Zeviani, M., Servidei, S., Gellera, C. *et al.* (1989) *Nature*, **339**, 309–11.

20 Analysis of giant deletions of human mitochondrial DNA in progressive external ophthalmoplegia

E.A. SCHON,[1,2] C.T. MORAES,[2] S. MITA,[1] H. NAKASE,[1] A. LOMBES,[1] S. SHANSKE,[1] E. ARNAUDO,[1] Y. KOGA,[1] M. ZEVIANI,[1] R. RIZZUTO,[1] A.F. MIRANDA,[1,3] E. BONILLA,[1] AND S. DiMAURO[1]
Departments of [1]Neurology, [2]Genetics and Development, and [3]Pathology, Columbia University, New York

20.1 INTRODUCTION

The 'mitochondrial myopathies' were initially defined mainly on the basis of morphological criteria, such as the appearance in muscle sections of 'ragged red fibres' (RRFs). RRFs are a morphological hallmark of proliferating mitochondria in muscle, and are seen in muscle sections stained with modified Gomori trichrome as red patches (Engel and Cunningham, 1963). In the last ten years, biochemical analyses have helped reclassify these diseases, mainly on the basis of the type of mitochondrial function which was impaired. These include defects in substrate transport and utilization, defects in the Krebs cycle, defects in oxidative phosphorylation, and defects in the respiratory chain (DiMauro *et al.*, 1987). The bulk of the recent work in this field has focused on the respiratory chain. Only now are we beginning to analyse these errors at the molecular genetic level.

The human mitochondrial genome is a 16 569 bp circle of double-stranded DNA (Anderson *et al.*, 1981). It contains genes encoding two ribosomal RNAs, 22 tRNAs, and 13 structural polypeptides, all of which are subunits of components of the respiratory chain. Of the 13 structural genes, seven encode subunits of complex I (reduced nicotinamide adenine dinucleotide (NADH)–coenzyme Q (CoQ) oxidoreductase), one encodes a subunit of complex III (CoQ–cytochrome c oxidoreductase), three encode subunits of complex IV (cytochrome c oxidase), and two encode subunits of complex V (ATP synthase). Each of these complexes also contains subunits encoded by nuclear genes, which are imported from the cytoplasm and assembled, together with the mitochondrial mtDNA-encoded subunits, into the holoenzyme located in the mitochondrial inner membrane.

Defects in mtDNA-encoded subunits should result in pedigrees exhibiting maternal inheritance, i.e. the disease should pass only through females, and essentially all children (both boys and girls) should inherit the error. It has often been difficult to confirm maternal inheritance and to distinguish it clearly from an autosomal dominant mode of inheritance.

Maternal inheritance of a mitochondrial myopathy has been demonstrated unequivocally in only two diseases. MERRF (myoclonic epilepsy with RRFs) is characterized by myoclonus, ataxia, hearing loss, weakness and generalized seizures. Muscle biopsies show abundant RRFs (Fukuhara *et al.*, 1980). However, although it is known to be maternally inherited (Rosing *et al.*, 1985; Wallace, 1987; Wallace *et al.*, 1988a), the exact genetic defect in mtDNA in MERRF has not yet been elucidated (Schon *et al.*, 1988a; Wallace *et al.*, 1988a). On the other hand, a second mitochondrial disease with maternal inheritance has now also been described – Leber's hereditary optic neuropathy (LHON), also called Leber's optic atrophy – in which almost all pedigrees share a single point mutation at codon 340 of the subunit 4 gene of complex I (Wallace *et al.*, 1988b; Singh *et al.*, 1989).

At least two other mitochondrial diseases, Kearns–Sayre syndrome (KSS) and ocular myopathy (OM), which are related myopathies characterized by progressive external ophthalmoplegia (PEO), could not be classified at all, because patients presenting with these disorders were almost all sporadic, with no apparent genetic component. KSS is a multisystem mitochondrial disorder defined by the presence of ophthalmoplegia, pigmentary retinopathy, onset before age 20, and at least one of the following: high cerebrospinal fluid (CSF) protein content, blockage in heart condition or ataxia (Pavlakis *et al.*, 1988). Morphologically, KSS patients display RRFs in muscle sections. KSS is ultimately fatal. OM shares with KSS ocular myopathy and often RRFs in muscle, but unlike KSS there is no systemic involvement and the disease is rarely fatal. Biochemically, both KSS and OM often show reduced respiratory chain enzyme activity, particularly that of cytochrome c oxidase (COX) (Johnson *et al.*, 1983; Muller-Hocker *et al.*, 1983; Byrne *et al.*, 1985).

Holt *et al.* (1988a) opened a new field of investigation when they found deletions of mtDNA in 9 of 25 patients with 'mitochondrial myopathies', defined by the appearance of morphologically abnormal organelles in muscle biopsy. However, they did not describe the clinical syndromes of the patients.

As described below, we have recently discovered that nearly all patients with PEO presenting with RRFs, including essentially all KSS patients and about half of all OM patients, harbour large-scale deletions of mtDNA in their skeletal muscle. The size and location of the deletions, and the number of deleted mtDNAs relative to the number of normal mitochondrial genomes, differed among patients, and did not appear to be correlated to the presentation or the severity of the disease phenotype (Zeviani *et al.*, 1988; Moraes *et al.*, 1989a; Schon *et al.*, 1988b, 1989) (see below).

Both mtDNA deletions (Holt *et al.*, 1988b; Lestienne and Ponsot, 1988; Rotig *et al.*, 1988; Johns *et al.*, 1989) and duplications (Poulton *et al.*, 1989) have now been observed by others. In addition, two reports have appeared demonstrating familial inheritance of mtDNA deletions in a mitochondrial myopathy with PEO (Ozawa *et al.*, 1988; Zeviani *et al.*, 1989). Interestingly, in these familial cases the deletions differed among family members, and in the family of Zeviani *et al.* (1989), in which the disease is inherited in an autosomal dominant manner, different deletions coexisted within the same muscle biopsies of different affected family members.

It is now clear that a new class of genetic disease is beginning to evolve: the mitochondrial disorders due to large-scale rearrangements of the human mitochondrial genome. For the first time, defined human pathologies have been associated with large-scale deletions of mtDNA, and this finding opens up new avenues for study of both the disease process itself and the basic biology of mitochondria (Harding, 1989).

Mitochondria contain multiple mtDNAs (Nass, 1969; Bogenhagen and Clayton, 1974), but it is not clear at this point whether the deleted mtDNAs are physically segregated from the wild-type genomes. However, all the data developed by us to date (see below) imply that genetic complementation – both within and between mitochondria – either is not occurring or is occurring at a level that does not affect the phenotype. This suggests that deleted genomes are segregated from wild-type genomes, at least within mitochondria, or that the mere presence of deleted genomes affects mitochondrial function adversely. If this is the case, we may be able to design a rational therapy to eliminate selectively mitochondria containing deleted genomes. Thus, studying genetic segregation and complementation is not only important on theoretical grounds but could also have practical implications.

These PEO patients present us with a rare opportunity to study questions about human mitochondrial biogenesis and genetics in a way that was heretofore difficult, if not impossible.

20.2 ANALYSIS OF PATIENTS FOR THE PRESENCE OF DELETED mtDNAs

In our initial investigations, we performed genomic Southern analysis of total DNA isolated from 7 patients with KSS, 10 disease controls, and 3 normals, using mtDNA as the hybridization probe (Zeviani *et al.*,

1988). Using the restriction enzyme *Pvu*II, which cuts only once in the mtDNA genome (and hence produces a single, 16.5 kb, linear molecule in normal mtDNA), the hybridization pattern in all seven patients with KSS was consistent with the presence of two populations of mtDNA: one corresponding to full-length mtDNA molecules (i.e. 16.5 kb long), and the other correspondingly to smaller mtDNAs (i.e. bands minus the deleted fragment migrating more rapidly). The size of the deletions ranged from 2.0 kb to 7.0 kb. None of the other (non-KSS) patients or normal controls showed any evidence of a deletion. The relative number of mutated genomes varied widely among patients, ranging from 45% to 75%, as measured by densitometry of the Southern autoradiograms.

Biochemical analysis of mitochondrial enzymes in muscle biopsies of six KSS patients showed a partial defect of COX activity on three patients; other enzymatic activities of the respiratory chain were normal, as was citrate synthase, a matrix enzyme.

In a follow-up study on 123 patients with all types of mitochondrial myopathies, deletions were found in 32 patients, all of whom had PEO (Moraes *et al.*, 1989a). We found no patients with mitochondrial myopathy who had deletions of mtDNA other than those with KSS or OM with RRFs. With a larger statistical base, we were able to show a significant correlation between the presence of deletions and reductions in respiratory chain activities.

20.3 MAPPING OF DELETIONS IN PEO

The deletions among the 32 patients ranged in size from 1.3 to 7.6 kb (Moraes *et al.*, 1989a). To map the site of each deletion, we performed Southern analysis with other restriction enzymes whose map locations on the mitochondrial genome are known (Anderson *et al.*, 1981). We showed that all the deletions were in regions of the mitochondrial genome containing structural components of the respiratory chain; no deletions were found in the ribosomal RNA genes, or in the region of either the origins of heavy- or light-strand replication, or of heavy- and light-strand transcription. Interestingly, of 29 mapped deletions, 12 (4 in patients with KSS and 8 in patients with OM) mapped to an identical location in the mtDNA, with deletion breakpoints about 5 kb apart, extending from the ATPase8 gene of complex V on the 'left', to the ND5 gene of complex I on the 'right'.

20.4 SEQUENCING OF DELETION BREAKPOINTS

Using the polymerase chain reaction (PCR) (Saiki *et al.*, 1985), we amplified selectively the region of the deleted genome spanning the deleted mtDNA in a number of patients with the 'common' deletion, and determined the exact site of the deletion breakpoint by DNA sequencing (Schon *et al.*, 1989). Of the 12 patients with the 'common' deletion, we sequenced mtDNA from three with KSS and three with OM. All six had the identical deletion breakpoint. The deletion was 4977 bp long, with the breakpoint on the 'left' side at nucleotide position 8483 within the ATPase8 gene, and on the 'right' side at position 13 460 within the ND5 gene. The deletion was flanked by a perfect 13 bp repeat; one repeat was found immediately prior to the 'left' deletion breakpoint, while the other 13 bp repeat was found (in normal, undeleted, mtDNA) at the extreme 3′ end of the deleted region, just prior to the 'right' deletion breakpoint. The 'fusion' gene thus created encodes an mRNA which is out of frame in the ND5 portion of the predicted transcript, resulting in a premature termination codon 12 nucleotides (four amino acids) beyond the deletion breakpoint: rather than encoding an ATPase8 protein of 68 amino acids (deduced molecular weight of 7.9 kDa), this mRNA, if translated, would encode a truncated protein 42 amino acids long (i.e. about 5 kDa). This 'common' deletion has now been reported in a patient with a totally different disease, Pearson's marrow/pancrease sydrome (Rotig *et al.*, 1988, 1989).

Because the deletion breakpoint in the three KSS patients was exactly the same as that of the three OM patients, we concluded that the size and/or location of the deletion per se had nothing to do with the phenotypic features that distinguish these two diseases. We had already noted previously that the relative number of deleted genomes was also not a distinguishing feature of KSS as compared to OM (Zeviani *et al.*, 1988; Moraes *et al.*, 1989a). Thus, the relatively 'mild' features of OM must be due to some other aspect of the pathogenesis of this disease: perhaps the key feature that differentiates the severity of the pathology in these two similar disorders is related to the extent to which the deleted genomes segregate during embryonic development in each disease.

The presence of a large-scale deletion flanked by direct repeats in the mtDNA implies that this deletion was caused by a homologous recombination event, and that there are as yet undefined mechanisms for rearranging human mtDNA. This work was the first documentation of direct repeats involved in the removal of large-scale regions of mammalian mitochondrial DNA. Moreover, the fact that 12 of 30 examined deletions in KSS/OM were flanked by this particular direct repeat implies that homologous recombination is a significant source of deletion in these diseases.

We have now sequenced 16 more different deletion breakpoints (Mita *et al.*, 1990). Broadly speaking, they fall into two classes: class I contains deletions that are precisely flanked by direct repeats (nine loci sequenced, including the 'common' deletion; Fig. 20.1(a)), while class II

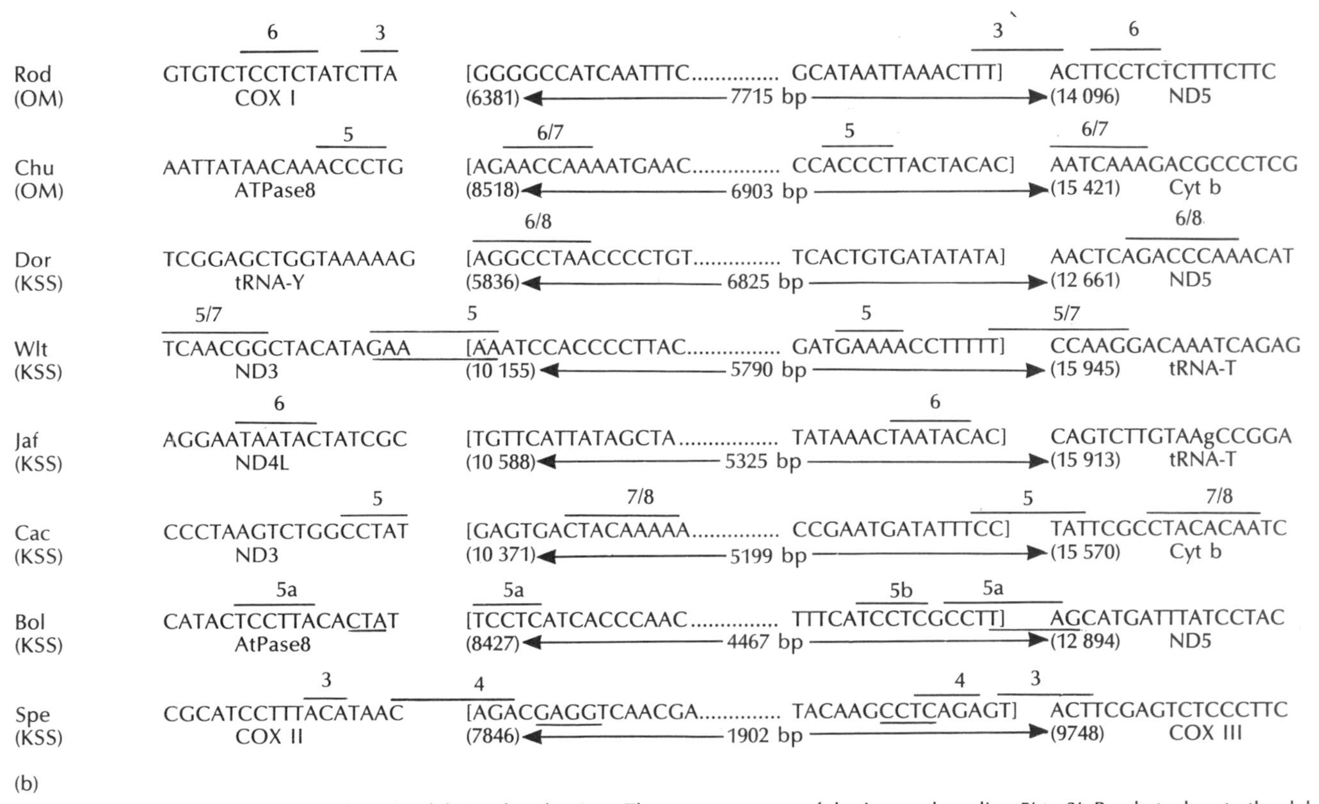

Fig. 20.1 DNA sequences surrounding the deletion breakpoints. The sequences are of the L-strand reading 5′ to 3′. Brackets denote the deletion breakpoints; mtDNA numbering (Anderson *et al.*, 1981) is of the first nucleotide at the 5′ end and of the first nucleotide immediately following the 3′ end of each deletion. The patient code is shown to the left of each deleted region; each patient is identified as having either KSS or OM. The interrupted genes are shown below the undeleted areas flanking the deletion. The length of the deletion is shown below the sequence. Repeated elements are overlined, showing the number of nucleotides in the repeat shown; inverted repeats are underlined. (a) Deletions flanked by direct repeats (b) Deletions not flanked by direct repeats. The lower-case ‘g’ in patient Jaf is a polymorphism relative to the published mtDNA sequence; 5a and 5b in patient Bol are two sets of 5 bp repeats.

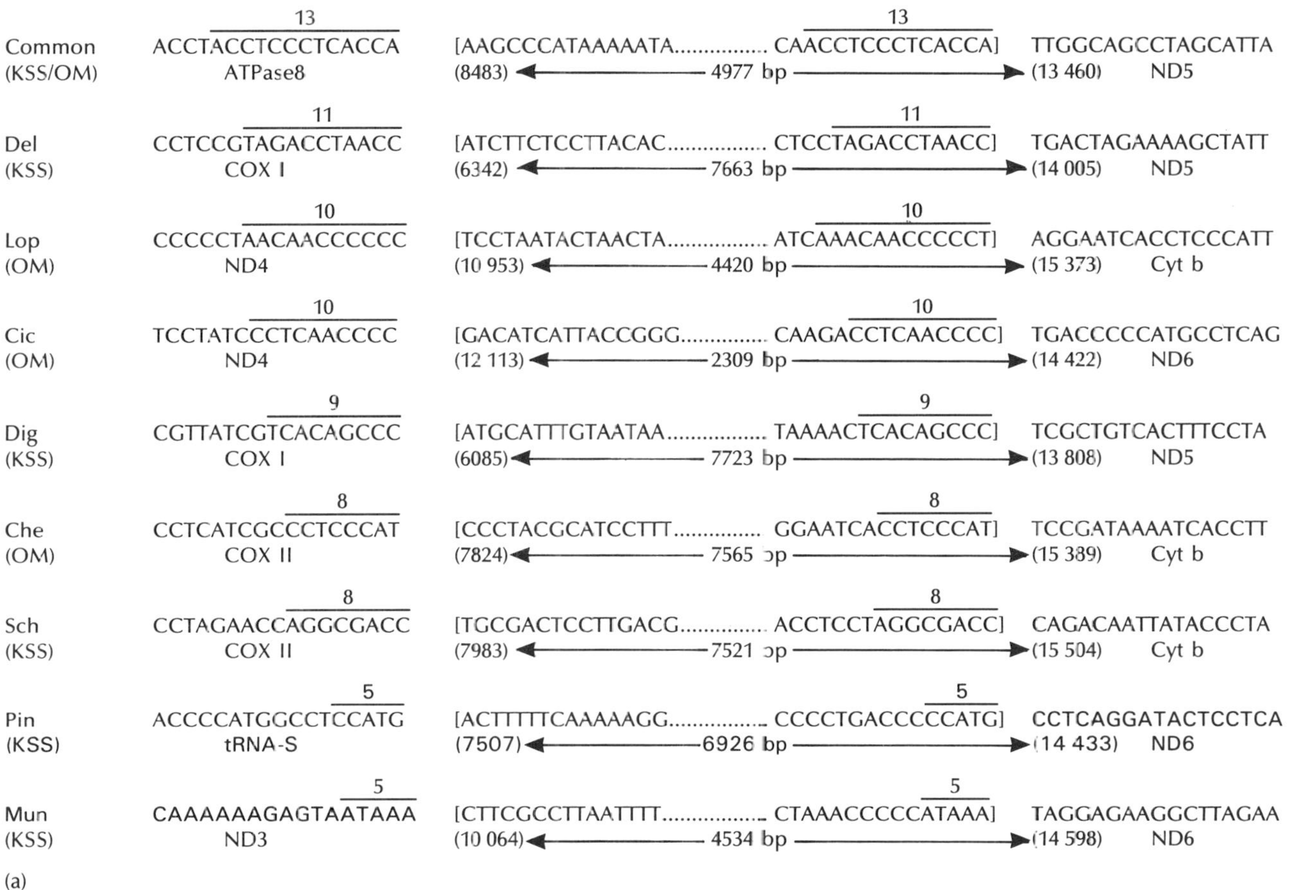

Common (KSS/OM)
13
ACCTACCTCCCTCACCA
ATPase8
[AAGCCCATAAAAATA................ CAACCTCCCTCACCA]
(8483) 4977 bp (13 460)
TTGGCAGCCTAGCATTA
ND5
Del (KSS)
11
CCTCCGTAGACCTAACC
COX I
[ATCTTCTCCTTACAC................ CTCCTAGACCTAACC]
(6342) 7663 bp (14 005)
TGACTAGAAAAGCTATT
ND5
Lop (OM)
10
CCCCCTAACAACCCCCC
ND4
[TCCTAATACTAACTA................ ATCAAACAACCCCCT]
(10 953) 4420 bp (15 373)
AGGAATCACCTCCCATT
Cyt b
Cic (OM)
10
TCCTATCCCTCAACCCC
ND4
[GACATCATTACCGGG.............. CAAGACCTCAACCCC]
(12 113) 2309 bp (14 422)
TGACCCCCATGCCTCAG
ND6
Dig (KSS)
9
CGTTATCGTCACAGCCC
COX I
[ATGCATTTGTAATAA................ TAAAACTCACAGCCC]
(6085) 7723 bp (13 808)
TCGCTGTCACTTTCCTA
ND5
Che (OM)
8
CCTCATCGCCCTCCCAT
COX II
[CCCTACGCATCCTTT.............. GGAATCACCTCCCAT]
(7824) 7565 bp (15 389)
TCCGATAAAATCACCTT
Cyt b
Sch (KSS)
8
CCTAGAACCAGGCGACC
COX II
[TGCGACTCCTTGACG.............. ACCTCCTAGGCGACC]
(7983) 7521 bp (15 504)
CAGACAATTATACCCTA
Cyt b
Pin (KSS)
5
ACCCCATGGCCTCCATG
tRNA-S
[ACTTTTTCAAAAAGG................ CCCCTGACCCCCATG]
(7507) 6926 bp (14 433)
CCTCAGGATACTCCTCA
ND6
Mun (KSS)
5
CAAAAAAGAGTAATAAA
ND3
[CTTCGCCTTAATTTT................ CTAAACCCCCATAAA]
(10 064) 4534 bp (14 598)
TAGGAGAAGGCTTAGAA
ND6
(a)

contains deletions that are not in class I (eight loci sequenced; Fig. 20.1(b).

It is not clear at this point whether the two classes of deletions reflect two distinct mechanisms of recombination – presumably homologous and non-homologous recombination respectively – or whether they are two manifestations of one underlying mechanism. Nevertheless, if the aetiology of KSS and OM is genetic, these results provide an important clue as to the potential nature of the error in these diseases, as the fundamental mutation is likely to reside in a nuclear gene associated with the cell's recombination machinery.

20.5 ANALYSIS OF HETEROPLASMY IN PEO

We have analysed heteroplasmy of mtDNA in clonal cultures from two KSS patients (Moraes *et al.*, 1989b). Surprisingly, many individual cellular clones (both myoblasts and fibroblasts) from these patients had only normal mtDNAs (at a level of detection in Southern analysis of $<1\%$ deleted genomes), while a minority of clones were heteroplasmic, i.e. they contained both wild-type and deleted mtDNAs. The heteroplasmic clones grew more slowly in tissue culture than did the homoplasmic clones with normal mtDNA, and only the heteroplasmic cells showed reduced COX activity.

We analysed multiple tissues from KSS patients, and found deleted mtDNAs, albeit in varying proportions, in all tissues examined (including rapidly dividing tissues, such as leucocytes and platelets). On the other hand, we detected deletions in OM patients only in muscle tissue (Moraes *et al.*, 1989a; Shanske *et al.*, 1990). Thus, analysis of heteroplasmy in various tissues lends credence to the hypothesis that the major difference in the severity and tissue specificity of the defect in OM as compared to KSS resides in the difference in tissue distribution of deleted mtDNAs between the two diseases.

20.6 TRANSCRIPTION AND TRANSLATION OF DELETED mtDNAs

We have found that mitochondria containing deleted mtDNAs are competent for transcription but are incompetent for translation (Nakase *et al.*, 1990).

Using mtDNA probes specific to the regions flanking the deletion in two KSS patients, we have shown by Northern analysis that the deleted genomes are transcribed. Specifically, we detected in both patients an aberrant hybridizing mRNA with a size consistent with a fusion mRNA derived from transcription across the deletion breakpoint. However, when we incorporated ^{35}S-Met into mitochondrial translation products (after inhibiting cellular translation with emetine), we found all the normal predicted mtDNA-encoded polypeptides, but *no* aberrant 'fusion' polypeptide, in fluorograms of translation products separated by sodium dodecyl suphate–polyacrylamide gel electrophoresis.

20.7 MORPHOLOGICAL ANALYSIS OF MUSCLE SECTIONS IN KSS

In order to establish a relationship between COX deficiency and abnormalities of muscle mtDNA at the single-fibre level, we have studied a patient with KSS harbouring the 'common' deletion (Mita *et al.*, 1989). Using a combination of in situ hybridization, cytochemistry and immunofluorescence, we showed that: (1) COX deficiency clearly correlates with a reduction in normal mtDNAs; (2) deleted mtDNAs were more abundant in COX-deficient fibres, especially RRFs; (3) focal COX deficiency is segmental in a given fibre; (4) the deleted mtDNAs are transcribed; and (5) mtDNA-encoded subunits *not* encompassed by the deletion, such as subunit II of COX, are not translated (antibodies to COX II immunofluoresce in COX-positive, but not in COX-negative fibres). These results help explain the pathogenesis of KSS at the cellular level.

20.8 IMPLICATIONS FOR PATHOGENESIS AND TREATMENT

Since mitochondria contain multiple genomes (Bogenhagen and Clayton, 1974), and since the PEO deletions are all believed to be clonal expansions of a single mutational event (Zeviani *et al.*, 1988), many progeny mitochondria presumably should contain both normal and deleted mtDNAs. Thus, a mitochondrion containing a deleted mtDNA might still be functional, since normal gene products may be provided by the other normal mtDNAs within the same organelle (i.e. genetic complementation). However, if segregation of deleted and wild-type genomes into separate organelles has occurred in these patients, or if the presence of deleted mtDNA molecules impairs the correct expression of the wild-type mtDNA, a decrease in mitochondrial function would be expected (absence of complementation).

At least five observations by our group imply that complementation is not occurring. (1) The PEO patients have observable deficiencies in respiratory chain enzyme activities. As stated above, if complementation were occurring, the wild-type genomes should have corrected or compensated for the error in these deletions. (2) The analysis of fibroblast clones from a patient with KSS demonstrated a decreased COX activity in heteroplasmic cells (cells containing both deleted and wild-type mtDNA) versus those cellular clones that contained exclusively wild-type mtDNA (Moraes *et al.*, 1989b). (3) Detailed Northern analysis of two KSS patients demonstrated that the deleted genomes were transcribed, and that aberrant 'fusion' mRNAs were synthesized from the region spanning the deletions, with sizes in agreement with those predicted from the deleted mtDNAs. Nevertheless, the incorporation of ^{35}S-Met into mitochondrial translation products in a heteroplasmic population of fibroblasts from one of these patients showed that all normal translation products of mtRNAs were synthesized, but *no* fusion protein derived from the fusion mRNA was observed (Nakase *et al.*, 1990). Thus, mitochondria containing deleted genomes seemed to be competent for transcription but incompetent for translation. (4) Of 28 patients with 17 different deletions, *every single one* had a deletion in mtDNA which eliminated at least one tRNA gene, implying that if a deletion leaves all the tRNA genes intact the phenotypic expression of PEO will not manifest itself. This, of course, is negative evidence, but is nonetheless supportive of the argument that intramitochondrial complementation does not occur in PEO. (5) A detailed morphological analysis of one KSS patient with the 'common' 5 kb deletion showed that the COX deficiency in this patient's muscle was not only segmental (i.e. normal in some fibres, deficient in others), but that RNAs *not* encompassed by the deletion were transcribed but were *not* translated, as measured by immunofluorescence (Mita *et al.*, 1989).

If deleted genomes are indeed segregated in PEO, with little or no complementation, we may be able to design a rational therapy for these diseases, based on the assumption that we could selectively 'kill off' the mitochondria containing deleted genomes.

20.9 CONCLUDING REMARKS

The discovery of large-scale deletions in human mitochondrial DNA associated with two poorly understood neuromuscular disorders is a breakthrough in terms of our understanding of the mitochondrial diseases. Aside from descriptive symptomatology, almost nothing was known about PEO prior to the discovery of giant deletions of mtDNA in these patients. We now have a way of classifying these diseases rationally, and in fact the presence of deleted mtDNA may now become a criterion for diagnosis of these disorders.

Now that we are able to categorize KSS and OM on the basis of specific clinical, biochemical and now, for the first time, genetic criteria, we can begin to asked truly focused questions that might help in an understanding of the aetiology, pathogenesis and ultimately treatment of these debilitating, and in the case of KSS, often fatal diseases.

Our observation of giant deletions in human mtDNA has already changed our thinking about the plasticity of the mammalian mitochondrial genome. Specifically, deletions, rearrangements and recombination in mtDNA in general had been observed previously only in lower eukaryotes and plants, but never in mammalian mtDNA, aside from one report of mtDNA deletions in mice (Boursot *et al.*, 1987). In this regard, we note that almost all cases of KSS and OM arise spontaneously; as opposed to MERRF and LHON, KSS and OM are mitochondrial diseases with no apparent maternal inheritance. Thus, although KSS and OM are genetic diseases of mitochondria, the primary heritable defect – if KSS and OM are indeed inherited diseases – may reside in the nuclear genome, because only the mtDNA-encoded respiratory chain genes are inherited maternally. The family of Zeviani *et al.* (1989), in which PEO was inherited in an autosomal dominant manner, supports this idea. The deletions in PEO may thus give us a way to analyse recombination in mammalian mtDNA and to search for components of the recombination machinery; these putative recombinases are almost certainly nucleus encoded because, as noted above, the mammalian mitochondrial genome encodes structural genes for just the respiratory chain.

ACKNOWLEDGEMENTS

This work was supported by grants form the NAtional Institutes of Health of the United States, the Muscular Dystrophy Association, the Aaron Diamond Foundation, and Dr and Mrs Libero Danesi, Milan, Italy.

REFERENCES

Anderson, S., Bankier, A.T., Barrell, B.G. *et al.* (1981) *Nature*, **290**, 457–65.
Bogenhagen, D. and Clayton, D.A. (1974) *J. Biol. Chem.*, **249**, 7991–5.
Boursot, P., Yonekawa, H. and Bonhomme, F. (1987) *Mon. Biol. Evol.*, **4**, 46–55.
Bryne, E., Dennett, X., Trounce, I. and Henderson, R. (1985) *J. Neurol. Sci.*, **71**, 257.
DiMauro, S., Servidei, S., Zeviani, M. *et al.* (1987) *Ann. Neurol.*, **22**, 498–506.
Engel, W.K. and Cunningham, G.G. (1963) *Neurology*, **13**, 919–23.
Fukuhara, N., Tokiguchi, S., Shirakawa, K. and Tsubaki, T. (1980) *J. Neurol. Sci.*, **47**, 117–33.
Harding, A.E. (1989) *N. Engl. J. Med.*, **320**, 1341–3.
Holt, I.J., Harding, A.E. and Morgan-Hughes, J.A. (1988a) *Nature*, **331**, 717–19.

Holt, I.J., Cooper, J.M., Morgan-Hughes, J.A. and Harding, A.E. (1988b) *Lancet*, **ii**, 1462.
Johns, D.R., Drachman, D.B., and Hurko, O. (1989) *Lancet*, **i**, 393–94.
Johnson, M.A., Turnbull, D.M., Dick, D.J. and Sheratt, H.S.A. (1983) *J. Neurol. Sci.*, **60**, 31.
Lestienne, P. and Ponsot, G. (1988) *Lancet*, **ii**, 885.
Mita, S., Schmidt, B., Schon, E.A. *et al.* (1989) *Proc. Natl Acad. Sci. USA*, **86**, 9509–13.
Mita, S., Rizzuto, R., Moraes, C.T. *et al.* (1990) *Nucl. Acids Res.*, **18**, 561–7.
Moraes, C.T., DiMauro, S., Zeviani, M. *et al.* (1989a) *N. Engl. J. Med.*, **320**, 1293–9.
Moraes, C.T., Schon, E.A., DiMauro, S. and Miranda, A.F. (1989b) *Biochem. Biophys. Res. Commun.*, **160**, 765–71.
Muller-Hocker, J., Pongratz, D. and Hubner, G. (1983) *Virchows Arch.*, **402**, 61.
Nakase, H., Moraes, C.T., Rizzuto, R. *et al.* (1990). *Am. J. Hum. Genet.*, **46**, 418–27.
Nass, M.M.K. (1969) *Science*, **165**, 25–35.
Ozawa, T., Toneda, M., Tanaka, M. *et al.* (1988) *Biochem. Biophys. Res. Commun.*, **154**, 1240–7.
Pavlakis, S.G., Rowland, L.P., DeVivo, D.C. *et al.* (1988) in *Contemporary Neurology* (ed. F. Plum), Davis, Philadelphia, pp. 95–123.
Poulton, J., Deadman, M.E. and Gardiner, R.M. (1989) *Lancet*, **i**, 236–40.
Rosing, H.S., Hopkins, L.C., Wallace, D.C. *et al.* (1985) *Ann. Neurol.*, **17**, 228–37.
Rotig, A., Colonna, M., Blanche, S. *et al.* (1988) *Lancet*, **ii**, 567–8.
Rotig, A., Colonna, M., Bonnefont, J.P. *et al.* (1989) *Lancet*, **i**, 902–3.
Saiki, R.K., Scharf, S., Faloona, F. *et al.* (1985) *Science*, **230**, 1350–4.
Schon, E.A., Zeviani, M., Sakoda, S. (1988a) *Aktuelle Aspekte Neuromuskulärer Erkrankungen: Therapie, Früherkennung, Genetik, Mitochondriopathien* (eds W. Mortier, R. Pothmann, and K. Kunze), Georg Thieme Verlag, Stuttgart, pp. 218–34.
Schon, E.A., Bonilla, E., Lombes, A. *et al.* (1988b) *Ann. NY Acad. Sci.*, **550**, 348–59.
Schon, E.A., Rizzuto, R., Moraes, C.T. *et al.* (1989) *Science*, **244**, 346–9.
Shanske, S., Moraes, C.T., Lombes, A. *et al.* (1990) *Neurology*, **40**, 24–8.
Singh, G., Lott, M.T. and Wallace, D.C. (1989) *N. Engl. J. Med.*, **320**, 1300–5.
Wallace, D.C. (1987) *Birth Defects*, Vol. 23, no. 3., Liss, New York, pp. 137–90.
Wallace, D.C., Zheng, X., Lott, M.T. *et al.* (1988a) *Cell*, **55**, 601–10.
Wallace, D.C., Singh, G., Lott, M.T. *et al.* (1988b) *Science*, **242**, 1427–30.
Zeviani, M., Moraes, C.T., DiMauro, S. *et al.* (1988) *Neurology*, **38**, 1339–46.
Zeviani, M., Servidei, S., Gellera, C. *et al.* (1989) *Nature*, **339**, 309–311.

21 Biochemical and genetic investigations in eleven cases of mitochondrial myopathies

C. MARSAC,[1] F. DEGOUL,[1] N.B. ROMERO,[2] J.L. VAYSSIERE,[4] P. LESTIENNE,[3] I. NELSON,[3] D. FRANCOIS[1] AND M. FARDEAU[2]
[1]INSERM U75, 156 rue de Vaugirard, 75015 Paris, France; [2]INSERM U153, 17 rue du Fer à Moulin, 75005 Paris, France; [3]INSERM U298, CHR, 49033 Angers, France; [4]Collège de France, Biologie Cellulaire, 11 place Marcelin Berthelot, 75231 Paris Cedex 05 France

21.1 INTRODUCTION

The mitochondrial myopathies are a clinically and biochemically heterogeneous group of human metabolic disorders with respiratory chain deficiencies associated with variable lactic acidaemia and mitochondrial proliferations in skeletal muscle fibres and in other affected cells (Morgan-Hughes *et al.*, 1977). To define the clinical phenotypes associated with biochemical respiratory chain defects, we performed polarography and enzymatic assays and mitochondrial DNA (mtDNA) analysis in 11 muscle samples from patients with mitochondrial myopathies: seven with Kearns-Sayre syndrome (KSS), two with chronic progressive ophthalmoplegia (CPO), one with stroke-like episodes and familial encephalomyopathy and the last one with seizures, weakness and maternal inheritance.

21.2 METHODS

About 1 g fresh quadriceps muscle was obtained under local anaesthesia and divided into several parts for histochemical studies, biochemical studies and DNA purification.

Histochemical studies were performed by N. Romero and M. Fardeau and presented by M. Fardeau in another communication.

In this chapter we present biochemical and genetic methods to diagnose mitochondrial respiratory chain defects and DNA mitochondrial deletions.

21.2.1 Biochemical methods

Preparation of mitochondria isolated from muscle for enzyme assays and oxidative studies was monitored as previously described by Morgan-Hughes *et al.* (1977).

Three mitochondrial enzyme activities were studied: cytochrome c oxidase (COX), succinate cytochrome c reductase (SCR) and citrate synthase (CSase). All of them were measured on muscle-isolated mitochondria using spectrophotometric methods previously reported (Morgan-Hughes *et al.*, 1977; Cooperstein and Lazarow, 1951).

Oxygen uptake was studied in a 0.5 ml reaction chamber at 25 °C on a Gilson 5/6 oxygraph. Respiratory substrates [5 mM pyruvate, 10 mM glutamate, 80 μM palmitoyl carnitine, each + 2.5 mM malate, 10 mM succinate, 2 mM ascorbate plus 50 μM tetra methyl phenylene diamine (TMPD)] were depleted by addition of 250 nmol ADP. State 3 respiration (ADP stimulated) and state 4 respiration (ADP consumed) were measured. Respiratory control ratios (RCR or state 3/state 4) were calculated.

21.2.2 Genetic methods

Southern blot analysis of mtDNA was performed (Southern, 1975).

The percentage of deleted mtDNA was calculated using densitometric scanning of autoradiography after Southern blot and correction by the molecular weight.

By using the polymerase chain reaction (Saiki *et al.*, 1988) the breakpoint of the deletion was amplified (in three cases), cloned and sequenced (in two cases). One μg of total DNA was amplified in a reaction mixture containing 10 mM Tris-HCl pH 8.5, 50 mM KCl, 2 mM $MgCl_2$, 0.01% (w/v) gelatin, 0.2 mM of each deoxyribo nucleotide triphosphates (dNTP), 10% dimethyl sulphoxide (v/v), 0.1% triton X-100, 30 pmol of each primer and 2.5 U Taq Polymerase (Promega). The amplification was performed with one cycle of 5 min denaturation at 90 °C, 2 min annealing at 50 °C, 4 min elongation at 70 °C, 34 cycles of 2 min at 90 °C, 2 min at 50 °C, and 4 min at 70 °C for elongation.

21.3 RESULTS

21.3.1 Clinical features

The summary of the clinical features is shown in Table 21.1.

21.3.2 Biochemical results

(a) Polarographic studies

The results of polarographic studies of muscle-isolated mitochondria from all patients, except case 1, with KSS are shown in table 21.2.

Table 21.1 Clinical features

7 KSS (cases 1–7)
- 6/7 pigmentary retinal degeneration
- 7/7 ptosis and ophthalmoplegia
- 7/7 heart block
- 3/7 cerebellar dysfunction
- 7/7 sporadic cases

2 CPO (cases 8 and 9)
- 2/2 ptosis and ophthalmoplegia
- 1/2 sporadic case, 1/2 autosomal dominant inheritance

1 case with dementia, deafness, recurrent strokes and pigmentary retinal degeneration (case 10)
1 case with seizures, weakness, maternal inheritance, deafness and ataxia (case 11)

Table 21.2 Polarographic studies

Patient	*Glutamate + malate*	*Pyruvate + malate*	*Palmitoyl carnitine + malate*	*Succinate + rotenone*	*TMPD + ascorbate*
Case 2	41	35	ND	40	325
RCR	1	6.2		2.4	
Case 3	148	105	85	98	
RCR	6.2	8	2.5	1.6	
Case 4	52	25	ND	80	418
RCR	4	3.8		2.4	
Case 5	109	67	32	64	
RCR	4.6	4.1	3.3	2	
Case 6	78	20	38	29	384
RCR	4.4	1	2.3	1.8	
Case 7	7	7	ND	56	
RCR	1	1		1	
Case 8	42	41	52	55	
RCR	2.2	1.9	3.2	1	
Case 9	237	182	ND	167	289
RCR	13	4.5		2.4	
Case 10	25	ND	ND	64	
RCR	1.2			1	
Case 11	32	22	56	120	186
RCR	1	1	2.3	2.6	
Controls	98 ± 17	96 ± 92	19 ± 18	96 ± 24	325 ± 129
RCR	5-10	5-10	5-10	2	1

ND, not determined.
Results are expressed in ng atoms O/min/mg protein at 25°C and refer to respiration in the presence of ADP.See Methods for details.

In all cases except one case with KSS (case 3) and another with CPO (case 9), complex I tested with reduced nicotinamide adenine dinucleotide-linked substrates was impaired. Oxygen uptake in the presence of succinate was lower than in controls in 4/11 cases (cases 2, 6, 7 and 8).

(b) Enzymatic assays

The mitochondrial activities of COX and SCR in isolated mitochondria were expressed per units of CSase and are shown in Table 21.3.

Table 21.3 Enzymatic assays in isolated mitochondria from quadriceps muscles

Patient	*COX/CSase*	*SCR/CSase*
Case 2	0.4	0.12
Case 3	1.55	0.17
Case 4	0.71	0.27
Case 5	0.47	0.135
Case 6	1.44	0.19
Case 7	0.64	0.15
Case 8	0.68	0.20
Case 9	2.3	0.17
Case 10	1	0.19
Case 11	0.8	0.16
Controls		
Mean	1.3	0.3
±SD	0.3	0.1
n	13	13

ND, not determined.

21.3.3 mtDNA studies

The biochemical characterization of the affected complexes are presented with the mtDNA studies and are shown in Table 21.4.

Biochemical and genetic correlations in patients with mtDNA deletions are shown in Tables 21.4 and 21.5. All deletions were large-scale deletions ranging in size from 2.5 to 8 kbp. The ratios of deleted mtDNA to normal mtDNA varied from 20 to 80% of the muscle mtDNA.

21.4 DISCUSSION

We have studied seven cases with KSS, two cases with CPO, one case with dementia, deafness and recurrent strokes and one case with

Table 21.4 mtDNA studies

Patient	*Affected complex*	*Length (kb)*	*Location*	*% deleted mtDNA*
Case 1	IV	4.7	CO1-ND5	75
Case 2	I-III-IV	8	CO1-cyt. b	66
Case 3	Normal	3	ND4-ND6	20
Case 4	I-IV	4.5	ND3-ND6	75
Case 5	I-III-IV	5	CO3-ND5	80
Case 6	I	4.977	CO3-ND5	28
Case 7	I-III-IV	4.7	ATPase8-ND5	66
Case 8	I-IV	4.5	CO1-ND4	47
Case 9	Normal	No deletion		
Case 10	I-IV	No deletion		
Case 11	I-IV	No deletion		

Table 21.5 Correlation between biochemical and genetic studies

Patient	*% COX deficiency*	*% deleted mtDNA*
Case 1	ND	75
Case 2	78	66
Case 3	10	20
Case 4	65	75
Case 5	67	80
Case 6	10	28
Case 7	ND	66
Case 8	63	47

ND, not determined.
COX activity was performed in fresh muscle mitochondria and COX deficiency is expressed as a percentage of the control value (1.3). Correlation coefficient was calculated between these two parameters and found to be 0.85.

seizures, weakness and maternal inheritance. These patients can be considered to have mitochondrial respiratory chain defects and also mitochondrial proliferations in their skeletal muscle fibres. We present here only biochemical and genetic results. All patients had respiratory chain deficiencies, affecting complex I in all cases except cases 3 and 9. Complex III deficiencies found by polarographic methods were confirmed by enzymatic assays of SCR in three cases (cases 2, 5 and 7) and Complex IV deficiency was found in all cases except in cases 3, 6 and 9. Severity of the clinical features seemed neither to depend on the genetic nor the biochemical defects. Deletions were found only in the first eight cases. Two patients (cases 5 and 6) presented with similar clinical features (except for the absence of the pigmentary retinal degeneration in case 6). Case 5 had complex I and IV deficiencies whilst case 6 had only complex I affected.

Amplification products of mtDNA were identical but the percentage of deleted mtDNA was different (Table 21.4). The two deletion junctions have been amplified by polymerase chain reaction (PCR). A lack of correlation was found between the percentage of complex I or III deficiency and the amount of deleted mtDNA. But we found a positive correlation between the percentage of COX deficiency and the percentage of deleted mtDNA (Table 21.5). When we looked for a correlation between the abnormal complex of the mitochondrial respiratory chain (found by polarographic methods and enzymatic assays) and the deleted reading frame of the mtDNA, we did not find a correlation in all cases, but only in cases 2 and 8. Case 6 had no defect in COX but deletion included a part of subunit 3 of the COX. Case 4 had a partial COX deficiency but deletion included only some subunits of complex I.

In three cases (cases 5, 6 and 7), the deleted junctions could be amplified by the PCR and sequenced in two cases (cases 6 and 7). Sequencing the DNA junction from the amplified deleted molecule gave two different results: in case 6, deletion at position 8571–13 237 produced a putative new open reading frame for a hybrid protein ND5–ATPase6. In the other case, deletion (8482–13 460) caused a frame shift. In case 6, the breakpoint was identical to those described in one-third of sequenced deletions (5, 6 and 7).

ACKNOWLEDGEMENTS

We thank Professors J. Clark and B. Kadenbach for their very precious collaboration. This work was financially supported by the Association Française contre les Myopathies (AFM) and by INSERM.

REFERENCES

Cooperstein, S.J. and Lazarow, A. (1951) A microspectrophotometric method for the determination of cytochrome oxidase. *J. Biol. Chem.*, **189**, 665–70.

Holt, I.J., Harding, A.E. and Morgan-Hughes, J.A. (1989) Deletions of muscle mitochondrial DNA in mitochondrial myopathies: sequence analysis and possible mechanisms. *Nucleic Acids Res.*, **17**, 4465–9.

Lestienne, P. and Ponsot, G. (1988) Kearns-Sayre syndrome with muscle mitochondrial DNA deletion. *Lancet*, **i**, 885.

Morgan-Hughes, J.A., Darveniza, P., Kahn, S.N. *et al.* (1977) A mitochondrial myopathy characterised by a deficiency in reducible cytochrome b. *Brain*, **100**, 617–40.

Rotig, A., Colonna, M., Bonnefont, J.P. *et al.* (1989) Mitochondrial DNA deletion in Pearson's marrow/pancreas syndrome. *Lancet*, **i**, 902–3.

Saiki, R., Gelfand, D.H., Stoffel, S. *et al.* (1988) Primer directed enzymatic amplification of DNA with thermostable DNA polymerase. *Science*, **239**, 487–91.

Schon, E.A., Rizzuto, R., Moraes, T. *et al.* (1989) A direct repeat is a hotspot for large-scale deletion of human mitochondrial DNA. *Science*, **244**, 346–9.

Southern, E. (1975) Detection of specific sequences among DNA fragments separated by gel electrophoresis. *J. Mol. Biol.*, **98**, 503–17.

22 Familiar cases of mitochondrial myopathies: mitochondrial DNA deletions and genetic analysis

N. BRESOLIN, I. MORONI, E. CIAFALONI, M. MOGGIO, G. MEOLA, A. GATTI, G. COMI AND G. SCARLATO
Istituto di Clinica Neurologica, Università degli Studi di Milano, Centro Dino Ferari, Via F.Sforza 35, 20122 Milan, Italy

22.1 INTRODUCTION

Mitochondrial myopathies are clinical heterogeneous disorders associated with abnormal mitochondria and impaired oxidative metabolism in muscles. Cerebral and cardiac symptoms may also be present, implying a more generalized biochemical defect. Recently, the concept that at least some of the mitochondrial myopathies may be genetic diseases due to an alteration of the mitochondrial genome (mtDNA) has emerged. This hypothesis is based on the notion that mtDNA encodes essential subunits of respiratory enzymes. In support of this idea, reports of large mtDNA deletions in muscle of a number of patients affected by various myopathies have appeared in the literature during recent years.

The mtDNA is inherited exclusively from the mother; males do not transmit this genomic element. The mtDNA codes for 13 essential polypeptides of oxidative phosphorylation (OXPHOS) as well as the mitochondrial rRNAs and tRNAs. The 13 polypeptides include seven (ND1, 2, 3, 4L, 4, 5 and 6) of the 25 subunits of complex I (rotenone-sensitive reduced nicotinamide adenine dinucleotide (NADH)–ubiquinone oxidoreductase), one (cytochrome b) of the approximately nine subunits of complex III (antimycin-sensitive ubiquinol–cytochrome c oxidoreductase), three (COI, II and III) of the 13 subunits of complex IV (cytochrome c oxidase) (COX)), and two (ATPase 6 and ATPase 8) of the 12 subunits of complex V (oligomycine-sensitive ATP synthase).

The familiar mtDNA mutation should be associated with defects in mitochondrial OXPHOS. The extent of the OXPHOS deficiency usually varies among individuals along the maternal lineage. The mtDNA is present in thousands of copies per cell. Cells containing a mixture of mutant and wild-type mtDNAs (heteroplasmy) randomly partition the two mtDNAs at cytokinesis such that the proportion of the two genomes drift along the cellular lineage (replicative segregation). The extent of the expression of mtDNA mutant phenotypes is proportional to the percentage of mutant mtDNAs inherited by the individual.

Different tissues should be affected sequentially as the respiratory capacity of individuals along the maternal lineage declines. This results from the different extents to which individual tissues rely on mitochondrial energy production (threshold effect).

In our laboratory, many sporadic and familiar cases with mitochondrial myopathies are genetically studied. In particular, total mtDNA is extracted from approximately 30 mg of muscle tissue obtained from biopsies, digested with the restriction enzymes Pvu II and Pst I, run on 0.8% agarose gel and transferred to nitrocellulose. Hybridization of Southern blots is carried out using probes encompassing the entire human mitochondrial genome. The enzyme PvuII linearizes the mitochondrial genome, cutting at the single sites. The enzyme PstI cuts twice, at positions 6910 and 9020. PvuII digestion results in a single DNA fragment migrating at 16.5 kb. PstI digestion results in two DNA fragments of 2.110 and 14.459 kb. The presence of additional faster-migrating bands indicates the existence of a deleted subpopulation of mtDNA molecules. mtDNA from patients carrying a deletion is further studied by digestion with additional restiction enzymes such as EcoRI, HindIII, XbaI, KpnI and BamHI to map the approximate limits of the deletions.

22.2 DESCRIPTION OF CLINICAL CASES

Myoclonic epilepsy associated with ragged red fibres (MERRF) is a rare disease of the central nervous system and skeletal muscle. We report a large pedigree (R. family) in which the proband displays the complete MERRF syndrome, including myoclonic epilepsy and mitochondrial myopathy (Fig. 22.1). Numerous family members all related through maternal lineage displayed manifestation of the disease. The maternal transmission of MERRF suggested that this disease might be the product of an mtDNA mutation. There was no known consanguineity in the R. family.

The first patient, R.E., the older of two brothers, showed at age 37, 'impotentia coeundi' and a computed tomography (CT) scan revealed a pituitary adenoma. At age 40 he presented proximal muscle weakness, rapidly progressing from the lower limbs to the upper ones. He had difficulties in walking, in climbing stairs and rising from sitting. Weakness of face muscles and limb girdle hypotrophy were present. He was admitted several times to our department and recently he could barely walk. He died at age 44 because of cardiorespiratory failure. Serum lactate level was 34.3 mg% (normal values 5.7–22 mg%), serum pyruvate

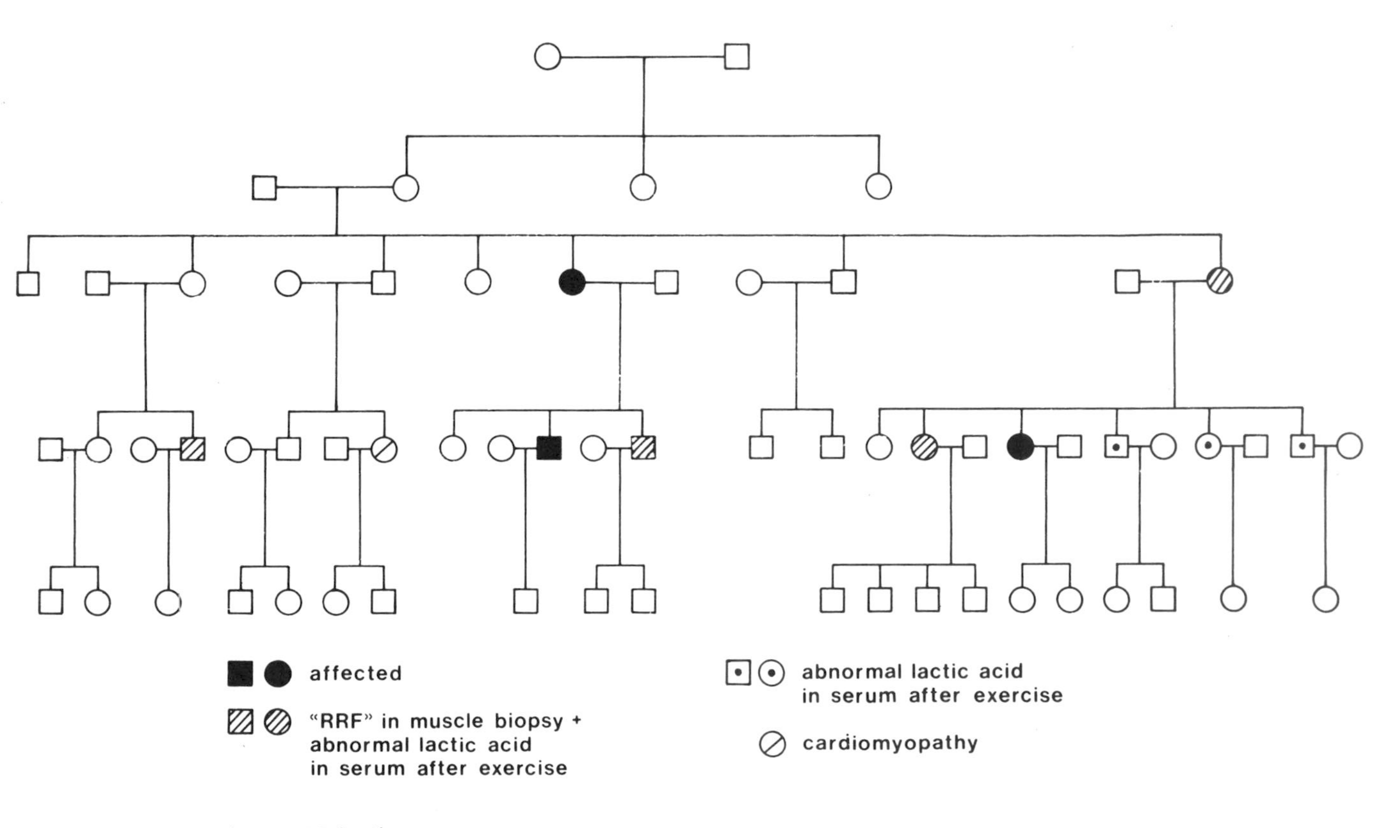

Fig. 22.1 Pedigree of a MERRF family.

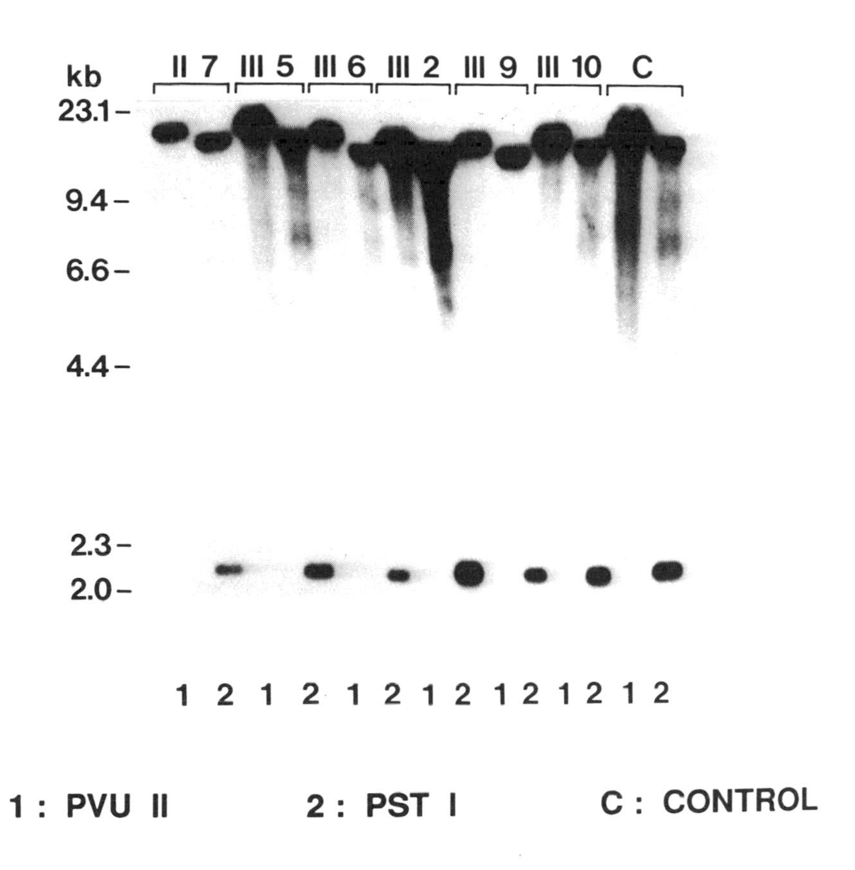

Fig 22.2 mtDNA of the R. family.

level was 1.19 mg% (n.v. 0.30–0.70 mg%), creatine kinase (CK) was 209 IU/ml, and lactate dehydrogenase (LDH) 345 mU/ml (n.v. up to 250 mU/ml). The patient underwent two muscle biopsies, which showed RRF, partial COX deficiency and paracristalline inclusions at electron microscopic examination. The mitochondrial enzyme activity on muscle homogenate presented an increase of citrate synthase (50%), and a decrease of COX (30%). Spectrophotometrically there was a decrease of the peaks corresponding to cytochromes aa3 (54%), b (54.6%) and c (78.8%) of the mitochondrial respiratory chain. ECG was normal before the last admission to hospital, and EEG showed minor abnormalities. He presented bilateral sensorineural hypoacusia and vestibular hyporeflexia. At EMG examination he presented a neurogenic pattern with distal prevalence. Genetic studies of mtDNA (Fig. 22.2) was negative in all patients while physiological aerobic and anaerobic balance studies, as well as NMR spectroscopy measurement, demonstrated a reduced efficiency of high-energy phosphate group synthesis and lower phosphocreatine resynthesis during exercise recovery. Autopsy showed cardiac infarction with left ventricular hypertrophy and cardiomegaly (610 g), atelectasis, brain oedema, and pituitary and thyroid adenoma. Heart and liver presented partial COX and succinate dehydrogenase (SDH) deficiency (Figs 22.3 and 22.4). At electron microscopic examination we confirmed the absence of COX activity in single muscle fibres. Figure 22.5 illustrates a fibre reacting near a fibre with unreacting mitochondria.

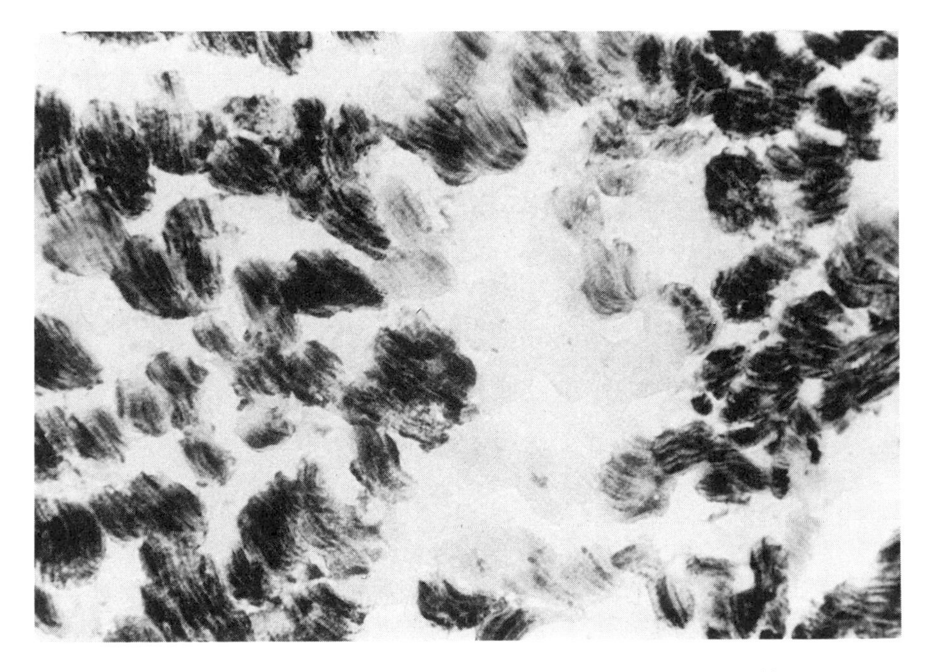

Fig. 22.3 COX: heart, × 250. Lack of COX activity is seen in many fibres.

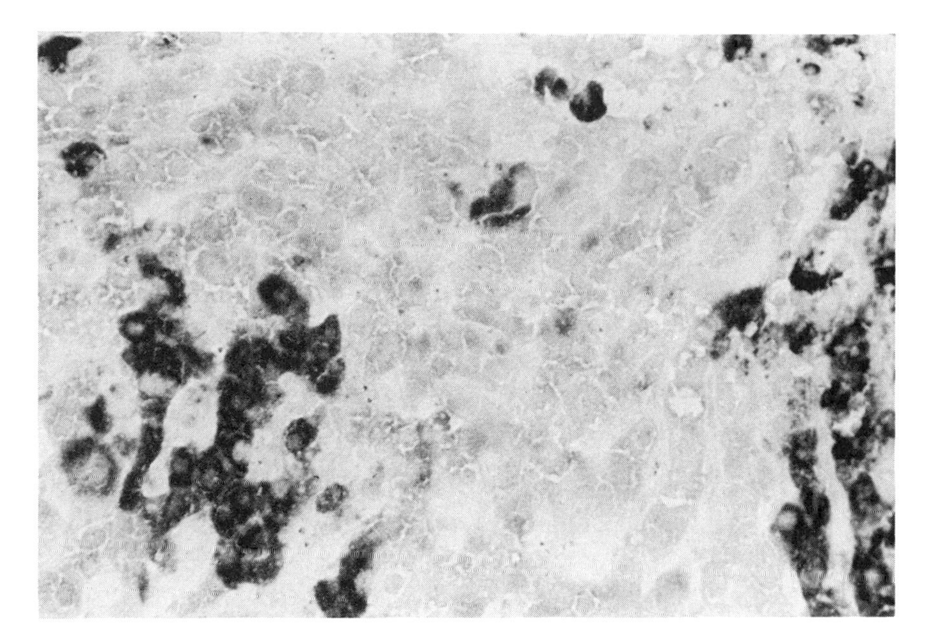

Fig. 22.4 COX: liver, × 250. Lack of COX activity is seen in many hepatocytes.

The second patient, R.M., is the brother who is now 40 years old. At age 37 he complained of muscle cramps and lower limb weakness. At neurological examination he presented a mild psychosis without focal neurological signs. Muscle strength was normal. Serum lactate level was 40.9 mg%, serum pyruvate level was 1.4 mg%, CK was 185 IU/ml and LDH was 341 mU/ml. Mitochondrial enzyme activity on muscle homogenate studied spectrophotometrically showed a reduction of cytochrome aa3 (59.6%). Cerebrospinal fluid (CSF) protein content was 81 mg% (n.v. 40); EEG, ECG and NMR scan were all normal. At EMG we found a myogenic pattern in the brachial biceps and a chronic neurogenic pattern in the vastum medialis. A muscle biopsy was performed showing many RRFs and partial COX, SDH deficiency and paracristalline inclusions at electron microscopic examination. Their mother died at age 66 because of a heart attack. She was never visited by us. It is known that she started having difficulties walking at age 40 and after a few years she could walk only with bilateral help. In 1980 she was admitted to hospital, where the doctors made first the diagnosis of 'ataxic polyneuropathy' and then of Ramsay Hunt syndrome. At neurological examination ophthalmoplegia was found in the upper gaze,

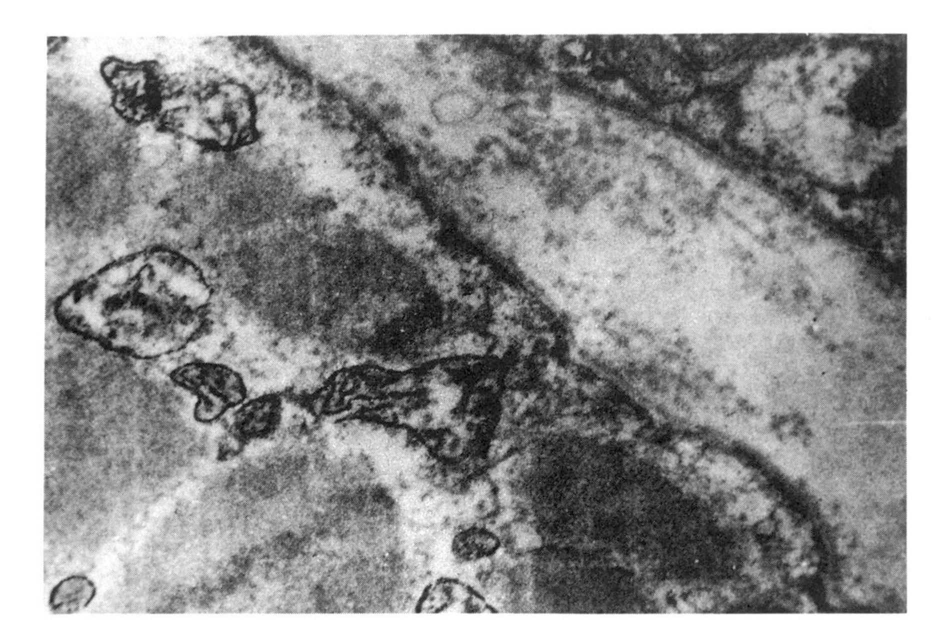

Fig. 22.5 EM: absence of COX activity in single muscle fibres.

ataxia, generalized muscle weakness, hypotrophia, hyporeflexia and myoclonic discharges.

The third patient, S.J., was their 38-year-old first cousin. She presented the complete MERRF syndrome with myoclonic epilepsy since age 28. At age 30 she developed muscle weakness, rapidly progressing from the lower limbs to the upper ones, and axial hypostenia. At neurological examination we found mild bilateral upward ophthalmoplegia, marked weakness of both girdles and bilateral hyporeflexia. EEG showed the typical myoclonic pattern. At ECG a Wolff–Parkinson–White syndrome and at echocardiography a mild impaired intraventricular septum systolic thickening were found. NMR and CT scan were normal as well as CSF proteins. She presented bilateral sensorineural hypoacusia, and bilateral vestibular hyporeflexia. EMG showed axonal polyneuropathy and a myopathic pattern in the proximal limb muscle. Serum lactate level was 27.7 mg%, serum pyruvate was 1.19 mg%, CK 374 IU/ml and LDH 663 MU/ml. Mitochondrial enzyme activity on muscle homogenate showed a decrease of succinate cit C reductase (30%) and of COX (10%). Reduced minus oxidized spectra obtained from isolated muscle mitochondria showed a reduction of cytochrome b (26.6%). The same myopathic pattern previously described was found in the muscle biopsy. Two sisters of this patient and their mother were clinically asymptomatic but muscle biopsy presented many RRFs and partial COX and SDH deficiency. Very recently we performed a muscle biopsy on the daughter (third generation) of another affected male cousin and the biopsy was normal, definitely confirming the maternal pattern of inheritance.

The second familiar case that we report is a 16-year-old patient (P.Y.) presenting palpebral ptosis, extrinsic ophthalmoparesis, deafness and short stature. Ptosis was present in several family members (not all related through the maternal lineage) spanning five generations (Fig. 22.6). In particular, one cousin presented ptosis without muscle weakness and with a normal muscle biopsy. Disturbance of cerebellar coordination was present at the upper limbs. Muscular strength was normal and no exercise intolerance was present despite a slight increase of blood lactate. Muscle biopsy was normal both at histological and ultrastructural examination. No RRFs were observed. Histochemically, several fibres showed lack of COX activity. Immunocytochemical and enzyme-linked immunosorbent assay (ELISA) studies using antibodies against human COX demonstrated the normal amount of the protein.

Determination of mitochondrial respiratory chain enzymes performed on isolated mitochondria showed normal activity. In muscle cultures normal fusion and differentiation was observed (Fig. 22.7). Southern analysis of mtDNA from a muscle biopsy of this patient revealed the existence of two mtDNA molecules (Fig. 22.8). The deleted mtDNA represented approximately 30% of the total mtDNA. The deletion spans 4.5 kb. More precise mapping of the limits of the deletion required digestion of mtDNA with accessory enzymes. We thought of obtaining mtDNA from cultures of the patient's muscle cells, grown from the original biopsy. Myoblast-enriched cultures were obtained after selective plating. Cells were allowed to fuse and differentiate into myotubes. Southern analysis of mtDNA from the two developmental stages revealed in both cell types an mtDNA population seemingly identical to that found in the original muscle biopsy. However, the deleted mtDNA was less abundant (with respect to the normal mtDNA population) in the cell cultures than in the original muscle. The DNA obtained from the culture was digested with the enzymes EcoRI, HindIII and XbaI (not shown), making it possible to map the limits of the deletion between the XbaI site located at nucleotide 8287 (retained) and the EcoRI site located at nucleotide 12 641 (deleted). The deleted region includes the gene coding for subunit III of COX. In this patient the lack of an overt muscle involvement may be due to the fact that the amount of mutated mtDNA does not reach the minimal threshold necessary for clinical manifestation of the genetic defect. Considering the normal biopsies obtained from

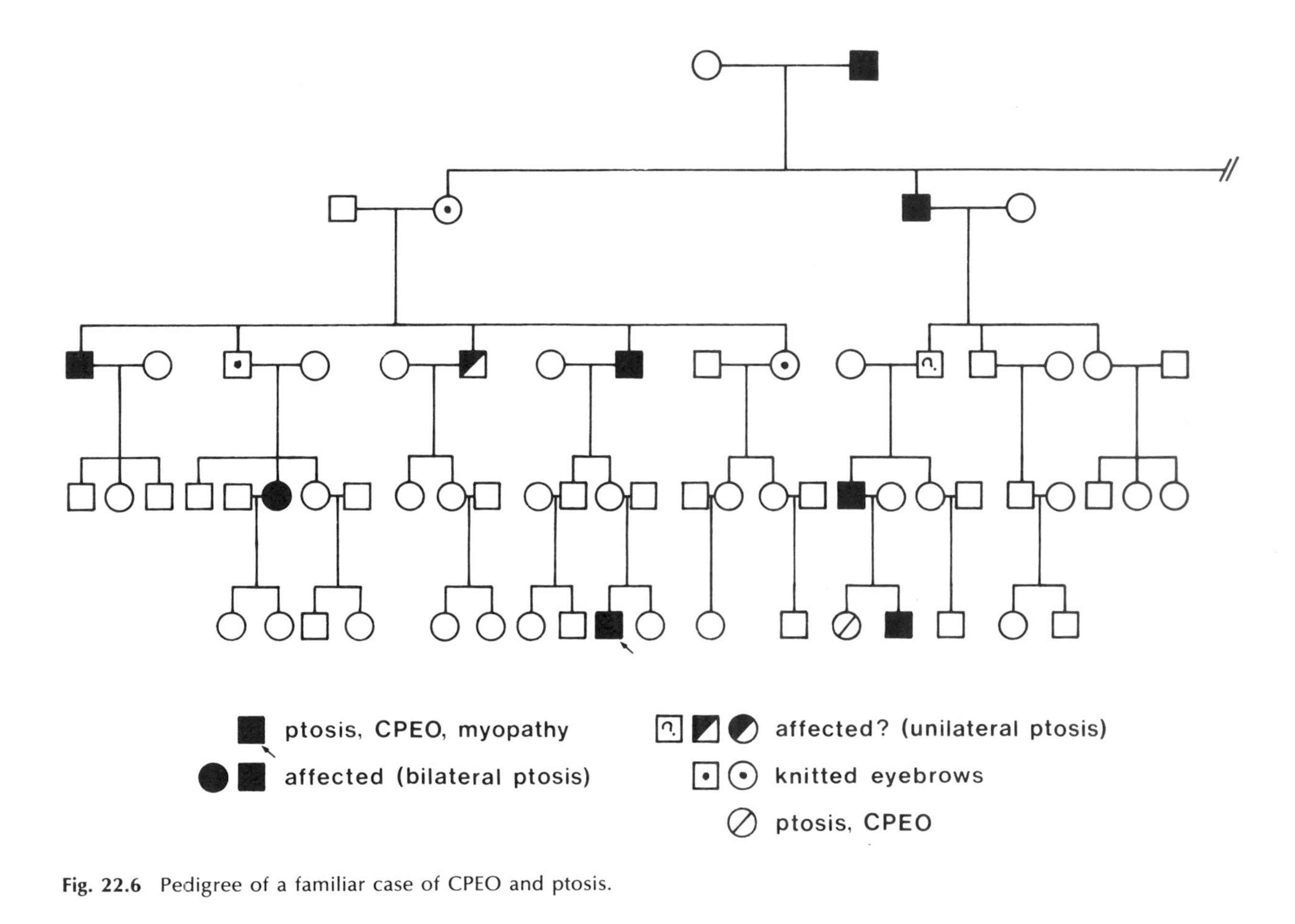

Fig. 22.6 Pedigree of a familiar case of CPEO and ptosis.

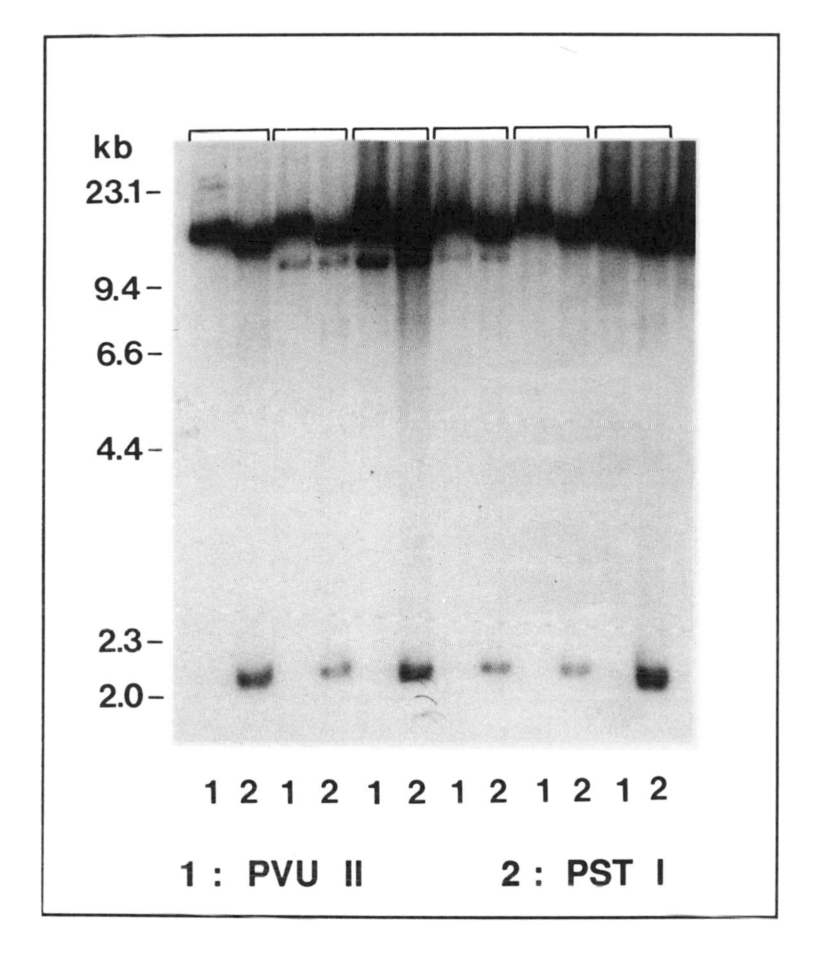

Fig. 22.7 mtDNA in muscle culture from the patient P.Y. CMT, control myotubes; MB, P.Y. myoblasts; MT, P.Y. myotubes; CL, P.Y. clones obtained from myotubes; CF, control fibroblasts; F, P.Y. fibroblasts.

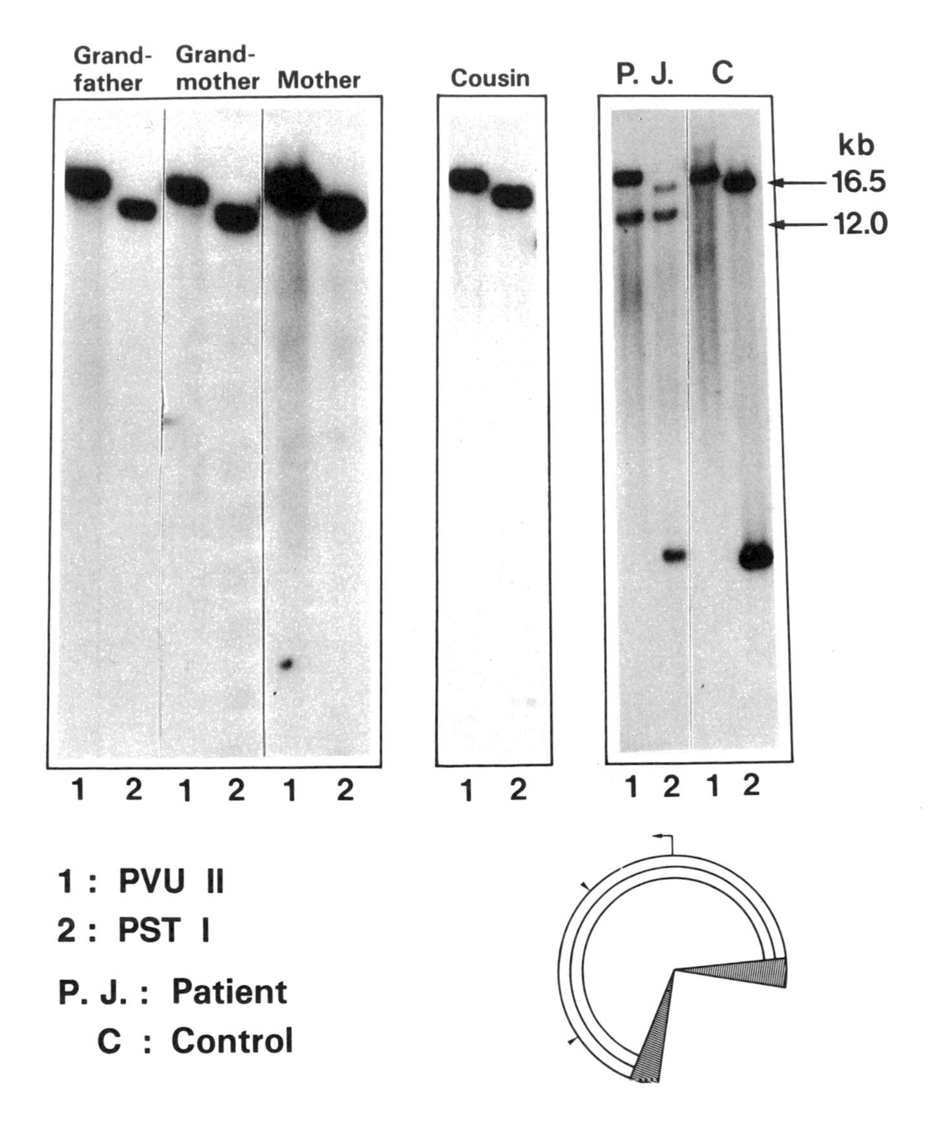
Grand-father
Grand-mother
Mother
Cousin
P. J.
C
kb
16.5
12.0
1 2 1 2 1 2
1 2
1 2 1 2
1 : PVU II
2 : PST I
P. J. : Patient
C : Control

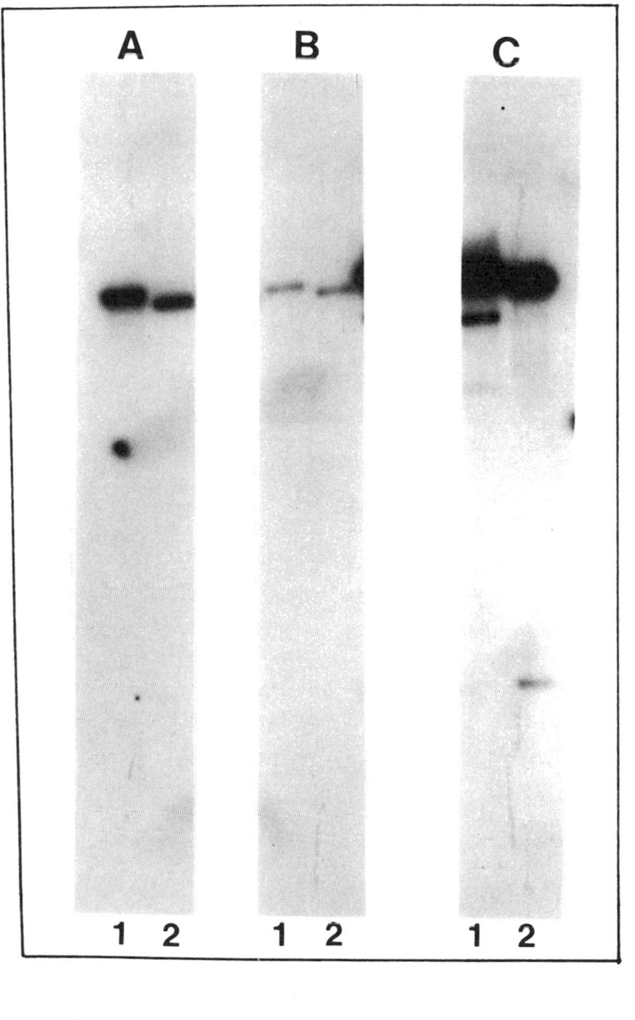

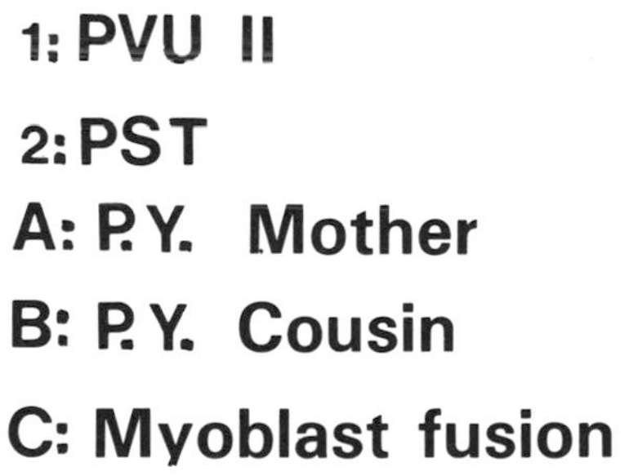

Fig. 22.9 mtDNA deletion in myoblast hybrid progeny.

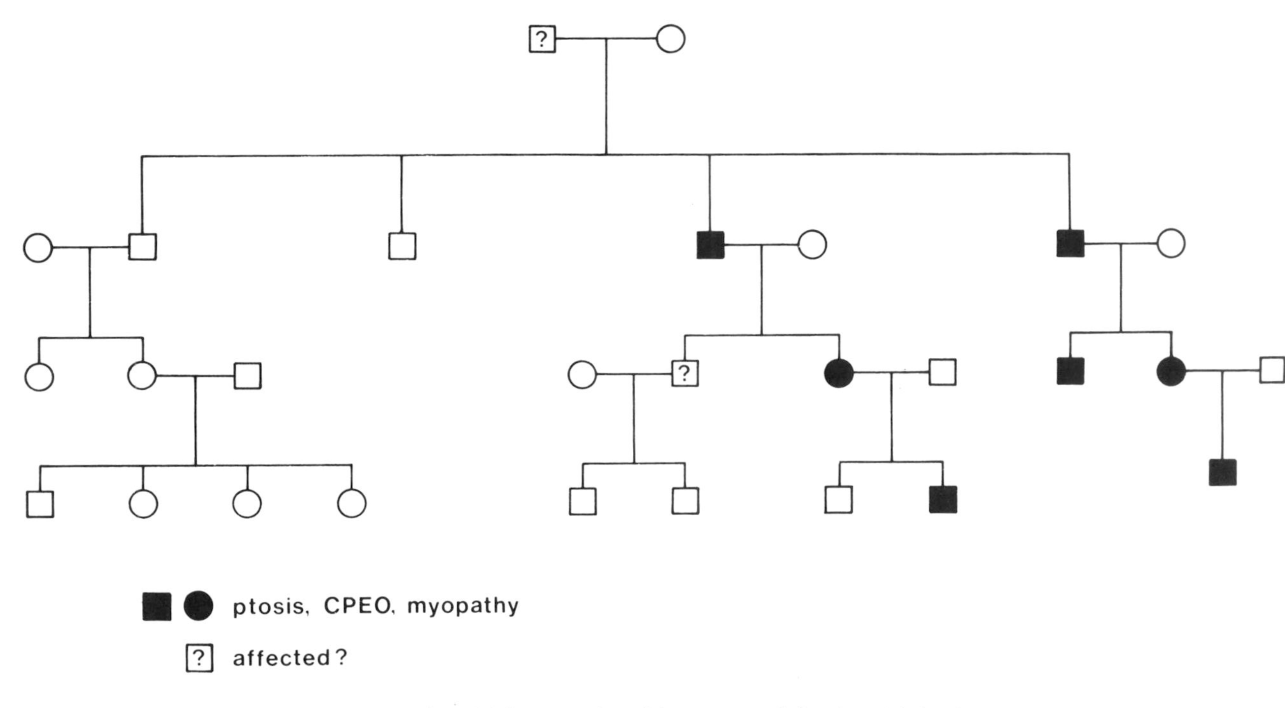

Fig. 22.10 Pedigree of a familiar mitochondrial myopathy with autosomal dominant inheritance

the patient's mother and from his cousin, we performed the fusion of myoblasts of these subjects. Preliminary results showed the presence of a new mtDNA deletion (Fig. 22.9).

A third family with mitochondrial myopathy and autosomal dominant inheritance is described (Fig. 22.10). One patient, a 61-year-old-man, presented severe ptosis and chronic progressive external ophthalmoplegia (CPEO), bilateral sensorineural hypoacusia and vestibular areflexia, short stature, prolapse of mitral valve, ventricular hypertrophia, tremor and severe muscle involvement. EMG showed the presence of an axonal polyneuropathy. Serum lactate level was high with respect to normal values, at rest and 5 min after exercise (respectively 24.2 mg%, and 35.6 mg%). Mitochondrial enzyme activity on muscle homogenate showed an increase of citrate synthase and a decrease of COX. Muscle biopsy showed the presence of RRFs, partial COX deficiency and variable degrees of neurogenic change. Analysis of mtDNA in muscle showed polymorphism with PvuII restriction enzyme. His father died because of the same disease, with identical clinical symptoms.

The sister of this patient, a 69-year-old woman, presented the same clinical characteristics. Serum lactate level was at the upper limit of normal values (22.4 mg%) at rest (she could not cycle because of severe muscle involvement). EMG showed an axonal polyneuropathy in all muscles checked. Analysis of muscle mtDNA showed polymorphism with PvuII restriction enzyme.

Her son presented severe ptosis and CPEO and mild generalized muscle weakness. Serum lactate level showed an increase at rest and after exercise (respectively 27.3 mg% and 38.4 mg%). A muscle biopsy showed the same characteristics as the first patient. The mitochondrial enzyme activity of muscle homogenate was normal. mtDNA analysis in muscle showed polymorphism with PvuII restriction enzyme. There was evidence of peripheral neuropathy at neurophysiological examination.

The second part of the pedigree of this family includes the grandfather's brother of the last patient, from whom was born an affected daughter who died because of the same disease. She was the mother of two sons: one was asymptomatic, while the other presented mild ptosis and CPEO, and a clear muscle involvement. Serum lactate level was normal at rest, but showed an increase 5 min after exercise (38.8 mg%). Mitochondrial enzyme activity of muscle homogenate was normal. Histopathological studies showed RRFs, partial COX deficiency and neurogenic changes. EMG showed mild neurogenic disturbance.

Genetic analysis on this family with an autosomal dominant pattern of inheritance is now under examination looking for mtDNA deletion and for possible interaction between nuclear encoded DNA and mtDNA.

FURTHER READING

Attardi, G. (1981) Organization and expression of the mammalian mitochondrial genome: a lesson in economy. *Trends Biochem. Sci.*, **6**, 86–9, 100–3.

Attardi, B. and Attardi, G. (1971) Expression of the mitochondrial genome in HeLa cells: properties of the discrete RNA component from the mitochondrial fraction. *J. Molec. Biol.*, **55**, 231.

Bresolin, N., Moggio, M., Bet, L. *et al.* (1987) Progressive cytochrome c oxidase deficiency in a case of Kearns–Sayre syndrome: morphological, immunological and biochemical studies in muscle biopsies and autopsy tissues. *Ann. Neurol.*, **21**, 564–72.

Di Mauro, S., Bonilla, E., Zeviani, M. *et al.* (1985) Mitochondrial myopathies. *Ann. Neurol.*, **17**, 521–38.

Hauswirth, W. and Lalpis, P.J. (1985) in *Achievement and Perspectives of Mitochondrial Research, Vol. II: Biogenesis* (eds E. Quagliarello, E.C. Slater, F. Palmieri *et al.*), Elsevier, New York, pp. 49–59.

Morgan-Hughes, J.A., Hayes, D.J., Clark, J.B. *et al.* (1982) Mitochondrial encephalomyopathies: biochemical studies in two cases revealing defect in the respiratory chain. *Brain*, **105**, 553–82.

Petty, R.H.K., Harding, A.E. and Morgan-Hughes, J.A. (1986) The clinical features of mitochondrial myopathies. *Brain*, **109**, 915–38.

Rosing, H.S., Hopkins, L.C., Wallace, D.C. *et al.* (1985) Maternally inherited mitochondrial myopathy and myoclonic epilepsy. *Ann. Neurol.*, **17**, 228–37.

Wallace, D.C. (1985) Mitochondrial genes and disease. *Hosp. Prac.*, 15 October, 77–92.

Wallace, D.C. (1987) Maternal genes: mitochondrial disease. in *Birth Defects*, vol. 23, no. 3, Liss, New York, pp. 137–90.

Wallace, D.C., Zheng X., Lott, M.T. *et al.* (1988) Familiar mitochondrial encephalomyopathy (MERRF): genetic, pathophysiological, and biochemical characterization of a mitochondrial DNA disease. *Cell*, **55**, 601–10.

Zeviani, M., Bonilla, E., De Vivo, D.C. and Di Mauro, S. (1989a) in *Neurologic Clinics* Vol. 7 (ed. W.G. Johnson), Saunders, Philadelphia, pp. 123–56.

Zeviani, M., Servidei, S., Gellera, C. *et al.* (1989b) An autosomal dominant disorder with multiple deletions of mitochondrial DNA starting at the D-Loop region. *Nature*, **339**, 309–11.

23 The correlation between pathology, biochemistry and molecular genetics in mitochondrial encephalomyopathies

A. OLDFORS,[1] E. HOLME,[2] B. KRISTIANSSON,[3] N.-G. LARSSON[2] AND M. TULINIUS[3]
[1]Department of Pathology, Gothenburg University, Sahlgren's Hospital, S-413 45 Gothenburg, Sweden; [2]Department of Clinical Chemistry, Gothenburg University, Sahlgren's Hospital, S-413 45 Gothenburg, Sweden; [3]Department of Paediatrics, Gothenburg University, East Hospital, S-416 85 Gothenburg, Sweden

23.1 INTRODUCTION

Most of the patients included in the present study are paediatric cases from various parts of Sweden but also from other Scandinavian countries. To discover patients with possible mitochondrial diseases, elevated levels of lactate in blood, in combination with myopathy or encephalopathy of unexplained type, have been the factors of most importance. In addition, patients without lactic acidosis but with clinical symptoms suggestive of mitochondrial disease have been thoroughly investigated with respect to mitochondrial dysfunction. In this way some 70 patients have been studied by clinical examination in addition to biochemical and morphological studies on muscle mitochondria. Among these 70 patients 37 patients had morphological and/or biochemical evidence of mitochondrial myopathy. We have not included primary carnitine deficiency or defects of fatty acid ß-oxidation in this material.

The patients may be divided into clinical entities such as Kearns-Sayre syndrome (KSS), mitochondrial myopathy, encephalopathy, lactic acidosis and stroke-like episodes (MELAS), myoclonus epilepsy and ragged red fibres (MERRF), Leber's hereditary optic neuropathy (LHON), Alpers' disease and Leigh's syndrome. The patient with LHON had in addition to blindness clinical and radiological evidence of involvement of basal ganglia. In the cases of Alpers' and Leigh's syndrome the diagnosis was based on clinical and radiological examinations and not on neuropathological investigation. In addition there are some cases with encephalomyopathies falling outside these

Table 23.1 Patients included in the study

Pat. no.	*Disease*	*Age (years)*	*Resp. chain defect*	*Light microscopy*	*Electron microscopy of mitochondria*	*Mol. genet. Southern blot*
1	KSS	12	Complex I	RRF	Small osmiophilic incl. + various	Deletion + heteroplasmy
2	KSS	17	Normal	RRF	Paracryst. incl. + various	Deletion + heteroplasmy
3	KSS	19	Normal	RRF	Paracryst. incl. + various	Deletion + heteroplasmy
4	MERRF	17	Complex I + IV	RRF	Paracryst incl. + vacuolated	MERRF point mutation
5	MERRF	10	Complex I + IV	RRF	Amorph. material	MERRF point mutation
6	MELAS	17	Complex I - III	RRF	Paracryst incl. + elongated	MELAS point mutation
7	MELAS	12	Complex I	RRF	Circular cristae Osmiophilic incl.	MELAS point mutation
8	MELAS	18	Not studied	RRF	Paracryst. incl. + various	Not studied
9	MELAS	40	Complex I	RRF	Sparse paracryst. incl. Filamentous material	MELAS point mutation
10	MELAS	20	Complex I	RRF	Elongated + vacuolated Osmiophilic incl.	MELAS point mutation
11	MELAS	13	Complex I	Minor myopathy	Normal	Normal
12	LHON	40	Complex I	Minor myopathy	Enlarged mitochondria Osmiophilic inclusions	LHON point mutation
13	Alpers	1	Complex I	Atrophy	Minor changes	Normal
14	Alpers	1	Complex III	Atrophy	Normal	Normal
15	Alpers	2	Complex I	Normal	Normal	Normal
16	Leigh	11	Complex I	Atrophy	Normal	Normal
17	Leigh	1/2	Complex I	Normal	Normal	Normal

18	Encephalomyopathy – cardiomyopathy	15	Complex I	Atrophy Necrotic fibres	Circ. cristae + osmiophilic incl.	Normal
19	Encephalomyopathy	11	Complex I	one RRF	Normal	Normal
20	Encephalomyopathy	40	Not studied	RRF	Paracryst. incl. + various	Not studied
21	Encephalomyopathy	4	Complex III	Normal	Normal	Normal
22	Encephalomyopathy	1/2	Complex III	Normal	Normal	Normal?
23	Encephalomyopathy	3	Complex I	Atrophy	Normal	Normal
24	Encephalomyopathy	3	Complex I	Normal	Normal	Normal
25	Encephalomyopathy	2	Complex IV	Normal	Normal	Normal
26	Encephalomyopathy	3	Complex I	Atrophy Lipid accum.	Not studied	Normal
27	Myopathy	1/2	Complex I + IV	RRF	Circular cristae + vacuolated	Normal
28	Myopathy	2	Complex III + IV	RRF	Circular cristae + vacuolated	Normal
29	Myopathy	14	Normal	RRF	Circular cristae + vacuolated	Normal
30	Myopathy	13	Complex I	RRF	Enlarged + osmiophilic incl.	Normal
31	Myopathy	41	Normal	RRF	Circular cristae	Normal
32	Myopathy + cardiomyopathy	1	Complex I + IV	RRF	Circular cristae – vacuolated	Normal
33	Myopathy + cardiomyopathy	1/2	Complex I + IV	RRF	Circular cristae Osmiophilic incl.	Normal
34	Myopathy = cardiomyopathy	3 days	Complex I + III + IV	RRF	Enlarged irregular Irregular cristae	Normal
35	CPEO	50	not studied	RRF	Paracryst incl. + various	Deletion + heteroplasmy
36	CPEO	33	Not studied	RRF	Paracryst. incl. + various	Normal
37	CPEO	70	Not studied	RRF	Not studied	Normal

entities. Two of these had attacks of unconsciousness. Twelve of the patients had no signs of encephalopathy but only myopathy, and some of these had cardiomyopathy. Three patients had chronic progressive external ophthalmoplegia (CPEO) without signs of encephalopathy.

23.2 MATERIALS AND METHODS

The function of the respiratory chain of freshly isolated muscle mitochondria was investigated by oximetric analysis (Scholte *et al.*, 1981). In addition most cases were studied by spectrophotometric assay of reduced nicotinamide adenine dinucleotide (NADH) ferricyanide reductase (Dooijewaard and Slater, 1976), succinate–cytochrome c reductase (Fischer *et al.*, 1985) and cytochrome c oxidase (COX) (Cooperstein and Lazarov, 1951).

The morphological studies included enzyme histochemical staining of succinate dehydrogenase to identify muscle fibres with abnormal accumulations of mitochondria. Enzyme histochemical demonstration of cytochrome c oxidase was performed at the light- and electron microscopic level (Seligman *et al.*, 1968). In addition electron microscopy was performed to identify any possible specific or typical mitochondrial change in each case. Predominant change could be, for example, circular arrangement of cristae. The occurrence or absence of paracrystalline inclusions was especially studied. However, often a mixture of various changes were observed. The MERRF point mutation at nucleotide 8344 (Shoffner *et al.*, 1990) was detected with PCR and direct sequency of mtDNA. The MELAS point mutation creates a novel restriction enzyme site for Apa I around nucleotide 3243 (Goto *et al.*, 1990). Cleavage of mtDNA with Apa I was used to detect the MELAS mutation.

Molecular genetic studies were performed by screening most of the patients with Southern blotting of skeletal muscle DNA after cleavage with the restriction enzyme Pvu 2 and hybridization with a mixture of radiolabelled mitochondrial DNA (mtDNA) fragments.

23.3 RESULTS AND DISCUSSION

A summary of the biochemical, morphological and molecular genetical studies are given in Table 23.1. In the cases with encephalomyopathy, complex I deficiency was the most prevalent biochemical defect. By oximetric analysis one of the KSS patients had a slightly reduced activity of the respiratory chain with pyruvate and glutamate as substrates, while analysis of the other KSS cases was normal. However, all the KSS patients had intermittently increased levels of lactate in blood. In the cases with myopathy without encephalopathy most of the cases had complex IV deficiency, but usually combined with other defects.

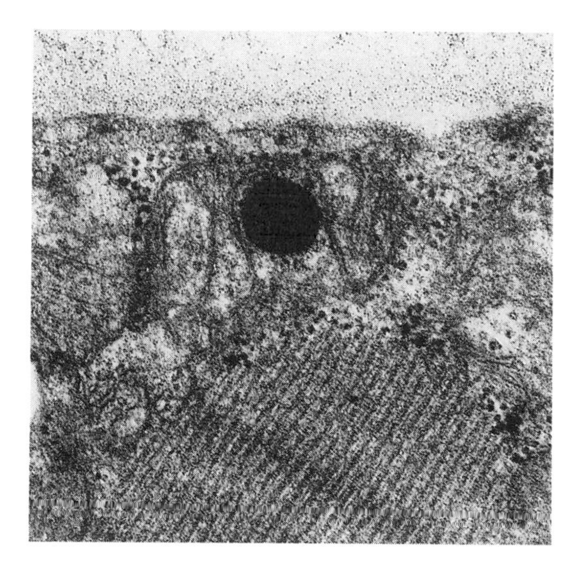

Fig. 23.1 Large osmiophilic inclusion in mitochondria of a patient with LHON.

Morphological examination revealed the presence of RRF in most cases of KSS, MERRF, and MELAS, but not the LHON case and not in the cases of Alpers or Leigh syndrome. It was obvious that there was a great variation in the predominant type of morphological change of the mitochondria even within the same disease entities. In many of the cases with complex I deficiency osmiophilic inclusions were frequently seen, but they were not specific for this type of respiratory chain dysfunction. In the case of LHON with an isolated defect at the level of complex I, osmiophilic inclusions were seen in many mitochondria (Fig. 23.1). Paracrystalline inclusions were often found in cases with RRF but not in all the cases of encephalomyopathy. In the paediatric cases of myopathy without encephalopathy paracrystalline inclusions were never observed, but enlarged mitochondria with circular cristae were a common finding in this group. Abnormal mitochondria were observed also in smooth muscle cells in a skin biopsy from one of the KSS patients (Fig. 23.2).

Southern blotting of mtDNA revealed obvious changes in five cases. The three patients with KSS all showed deletions and heteroplasmy.

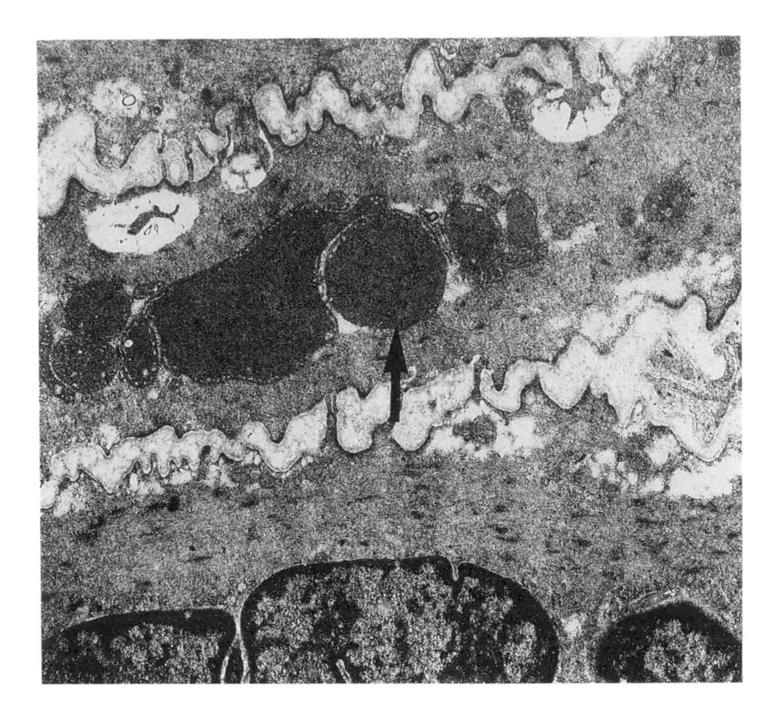

Fig. 23.2 Abnormal mitochondria (arrow) in smooth muscle cells of the skin in a patient with KSS. The same type of abnormality was also seen in skeletal muscle mitochondria in this patient.

The size of the deletions varied from 4.9 to 7.6 kb. The deleted parts did not include the origins of replication or the D-loop region needed for transcription. None of the patients showed a deletion identical to the common deletion found in one-third of the cases with CPEO or KSS (Schon *et al.* 1989; Holt *et al.*, 1989; Moraes *et al.*, 1989). Deleted mtDNA was also found but to a lesser amount also in cultured fibroblasts but not in lymphocytes. One of the patients with CPEO had mtDNA deletion and heteroplasmy. The patient with LHON showed loss of an SfaNI restriction enzyme site in mtDNA. This point mutation has been reported in a half of the cases with LHON (Wallace *et al.*, 1988). Both patients with MERRF and four of the patients with MELAS had heteroplasmy with point mutations of mtDNA which have been associated with these syndromes (Shoffner *et al.*, 1990; Goto *et al.*, 1990). The MELAS patient without RRF (patient 11) did not have the MELAS point mutation.

Several paediatric cases with myopathy in our material showed deficiency of COX by biochemical investigation of isolated mitochondria

Table 23.2 Patients with COX deficiency in isolated muscle mitochondria

Pat. no.	*Disease*	*Age (years)*	*Resp. chain defect*	*Light microscopy*	*Electron microscopy of mitochondria*	*COX deficiency histochemistry*
5	MERRF	10	Complex I + IV	RRF	Amorph. material	Deficient in 50% progressive
25	Encephalomyopathy	2	Complex IV	Normal	Normal	Deficient in 90%
27	Myopathy	1/2	Complex I + IV	RRF	Circular cristae, vacuolated	Deficient in 95%
28	Myopathy	2	Complex III + IV	RRF	Circular cristae, vacuolated	Deficient in 50%
32	Myopathy + cardiomyopathy	1	Complex I + IV	RRF	Circular cristae, vacuolated	Deficient in 100%
33	Myopathy + cardiomyopathy	1/2	Complex I + IV	RRF	Circular cristae Osmiophilic incl.	Deficient in RRF
34	Myopathy + cardiomyopathy	3 days	Complex I + III + IV	RRF	Enlarged, irregular cristae	Deficient in 100%

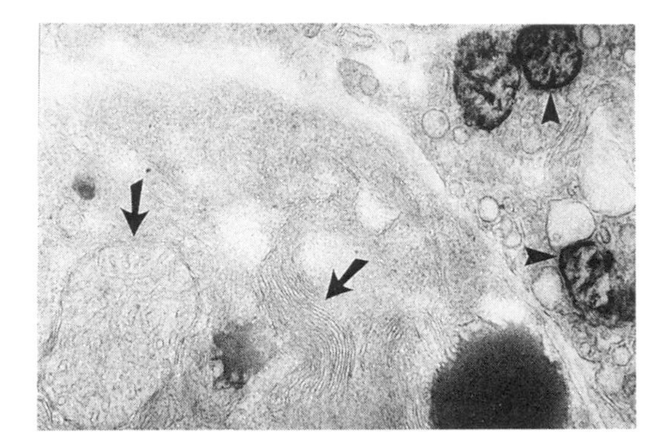

Fig. 23.3 Ultrastructural enzyme-histochemical staining of COX in a patient with severe COX deficiency. Activity is present in mitochondria of endothelial cells (arrowhead) but not in muscle fibres (arrow). From Oldfors *et al.* (1989).

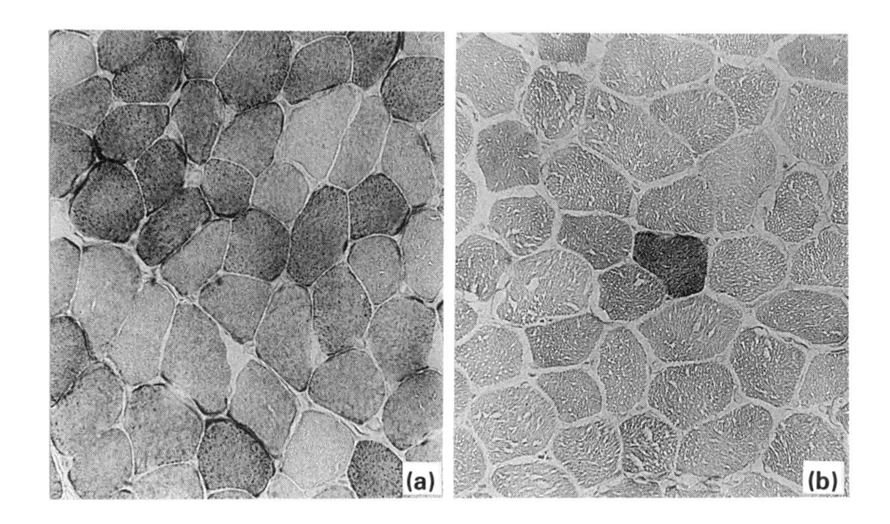

Fig. 23.4 Enzyme histochemical staining of COX in a patient with progressive COX deficiency. (a) 10 years of age; (b) 13 years of age.

Table 23.3 Enzyme histochemical activity of COX in RFF[a]

Pat. no.	Disease	Age (years)	Resp. chain defect	Light microscopy	Electron microscopy of mitochondria	COX deficiency histochemistry
1	KSS	12	Complex I	RRF	Small osmiophilic incl. + various	Deficient in RRF
2	KSS	17	Normal	RRF	Paracryst. incl. + various	Deficient in RRF
3	KSS	19	Normal	RRF	Paracryst incl + various	Deficient in RRF
4	MERRF	17	Complex I + IV	RRF	Paracryst incl. + vacuolated	High activity in RRF
6	MELAS	17	Complex 1 + III	RRF	Paracryst incl. + elongated	Deficient in 50% of RRF
7	MELAS	12	Complex I	RRF	Circular cristae Osmiophilic incl.	High activity in RRF
8	MELAS	18	Not studied	RRF	Paracryst incl + various	Deficient in RRF
9	MELAS	40	Complex I	RRF	Sparse paracryst. incl. Filamentous material	High activity in RRF
10	MELAS	20	Complex I	RRF	Elongated, vacuolated Osmiophilic incl.	High activity in RRF
20	Encephalomyopathy	40	Not studied	RRF	Paracryst. incl. + various	Deficient in RRF
29	Myopathy	14	Normal	RRF	Circular cristae + vacuolated	Deficient in RRF
30	Myopathy	13	Complex I	RRF	Enlarged + osmiophilic incl.	High activity in RRF
31	Myopathy	41	Normal	RRF	Circular cristae	Deficient in RRF
35	CPEO	50	Not studied	RRF	Paracryst. incl. + various	Deficient in RRF
36	CPEO	33	Not studied	RRF	Paracryst.incl. + various	Deficient in RRF
37	CPEO	70	Not studied	RRF	Not studied	Deficient in 50% of RRF

[a]Patients with COX deficiency in isolated mitochondria (Table 23.2) are excluded from this table.

(Table 23.2). There was a good correlation between the oximetric and spectrophotometric results and the histochemical staining of COX. All patients with low biochemical activity of COX showed many fibres with absence of staining of the enzyme. However, spectrophotometric analysis showed that most patients with COX deficiency also had other respiratory chain defects. Ultrastructural enzyme histochemical staining in two patients with severe COX deficiency revealed absent activity in muscle fibre mitochondria and normal activity in the mitochondria of adjacent endothelial cells (Fig. 23.3). This finding indicates a cell-type specific deficiency, possibly due to different isoenzymes or to mtDNA mutations and heteroplasmy affecting various cell types differently. In one case of MERRF progressive muscle weakness was noted and muscle biopsies were studied at two different occasions three years apart. Oximetric analysis of the mitochondria in the first biopsy indicated a respiratory chain defect at the level of complex I. The activity at the level of complex IV was only slightly reduced. The enzyme histochemical staining of COX showed scattered fibres with absent activity (Fig. 23.4 (a)). Three years later most muscle fibres showed deficient staining (Fig. 23.4 (b)) and oximetric analysis disclosed low activity also at the level of comlex IV. The spectrophotometric findings indicated a combined defect of complex I and COX. This patient is thus an example of progressive COX deficiency. Progressive deficiency of COX has previously been reported in a patient with KSS (Bresolin *et al.*, 1987). Table 23.3 illustrates the occurrence of COX deficiency in muscle fibres in patients with RRF and normal biochemical activity of COX. For example, all the cases of KSS showed absent staining of COX in the RRF. This was also seen in other diseases. Many cases, however, showed high activity of COX in RRF as, for example, in some of the MELAS cases. In two cases with RRF we have observed a mixture of RRF with and without COX activity. By ultrastructural enzyme histochemistry we have compared the morphology and enzyme activity in these cases. In RRF with high activity the abnormal and enlarged mitochondria had many cristae with apparently normal morphological appearance, and normal COX activity. In the same patient the RRF with low activity of COX showed absence of activity in the interior of the mitochondria, and morphologically abnormal cristae (Fig. 23.5). Thus there is a correlation between the morphological abnormality of cristae and COX deficiency in certain cases, with COX deficiency in only some of the RRF.

By immunocytochemical staining with antisera raised against COX (bovine heart), we have not seen complete absence of immunoreactive material in muscle fibres in any of the cases with generalized COX deficiency. The presence of immunoreactive material indicates a defect or inactivated enzyme in most cases of COX deficiency.

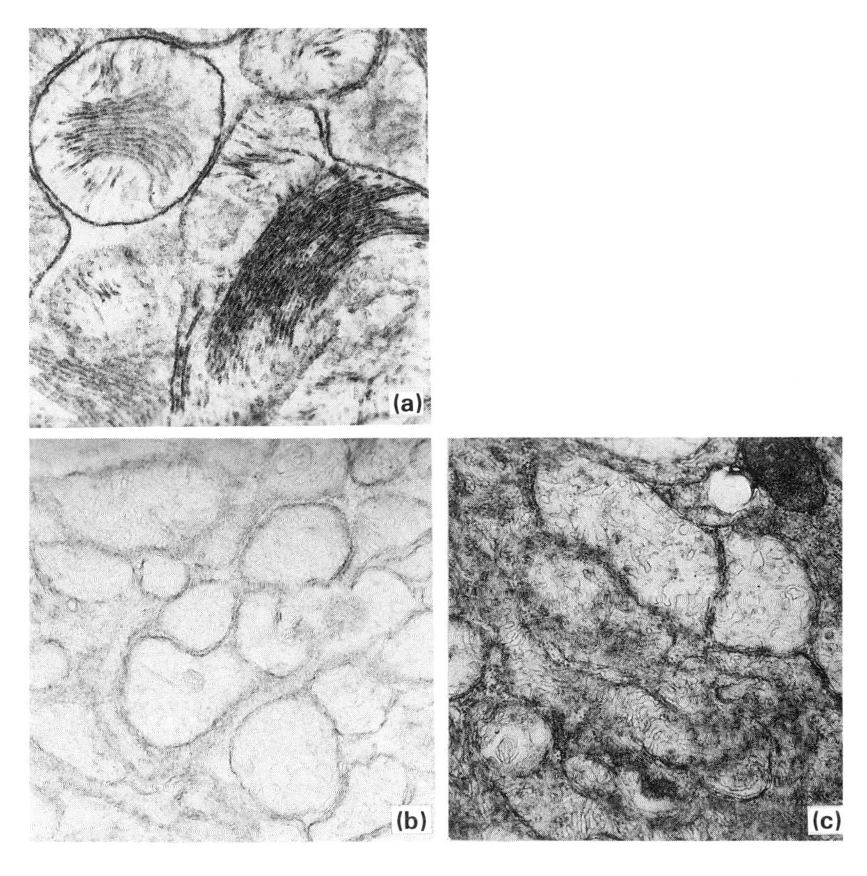

Fig. 23.5 Ultrastructural enzyme histochemical staining of COX in a patient with a mixture of RRF with and without COX activity. (a) RRF with high activity of COX. Although the mitochondria are markedly enlarged they show morphologically normal cristae with normal activity of COX. (b) RRF with low activity of COX. The abnormal mitochondria show low COX activity. After contrasting the section (c) the abnormal cristae of these mitochondria are demonstrated.

ACKNOWLEDGEMENT

This study was supported by grants from the Swedish Medical Research Council, project no. 07122.

REFERENCES

Bresolin, N., Moggio, M., Bet, L. *et al.* (1987) *Ann. Neurol.*, **21**, 564–72.
Cooperstein, S.J. and Lazarov, A. (1951) *J. Biol. Chem.*, **18**, 665–70.

Dooijewaard, G. and Slater, E.C. (1976) *Biochim. Biophys. Acta,* **440**, 1–15.
Fischer, J.C., Ruitenbeek, W., Berden, J.A. *et al.* (1985) *Clin. Chim. Acta,* **153**, 23–36.
Goto, Y.-l, Nonaka, I. and Horai, S. (1990) *Nature*, **348**, 651–3.
Holt, I.J., Harding, A.E. and Morgan-Hughes, J.A. (1989) *Nucleic Acids Res.* **17**, 4465–9.
Moraes, C.T., DiMauro, S., Zeviani, M. *et al.* (1989) *N. Engl. J. Med.* **320**, 1293–9.
Oldfors, A., Sommerland, H., Holme, E. *et al.* (1989) *Acta Neuropathol.*, **77**, 267–75.
Scholte, H.,R., Busch, H.F.M. and Luyt-Houwen, I.E.M. (1981) in *Mitochondria and Muscular Diseases* (eds H.F.M. Busch, F.G.I. Jennekens and H.R. Scholte) Beetserzwaag, Netherlands, Mefar, pp. 133–45.
Schon, E.A., Rizzuto, R., Moraes, C.T. *et al.* (1989) *Science*, **244**, 346–9.
Seligman, A.M., Karnovsky, M.J., Wasserkrug, H.L. and Hanker, J.S. (1968) *J. Cell. Biol.*, **38**, 1–140
Shoffner, J.M., Loh, M.T., Lezza, A.M. *et al.* (1990) *Cell*, **61**, 931–7.
Wallace, D.C., Singh, G., Loh, M.T. *et al.* (1988) *Science*, **242**, 1427–30.

24 Various clinical presentation of mitochondriopathies: clinical and therapeutic considerations

C. ANGELINI, A. MARTINUZZI, M. FANIN, M. ROSA, R. CARROZZO AND L. VERGANI
Regional Neuromuscular Centre, Neurological Clinic, University of Padova, Italy

24.1 INTRODUCTION

Mitochondriopathies constitute a heterogeneous clinical group of disorders (DiMauro *et al.*, 1985; Morgan Hughes, 1986). Their diagnosis is provided by clinical criteria together with mitochondrial DNA (mtDNA) studies and enzymic studies in muscle.

24.2 MORPHOLOGICAL CRITERIA

24.2.1 Histochemical approach

Mitochondrial alterations are more prominent in muscle, but they may be expressed in other organs, i.e. liver, brain (Spiro *et al.*, 1970), heart (Sengers *et al.*, 1976), kidney and fibroblasts (Treem *et al.*, 1988) if the disorder is multisystemic.

The most used histochemical criteria are:

1. 'ragged red fibres' (RRFs) and their variants;
2. 'subsarcolemmal accumulation of mitochondria in rims;
3. fibres with lipid droplets (they are usually seen in defects of fatty acid utilization, but also in respiratory chain defects);
4. fibres negative for oxidative enzymes, i.e. cytochrome c oxidase (COX), succinate dehydrogenase.

24.2.2 Ultrastructural findings

The ultrastructural mitochondrial changes are not aspecific, but useful when relevant for a diagnosis. They consist of the occurrence of megaconia, pleoconia (Shy *et al.*, 1966) or light cores (Bender and Engel, 1976) and paracrystalline structures within mitochondria. It is assumed that these are responses to the metabolic state such as:

1. the increase in number and size of mitochondria may be an attempt to compensate for a defective enzyme;
2. swelling and morphological alteration may result from fatty acid accumulation;
3. intramitochondrial 'paracrystalline structures' may derive from a conformational change of the inner membrane itself (Morgan Hughes, 1982).

24.3 BIOCHEMICAL CRITERIA

24.3.1 Laboratory and enzymic dat a

The definition of a mitochondriopathy (Angelini, 1987) relies on clinical chemistry and tissue biochemical measurements. The diagnostic work-up includes: plasma lactate and pyruvate levels after a standardized effort, acid–base assessment, creatine kinase isoenzymes and carnitine fractions. In the urine ketone bodies, organic acids and carnitine fraction should be measured. A useful diagnostic technique used in our laboratory is the study of mitochondrial enzymes in whole homogenate, mitochondria pellets or thrombocytes. The isolation of mitochondria can be performed using a fresh biopsy (1–2 g) for polarographic purposes or a smaller amount of frozen tissue (Morgan Hughes, 1986). Since often the mitochondrial defect is not restricted to muscles we use lymphocytes or cultured fibroblasts to evaluate DNA deletions, oxidative enzymes and the utilization of labelled substrates for $^{14}CO_2$ production. Antibody against enzymes such as cytochrome c oxidase subunits, electron transfer and flavoprotein (ETF) are useful techniques for arriving at a molecular diagnosis.

24.3.2 Molecular DNA polymorphism

Mitochondrial DNA deletions occur both in progressive external ophthalmoplegia and Kearns–Sayre syndrome (Moraes *et al.*, 1989). Mitochondrial respiratory complexes have both nuclear and mtDNA-encoded proteins (DiMauro *et al.*, 1985), and many pleiotropic alterations and partial defects are expressed in mitochondrial encephalomyopathies. In fact the presence of two different sets of genes implies in the assembly of the complexes a synchronous interaction. Besides deletion and mutation, enzyme activity defects may be due to deformation of quaternary structure of a complex or to abnormal interaction with a coenzyme, or to errors in the transport and assembly of proteins inside the mitochondria. Some defects may be expressed during mitochondrial biogenesis and then disappear (such as the reversible complex IV defect).

24.4 CLINICAL POLYMORPHISM

The evolving concept of mitochondrial disorders now includes several clinical subtypes.

24.4.1 Myopathic type of mitochondriopathy

(a) Congenital or childhood-onset myopathy (DiMauro, 1979)

These are characterized by shoulder and pelvic girdle weakness, present either at birth or during infancy. Episodes of severe paresis and salt craving lasting several hours have been described in the original case of 'pleoconial myopathy' of Shy *et al.* (1966).

(b) A benign congenital myopathy

With mitochondria lipid glycogen accumulation this presents as feeding and respiratory problems, severe weakness, hepatomegaly and macroglossia (Jerusalem *et al.*, 1973).

(c) Proximal neck and limb muscle weakness

This is characteristic of muscle carnitine deficiency with lipid storage myopathies and may be variably responsive to medium-chain triglycerides, steroids and carnitine administration. Primary carnitine deficiency in some cases seems due to a generalized transport defect (Treem *et al.*, 1988).

(d) Limb girdle myopathy with RRF

This form may become overt during adolescence or adult age. Proximal muscles are weak but characteristically the whole leg may appear hypertrophic. Progression is generally slow or fluctuating.

(e) 'Myoglobinuria'

This can be either spontaneous or triggered by cold, infection and starvation. It is seen in carnitine palmitoyltransferase deficiency and long-chain acyl-CoA dehydrogenase defects (Angelini, 1987).

(f) Exercise intolerance

This occurs as limb weakness and excessive lactate response. In these patients even moderate exertion causes stiffness, tenderness, pain and weakness of exercising muscles. The symptoms may begin in childhood, and though exercise intolerance is persistent the course fluctuates. Complex I defect has been described with this syndrome (Morgan Hughes, 1986).

(g) Non-thyroidal hypermetabolism (DiMauro, 1979; Luft *et al.*, 1962)

Two patients with fever, heat intolerance and profuse sweating, tachypnoea, and dyspnoea at rest are described with the characteristic of 'Luft's disease' due to uncoupling of mitochondria.

24.4.2 Cardiomyopathic type mitochondriopathy

First described in two patients – an 11-year-old girl and a 15-year-old boy, the clinical disorder was characterized by mild non-progressive myopathy, short stature and cardiomegaly (Bender and Engel, 1976; Sengers *et al.*, 1976); characteristic morphological features were many light-cored osmiophilic inclusions in the first patient (attributed to calcium) and lipid accumulation in the second. A similar disorder was designated as 'Sengers' syndrome'. Overt hypertrophic cardiomegaly has often been associated with a variety of mitochondrial defects

Table 24.1 Metabolic cardiomyopathy with mitochondriopathy

	Type	*Defect*
(a)	Hypertrophic	
	Obstructive	-
	Non-obstructive	Long-chain acyl CoA dehydrogenase
		ETF
		Complex II
	(Histiocytoid cardiac lipidosis)	Complex III
		Complex IV
(b)	Dilatative	Kearns-Sayre syndrome (mtDNA deletion)
(c)	Constrictive	
	Restrictive	Systemic carnitine deficiency
	(Endocardial fibroelastosis)	(Carnitine transport)
	Obliterative	-
(d)	With conduction defect	Kearns-Sayre syndrome (mtDNA deletion)

(Table 24.1). In two brothers of 25 and 19 years of age respectively we diagnosed an asymmetrical hypertrophic cardiomyopathy by echocardiography; both showed a broad base gait, proximal limb muscle weakness and abnormal EMG. On muscle biopsy the most prominent findings were fibres with lipid droplets, 'core-like areas' and increased mitochondrial rims, except for succinate dehydrogenase stain which was absent in extrafusal fibres but not in intrafusal fibres.

Enzymic studies revealed reduced activity of complex II enzymes found also in platelets.

24.4.3 Generalized fatal infantile type mitochondriopathy

The fatal infantile mitochondriopathy with lactic acidosis is associated with multiple defects of the respiratory chain, although often it is due to a defect of both complexes I and IV (Leigh's disease) or to isolated complex IV defect. Given the complexity of COX (complex IV) it is not surprising that COX deficiency should be characterized by remarkable clinical heterogeneity (DiMauro *et al.*, 1985). The most common picture is a myopathy associated with renal dysfunction. In a child of 11 months with congenital cataracts, glaucoma, marked hypotonia and respiratory insufficiency we found amino aciduria and low plasma carnitine and muscle COX deficiency. In a 10-year-old boy, with a childhood encephalomyopathy (Angelini *et al.*, 1986) characterized

Table 24.2 Encephalomyopathic types

Clinical syndrome	*mtDNA (tRNDefet*
MELAS (Mitochondrial encephalomyopathy, lactic acidosis, stroke-like episodes)	mtDNA ($tRNA^{leu}$)
MERRF (Myoclonus epilepsy, ragged red fibres)	mtDNA ($tRNA^{lys}$)
Ataxia (Severe psychomotor retardation, lactic acidosis)	PDH complex
Dementia (Optic atrophy, dystonic rigidity, ophthalmoplegia)	Complex 1
Ataxia (Mental impairment, myoclonic jerks, short stature)	Complex II
Myoclonus, ataxia (Proximal weakness, areflexia, Babinski, dementia, ophthalmoplegia)	Complex III
Ataxia (Muscle wasting, mental impairment)	Complex IV

by ataxia, muscle wasting and mental impairment who died of respiratory insufficiency and bulbar dysfunction we found a generalized COX defect in all tissues and in cultured fibroblasts.

24.4.4 Encephalomyopathic types

Several syndromes have been associated with neurological impairment (Table 24.2).

In a family we have observed the association of MELAS, ocular myopathy and mitochondrial dementia suggesting a spectrum of clinical entities rather than various discrete syndromes.

24.5 SYNDROMES WITH PROGRESSIVE EXTERNAL OPHTHALMOPLEGIA

A major breakthrough has been the demonstration of deletions of mitochondrial DNA in most cases of Kearns–Sayre syndrome (Moraes *et al.*, 1989). In six patients with progressive external ophthalmoplegia (PEO) and variable heart involvement we have found DNA deletions (Table 24.3). Major difficulties have arisen in the nosographic classification of these two syndromes because of the overlapping of clinical signs of multisystem involvement such as growth retardation, atypical pigmentary retinopathy, sensorineural deafness, cerebellar ataxia, heart block and peripheral neuropathy (Berenberg *et al.*, 1977). No clearcut enzymic change has been found but COX-negative fibres appear in the advanced

Table 24.3 Clinical signs and mtDNA deletion in 20 KSS and PEO cases

	KSS	*PEO*
No. of cases	*7*	*13*
Sex	4M, 3F	5M, 8F
Onset (years)	3–19	19–49
Family history	2	2
Ophthalmoplegia	7	8
Ptosis	6	13
Retinitis pigmentosa	3	3
Heart block	2	3
Mitral prolapse	2	2
Short stature	4	0
Hearing loss	1	2
Muscle weakness	5	7
Endocrinopathy	1	2
Ragged red fibres	5	13
Lactic acidosis	5/5	9/11
mtDNA deletion	3/4	6/6

stage of the disease. Heart involvement has been documented in one of our patients by cardiac biopsy (Charles *et al.*, 1981). We favour the 'lumping hypothesis' (Berenberg *et al.*, 1977) in PEO and Kearns–Sayre syndrome and attribute the spectrum of clinical symptomatology to the quantity of heteroplasmic mitochondria with mtDNA deletions.

24.6 REASONS FOR CLINICAL SUBTYPES AND THERAPEUTIC CONSIDERATIONS

The clinical subtypes may originate:

1. since isoenzymes of mitochondrial enzymes are expressed in various tissues;
2. from mitochondrial biogenesis;
3. from heteroplasmic populations of mitochondria.

A striking modulation of mitochondrial function is the adaptation to tissue metabolic requests. There are tissues with low mitochondrial activity and tissues with high energetic requirements. Disease states may be revealed with increasing age or energy crisis. If the defect is functionally decompensated it leads to a progressive fatal syndrome where bioenergetic failure is the result of metabolic disturbance. It is conceivable that therapeutic intervention has to be early and multifactorial. In lipid storage myopathies with organic acidurias, MCT diet or carnitine replacement therapy removes toxic acyl groups in mitochondria, enhances lipid metabolism and decreases the ratio of CoA : acyl CoA in tissues (Chapoy *et al.*, 1980). The use of mobile carriers such as coenzyme Q is of benefit in a subgroup of patients with the Kearns–Sayre or ophthalmoplegia plus phenotype, even when mtDNA is deleted. Vitamin K_3 and ascorbate have been used as a 'shuttle' in a case of complex III defect (DiMauro *et al.*, 1985). Genetic therapy is still a challenge: in the future the difficulty in allowing the penetration of mtDNA through the double mitochondrial membrane may be overcome by biological ballistic technology (Johnston *et al.*, 1988).

ACKNOWLEDGEMENT

We thank Dr S. Shanske, Department of Neurology, Columbia University, New York, for the study of mtDNA deletion in our patients.

REFERENCES

Angelini, C. (1987) Metabolic myopathies. *Curr. Neurol.* **7**, 31–49.
Angelini, C., Bresolin, N., Pegolo, G. *et al.* (1986) Childhood encephalo-

myopathy with cytochrome C oxidase deficiency, ataxia, muscle wasting and mental impairment. *Neurology*, **36**, 1048–62.

Bender, A.N. and Engel, W.K. (1976) Light-cored dense particles in mitochondria of a patient with skeletal muscle and myocardial disease. *J. Neuropathol.*, **35**, 46–52.

Berenberg, R.A., Pellock, J.M., DiMauro, S. *et al.* (1977) Lumping or splitting? 'Ophthalmoplegia-plus' or Kearns–Sayre syndrome? *Ann. Neurol.*, **1**, 37–54.

Chapoy, P.R., Angelini, C., Brown, W.J. *et al.* (1980) Systemic carnitine deficiency: a treatable inherited lipid storage disease presenting as Reye's syndrome. *N. Engl. J. Med.*, **303**, 1389–94.

Charles, R., Holt, S., Kay, J.M. *et al.* (1981) Myocardial ultrastructure and the development of atrioventricular block in Kearns–Sayre syndrome. *Circulation*, **63**, 214–19.

Cederbaum, S., Blass, J.P., Minkoff, N. *et al.* (1976) Sensitivity to carbohydrate in a patient with familial intermittent lactic acidosis and pyruvate dehydrogenase deficiency. *Pediatr. Res.*, **10**, 713–20.

DiMauro, S. (1979) in *Handbook of Clinical Neurology: Diseases of Muscle.*, Vol. 41 (eds P.J. Vinken and G.W. Bruyn), North-Holland, Amsterdam, pp. 175–234.

DiMauro, S., Bonilla, E., Zeviani, M. *et al.* (1985) Mitochondrial myopathies. *Ann. Neurol.*, **17**, 521–38.

Jerusalem, F., Angelini, C. and Engel, A.G. (1973) Mitochondria lipid glycogen (MLG) disease of muscle. *Arch. Neurol.*, **29**, 162–9.

Johnston, S.A., Anziano, P.Q., Shark, K., Sanford, J.C. and Butow, R.A. (1988) Mitochondrial transformation in yeast by bombardment with microprojectiles. *Nature*, **240**, 1538–41.

Luft, R., Ikkos, D., Palmieri, G. *et al.* (1962) A case of severe hypermetabolism of nonthyroid origin with a defect in the maintenance of mitochondrial respiratory control: a correlated clinical biochemical and morphological study. *J. Clin. Invest.*, **41**, 1776–804.

Moraes, C.T., Di Mauro, S., Zeviani, M. *et al.* (1989) Mitochondrial DNA deletions in progressive external ophthalmoplegia and Kearns–Sayre syndrome. *N. Engl. J. Med.*, **320**, 1293–9.

Morgan Hughes, J.A. (1982) in *Advances in Clinical Neurology*, Vol. 3 (eds W.B. Matthews and G.H. Glaser), Churchill Livingstone, Edinburgh, pp. 1–46.

Morgan Hughes, J.A. (1986) in *The Mitochondrial Myopathies in Myology: Basic and Clinical* (eds A.G. Engel and B.Q. Banker), McGraw Hill, Toronto, pp. 1709–13.

Sengers, R.C.A., Stadhouders, A.M., Jaspar, H.H.J. *et al.* (1976) Cardiomyopathy and short stature associated with mitochondrial and/or lipid storage myopathy of skeletal muscle. *Neuropediatrics*, **7**, 196–208.

Shy, G.M., Gonatas, N.K. and Perez M. (1966) Two childhood myopathies with abnormal mitochondria: 1 megaconial myopathy, II pleoconial myopathy. *Brain*, **89**, 133–58.

Spiro, A.J., Moore, C.L., Prineas, J.W. *et al.* (1970) A cytochrome related inherited disorder of central nervous system and muscle. *Arch. Neurol.* **23**, 108–12.

Treem, W.R., Stanley, C.A., Finegold, D.N. *et al.* (1988) Primary carnitine deficiency due to a failure of carnitine transport in kidney, muscle and fibroblasts. *N. Engl. J. Med.*, **319**, 1331–6.

PART THREE

Pharmacological Agents and Toxic Insults

25 Voltage-gated sodium channels as target for pharmacological agents

M.R. CARRATÙ AND D. MITOLO-CHIEPPA
Institute of Pharmacology, Medical Faculty,
University of Bari, I-70124 Bari, Italy

25.1 INTRODUCTION

Chemical treatments and neurotoxins together provide important information about the mechanisms of gating. Well-investigated examples include the peptide toxins in scorpion venoms and in nematocysts of coelenterate tentacles, the alkaloidal toxins secreted by tropical frogs, and other lipid-soluble, insectidal substances of many plant leaves. They act on sodium channels by increasing the probability that a channel opens or remains open. They cause pain and death by promoting repetitive firing or constant depolarization of nerve and muscle and by inducing arrhythmias. In addition, some simple chemical treatments and even just changing the ionic content of the bathing medium can be used to change gating.

In the Hodgkin–Huxley (HH) model (Hodgkin and Huxley, 1952) the time course of macroscopic current in sodium channels is described in terms of a rapid 'activation' process which controls the rate and voltage dependence of sodium permeability increase following depolarization, and a slower, independent 'inactivation' process which controls the rate and voltage dependence of the subsequent return of the sodium permeability to the resting level during a maintained depolarization. Though the underlying mechanisms are actually neither as simple nor as separable as the HH model suggests, however, the operational definitions of activation and inactivation introduced by Hodgkin and Huxley proved useful in describing the effects of toxins and pharmacological agents that alter sodium channel function.

The voltage-clamp method measures three properties of sodium channels: voltage-dependent activation, voltage-dependent inactivation and selective ion transport. In the sections below, one class of gating modification is mainly considered: prevention or slowing down of inactivation.

25.2 PREVENTION OR SLOWING OF SODIUM INACTIVATION BY ENZYMES, REACTIVE REAGENTS AND NEUROTOXINS

The inactivation process is easily impaired by chemical agents and enzymes acting from the axoplasmic side of the membrane. Internal treatment with pronase or some other proteolytic enzyme is a classical example (Armstrong *et al.*, 1973; Armstrong and Bezanilla, 1973; Rojas and Rudy, 1976; Carbone, 1982). In the presence of pronase, which is a mixture of proteolytic enzymes, sodium currents activate normally but fail to inactivate, so that during a long depolarizing pulse the sodium channels remain open.

A similar loss or slowing of sodium inactivation, often accompanied by a decrease of sodium current, is seen when excitable cells are treated with a variety of reactive reagents (Eaton *et al.*, 1978; Horn *et al.*, 1980; Oxford *et al.*, 1978; Stämpfli, 1974); in addition, high internal pH (>9.5) reversibly stops inactivation in squid giant axon (Brodwick and Eaton, 1978). These reagents seem to act on a gating process whose kinetics correspond qualitatively to those of inactivation in the classical HH model, while leaving activation intact. None of the reactive reagents or enzymes is completely specific for one chemical group, and their chemical actions on sodium channels have not been determined; it has been proposed that the target is a protein which is vulnerable at some intracellularly accessible arginyl and tyrosyl residues (Eaton *et al.*, 1978).

Most of the anemone and scorpion toxins bind reversibly and competitively to a common external receptor on the sodium channel; they slow the rate of inactivation so much that the duration of single action potentials is increased to several seconds or even minutes (Bergman *et al.*, 1976; Romey *et al.*, 1976; Catterall, 1980). The decay of sodium current in a voltage-clamp step acquires a multi-exponential character, extending for seconds; the rate and voltage dependence of activation is little changed and peak current amplitudes typically decrease a little or stay the same. Though the major scorpion and anemone toxins intensify excitation by a common slowing of sodium inactivation, other scorpion toxins intensify excitation acting through different mechanisms (Carbone *et al.*, 1982; Cahalan, 1975; Wang and Strichartz, 1983).

25.3 ENDOGENOUS POLYPEPTIDES CHANGE MANY PROPERTIES OF SODIUM CHANNELS

We have seen that anemone and scorpion toxins slow inactivation of the sodium channel; endogenous polypeptides are able to modify the

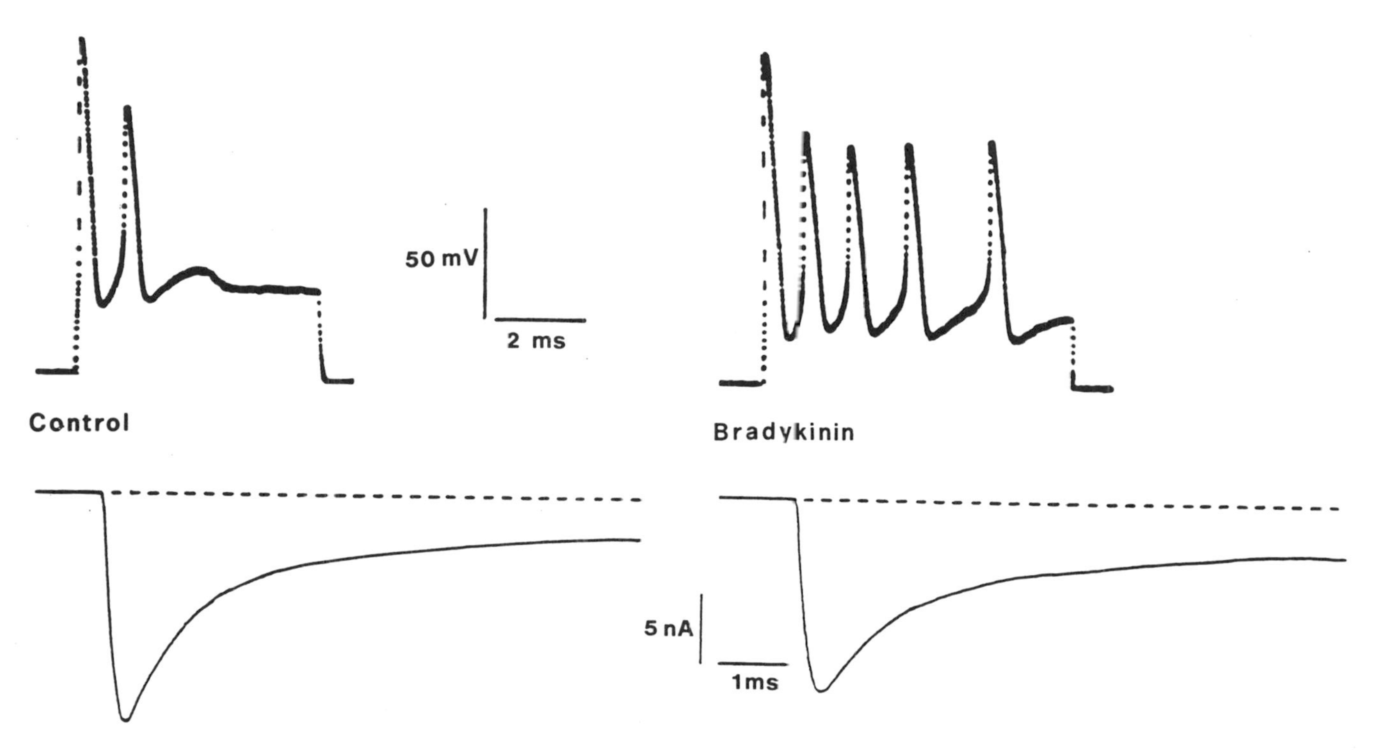

Fig. 25.1 Upper panel: action potentials recorded from a node of Ranvier from a frog myelinated nerve fibre before and after treatment with 10 μM bradykinin (BK). Action potentials are elicited by applying a 25 ms depolarizing current across the node. Lower panel: traces of sodium current from the same fibre under voltage-clamp before and after treatment with 10 μM BK. Potassium currents are eliminated by cutting the ends of the fibre in 80 mM CsCl + 10 mM NaCl and by using 10 mM tetraethylammonium outside. Sodium current was recorded during depolarizations to 0 mV preceded by 50 ms hyperpolarizations to −120 mV. From Carratù and Mitolo-Chieppa, (1989).

voltage-gated sodium channels as well. Among these, bradykinin changes many properties of sodium channels and causes repetitive discharges in nerves (Fig. 25.1) (Carratù and Mitolo-Chieppa, 1989).

Inactivation of sodium channels is slowed, so sodium currents are maintained rather than just transient and peak macroscopic sodium conductance is smaller than in controls. In the presence of bradykinin the decay of sodium current in a voltage-clamp step acquires a multi-exponential character and three phases of inactivation can be distinguished: fast, slow and late. In this respect, the effect of bradykinin is reminiscent of a derivative of phencyclidine – ketamine – used to induce general anaesthesia (Benoit *et al.*, 1987). Finally, like the pentapeptide leu- and met-enkephalins, internal bradykinin produces frequency-dependent block of sodium current (Carratù *et al.*, 1982). The voltage-dependent properties of the 'late' current are quite different from those of the transient current; the voltage at which the current activates and its reversal potential are more negative, and different voltage properties were supposed to reflect the existence of a discrete population of sodium channels (Dubois and Bergman, 1975); however, there are convincing arguments substantiating the interconversion hypothesis.

25.4 SODIUM INACTIVATION AND TRANSITION METAL IONS

Sodium inactivation acquires a multi-exponential character in the presence of transition metal ions as well. Conventional voltage-clamp analysis (Nonner, 1969) in frog's myelinated nerve fibre has shown that $FeCl_2$ (10–100 μM) slows inactivation and produces a non-inactivatable sodium current accounting for the repetitive firing (Figure 25.2).

Neither macroscopic peak sodium conductance nor activation gating processes are modified. The effect of $FeCl_2$ cannot be easily explained in terms of simple changes of a negative surface potential; the simple 'surface potential theory' would predict that all parameters of one gating function would be shifted equally. This is not always the case. In frog nerve, Ni^{2+} ions slow activation and inactivation of sodium channels, and, in squid axon, Zn^{2+} ions slow inactivation (Hille, 1968; Conti *et al.*, 1976; Gilly and Armstrong, 1982). The stronger action of transition metals is interpreted by assuming that they bind to surface groups with a higher affinity.

Therefore, when the external ion acts primarily by binding, it causes more than a simple shift and it is not surprising that more than an electrostatic effect on the channel protein would occur. However, the need to postulate a binding theory does not argue against the importance of counterion-dependent surface potentials.

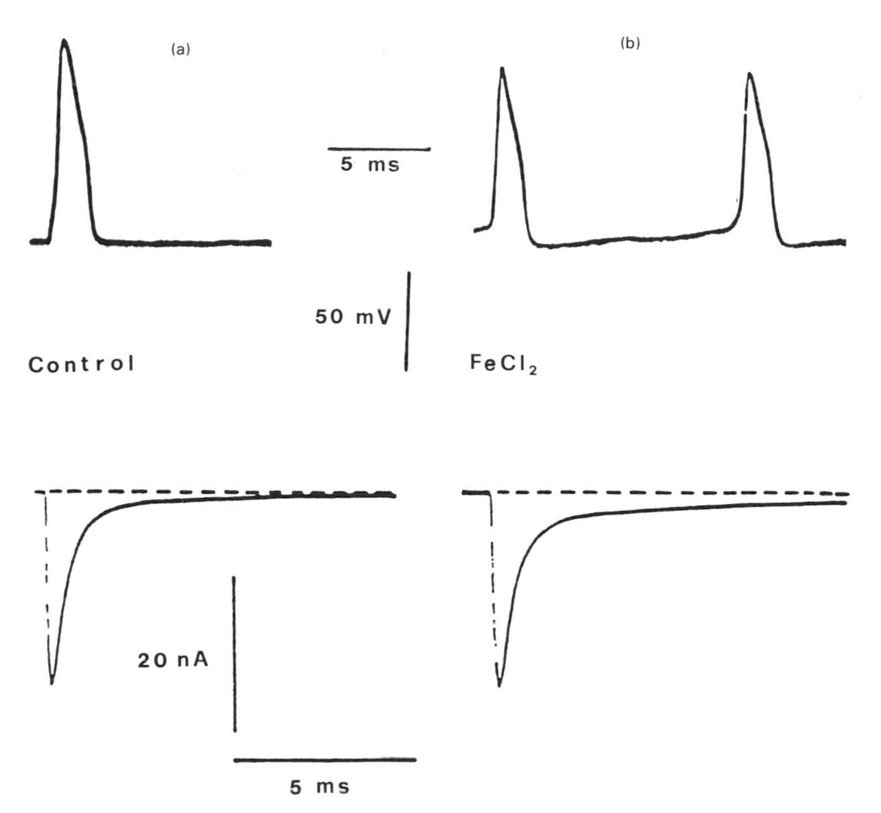

Fig. 25.2 Upper panel: (a) Control action potential in Ringer's solution, evoked by a 0.5 ms depolarizing stimulus. (b) Spontaneous action potentials recorded 5 min after the addition of 100 μM $FeCl_2$. Lower panel: traces of sodium current recorded from the same fibre under voltage-clamp before and after treatment with 100 μM $FeCl_2$. Potassium currents are blocked by cutting the ends of the fibre in 80 mM CsCl + 10 mM NaCl and by using 10 mM tetraethylammonium outside. Sodium current was recorded during depolarizations to 0 mV preceded by 50 ms hyperpolarizations to −120 mV.

25.5 DOES 'LATE' SODIUM CURRENT RELATE TO A NEW STATE OF SODIUM CHANNEL?

We have seen that a 'late' sodium current may be induced under several conditions: peptide neurotoxins, endogenous polypeptides, general anaesthetics and ions. Mostly, this current reflects a slowing of sodium inactivation and it relates to molecular rearrangements of the sodium channel which may exist in a new state. Different models have been proposed to explain the effects of sodium inactivation gate modifiers: a reopening of a fraction of inactivated channels; an unmasking of pre-existent silent channels into non-inactivatable ones; quantitative changes in rate constants of channel inactivation. According to the model proposed by Schmidtmayer (1985) and consisting of quantitative changes in rate constants of inactivation, channels would have only one open state and, consequently, peak and maintained currents should reverse at the same voltage. The observation that peak and maintained currents reverse at different potentials indicates that this model cannot explain the effects of the pharmacological agents considered. However, sometimes maintained current reverses at the same voltage as the peak current (Benoit *et al.*, 1987); this indicates that the properties of sodium channels are differentially modified by various drugs. The observation that in the presence of some drugs the initial amplitudes of fast and slow inactivation phases and maintained current change simultaneously is in favour of the interconversion hypothesis which consists of a transformation of fast channels into slow and late channels. The fast and slow channels correspond to fast and slow inactivation phases of the peak current, respectively. From a molecular point of view, this can be explained if one assumes that emergence of the late channels is controlled by the environment of the channel, and only about 30% of channels have an environment favourable to their transformation into the late form. Modification of membrane lipids is a tentative explanation for the effects observed; on the other hand, the reagents which inhibit sodium inactivation are able to modify double bonds. In addition, drugs acting by specific binding with channel protein produce additional effects which are independent and specific of their properties: increase of inactivation rate constant, negative shift of the activation–voltage relationship and decrease of the ionic selectivity of late channels.

25.6 CONCLUDING REMARKS

We have seen many classes of sodium channel gating modifiers: enzymes and chemical treatments acting from the cytoplasmic side of the membrane; some peptide toxins, endogenous polypeptides and transition metal ions acting from outside. The vulnerability of inactivation gating to internal chemical attack shows that it depends on a protein domain accessible from the cytoplasmic side; nevertheless, the inactivation-slowing action of external peptides and drugs suggests that the external channel face also has domains coupled to the inactivation gating mechanism. The slowing or elimination of inactivation without effects on activation means that these same domains are not involved in activation gating. More recently, structure–function relationships of the voltage-gated sodium channel expressed in *Xenopus* oocytes have been investigated by the combined use of site-directed mutagenesis and patch-clamp recording (Stühmer *et al.*, 1989); this functional analysis of mutated sodium channels provides evidence that there is a well-defined region important for the inactivation of the sodium channel and this region is likely to contain the sites at which intracellular proteolytic digestion exerts its effect. This region, distinct from that involved in the voltage-dependent activation and located on the cytoplasmic side of the membrane, contains a cluster of conserved positively charged residues (mainly lysine) and has been proposed to be involved in the inactivation of the sodium channel.

REFERENCES

Armstrong, C.M. and Bezanilla, F. (1973) *Nature*, **242**, 459–61.

Armstrong, C.M., Bezanilla, F. and Rojas, E. (1973) *J. Gen. Physiol.*, **62**, 375–91.

Benoit, E., Carratù, M.R., Dubois, J.M. and Mitolo-Chieppa, D. (1987) *Br. J. Pharmacol.*, **87**, 291–7.

Bergman, C., Dubois, J.M., Rojas, E. and Rathmayer, W. (1976) *Biochem. Biophys. Acta* **455**, 173–84.

Brodwick, M.S. and Eaton, D.C. (1978) *Science*, **200**, 1494–6.

Cahalan, M.D. (1975) *J. Physiol.*, **244**, 511–34.

Carbone, E. (1982) *Biochem. Biophys. Acta*, **693**, 188–94.

Carbone, E., Wanke, E., Prestipino, G. *et al.* (1982) *Nature*, **296**, 90 1.

Carratù, M.R. and Mitolo-Chieppa, D. (1989) *Experientia*, **45**, 346–9.

Carratù, M.R., Dubois, J.M. and Mitolo-Chieppa, D. (1982) *Neuropharmacology*, **21**, 619–23.

Catterall, W.A. (1980) *Annu. Rev. Pharmacol. Toxicol.*, **20**, 15–43.

Conti, F., Hille, B., Neumcke, B. *et al.* (1976) *J. Physiol.*, **262**, 729–42.

Dubois, J.M. and Bergman, C. (1975) *Pflügers Arch.*, **357**, 145–8.

Eaton, D.C., Brodwick, M.S., Oxford, G.S. and Rudy, B. (1978) *Nature*, 271, 473–6.

Gilly, W.F. and Armstrong, C.M. (1982) *J. Gen. Physiol.*, **79**, 935–64.

Hille, B. (1968) *J. Gen. Physiol.*, **51**, 221–36.

Hodgkin, A.L. and Huxley, A.F. (1952) *J. Physiol.*, **117**, 500–44.

Horn, R., Brodwick, M.S. and Eaton, D.C. (1980) *Am. J. Physiol.*, **238**, C127–C132.

Nonner, W. (1969) *Pflügers Arch.*, **309**, 176–92.

Oxford, G.S., Wu, C.H. and Narahashi, T. (1978) *J. Gen. Physiol.*, **71**, 227–47.

Rojas, E. and Rudy, B. (1976) *J. Physiol.*, **262**, 501–31.

Romey, G., Abita, J.P., Schweitz, H. *et al.* (1976) *Proc. Natl Acad. Sci. USA*, **73**, 4055–9.

Schmidtmayer, J. (1985) *Pflügers Arch.*, **404**, 21–8.

Stämpfli, R. (1974) *Experientia*, **30**, 505–8.

Stühmer, W., Conti, F., Suzuki, H. *et al.* (1989) *Nature*, **339**, 597–603.

Wang, G.K. and Strichartz, G.R. (1983) *Mol. Pharmacol.*, **23**, 519–33.

26 A K^+ channel opener in vascular smooth muscle: pharmacology and mechanism of action of cromakalim

U. QUAST
Preclinical Research, Sandoz Ltd, CH 4002 Basel, Switzerland

26.1 INTRODUCTION

Membrane K^+ channels determine the resting membrane potential of the cell; they influence parameters as different as cell excitability and cell volume. Opening of K^+ channels will bring the membrane potential closer to the Nernst potential for K^+ (usually −80 to −90 mV) and oppose depolarization. Many neurotransmitters, neuropeptides and hormones open membrane K^+ channels, generally via activation of a G-protein. This review concentrates on the in vitro activity of cromakalim (BRL 34915), the representative of a new class of drugs which open K^+ channels, preferentially in smooth muscle (but also in other tissues). These K^+ channel openers (KCOs) constitute a chemically diverse group of compounds, including nicorandil, cromakalim, pinacidil, diazoxide, minoxidil sulphate and RP 34915 (see Fig. 26.1 for structures). Therapeutically, these agents are used or tested in hypertension, asthma, angina pectoris, cardiac insufficiency, peripheral arterial disease, irritable bladder and even male pattern baldness.

The coronary vasodilator nicorandil was the first drug shown to hyperpolarize smooth muscle by a presumed opening of membrane K^+ channels (Furukawa *et al.*, 1981). In fact, nicorandil is a mixed vasodilator, since the nitro group within its structure enables it to activate soluble guanylate cyclase, thereby increasing the concentration of cGMP in the cell (Holzmann, 1983), an effect also leading to relaxation of smooth muscle.

The discovery by A. Weston and colleagues that the vasorelaxant effects of cromakalim are linked to a hyperpolarization of the vascular smooth muscle membrane induced by the opening of K^+ channels (Hamilton *et al.*, 1986; Weir and Weston, 1986b) has boosted the interest in this mechanism of action. In comparison to nicorandil, cromakalim is

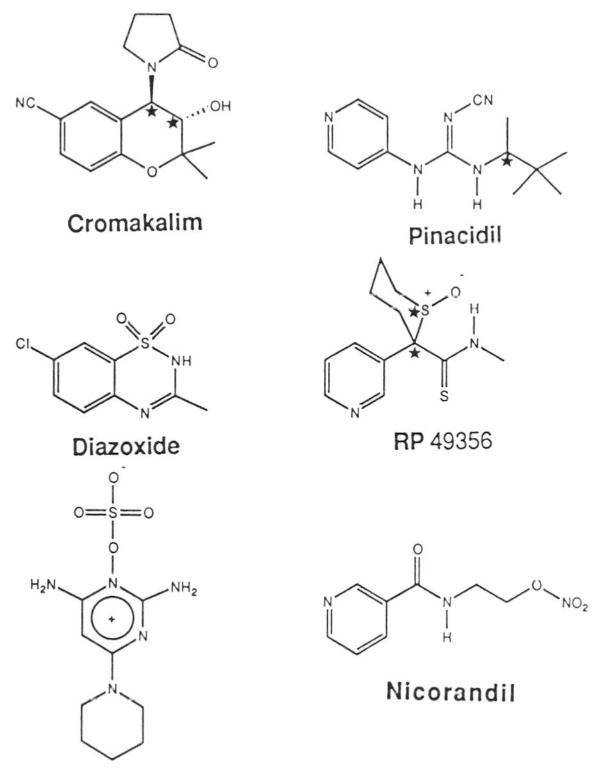

Fig. 26.1 Chemical structures of K^+ channel openers (a) and the insulinotropic K_{ATP} blockers glibenclamide and AZ-DF 265 (b). Asterisks indicate chiral centres; the effects of cromakalim, pinacidil and AZ-DF 265 have been shown to be stereospecific. See Cook and Quast (1990) for references; modified after Quast and Cook (1989).

approximately 100 times more potent in relaxing smooth muscle, and its mechanism of action appears to lie exclusively in the increase of the membrane permeability for K^+ (P_k) presumably by opening of K^+ channels.

In this review, the pharmacological properties of cromakalim, the prototype of this class of drugs, are described. Evidence for and against K^+ channel opening being solely responsible for its smooth muscle relaxant effects is reviewed. The attempts to identify the nature of the channel opened by cromakalim are discussed and the review is concluded by a consideration of the possible ways in which membrane hyperpolarization may lead to a modulation of smooth muscle tone.

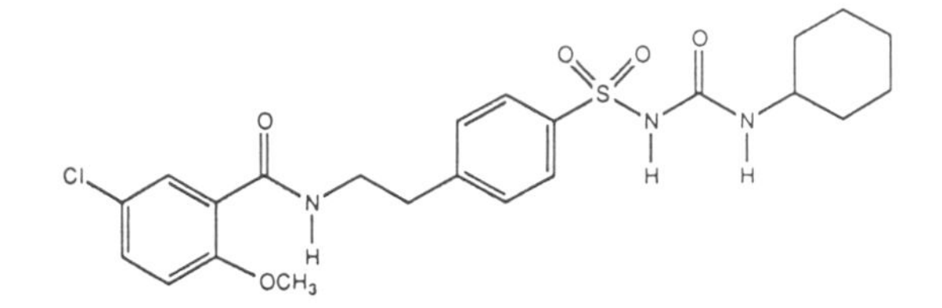

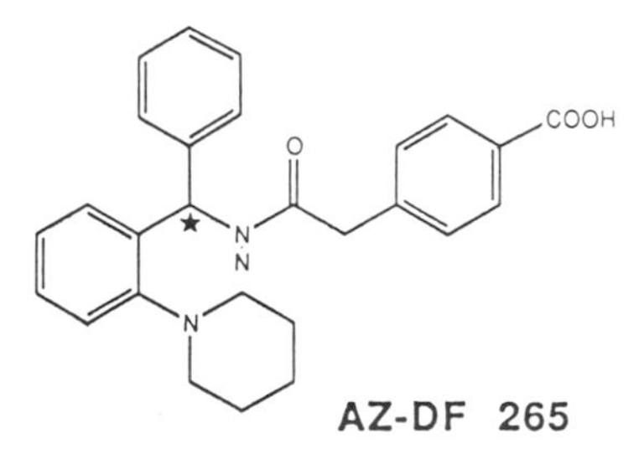

(b)

Fig. 26.1 (b)

26.2 PHARMACOLOGICAL PROPERTIES OF CROMAKALIM

26.2.1 Evidence for K+ channel opening and tracer flux studies

The suggestion that cromakalim opens membrane K^+ channels was based on its ability to hyperpolarize various smooth muscle preparations and/or to promote $^{86}Rb^+$ efflux, the latter as a measure of P_k (vascular tissue: Hamilton *et al.*, 1986; Weir and Weston, 1986b; Coldwell and Howlett, 1987; Quast, 1987; Hof *et al.*, 1988; Cook *et al.*, 1988b; guinea-pig trachea: Allen *et al.*, 1986; guinea-pig taenia caeci: Weir and Weston 1986a). A typical experiment in rat portal vein is shown in Fig. 26.2. Opening of membrane K^+ channels would be expected to shift the membrane potential towards the Nernst potential for K^+ and therefore lead to a hyperpolarization of the cell. Since stimulation of $^{86}Rb^+$ efflux and hyperpolarization are generally observed at similar concentrations (>0.1 μM), they would indeed appear to be linked.

In most tracer flux studies with cromakalim, $^{86}Rb^+$ was chosen as a convenient substitute for $^{42}K^+$ and/or $^{43}K^+$, due to its longer half-life. In the guinea-pig portal vein, Quast and Baumlin (1988) showed in double labelling experiments that using $^{42}K^+$ allowed the detection of an effect of cromakalim on flux at lower concentrations than with

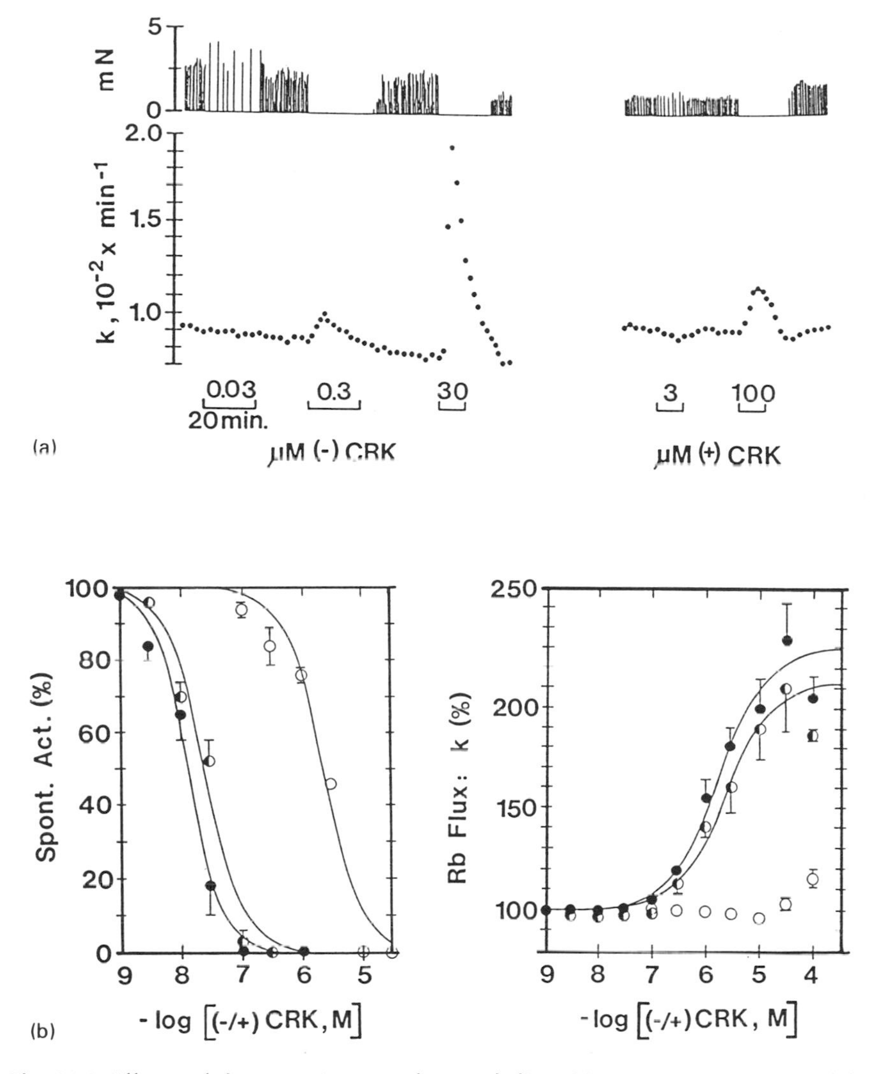

Fig. 26.2 Effects of the enantiomers of cromakalim (CRK) on spontaneous activity and $^{86}Rb^{+}$ efflux from the rat portal vein. Inhibition of spontaneous activity. (a) prominent effect on the frequency of the contractions and stimulation of $^{86}RB^{+}$ efflux (middle trace; quantified by the rate of efflux, *k*) are stereospecific. (b) Concentration dependence of these effects. ●, (−)-enantiomer of CRK; ○, (+)-enantiomer, ◐ (racemic) cromakalim. Modified after Hof *et al.* (1988).

$^{86}Rb^+$ (20 nM as opposed to 60 nM). In both the rat aorta and guinea-pig portal vein, the $^{42}K^+$ efflux stimulated by cromakalim below 0.1 μM was about threefold larger than that of Rb^+, whereas at concentrations above 1 μM this ratio was only 1.5-fold (Fig. 26.3). This suggested heterogeneity in the K^+ channels opened by different

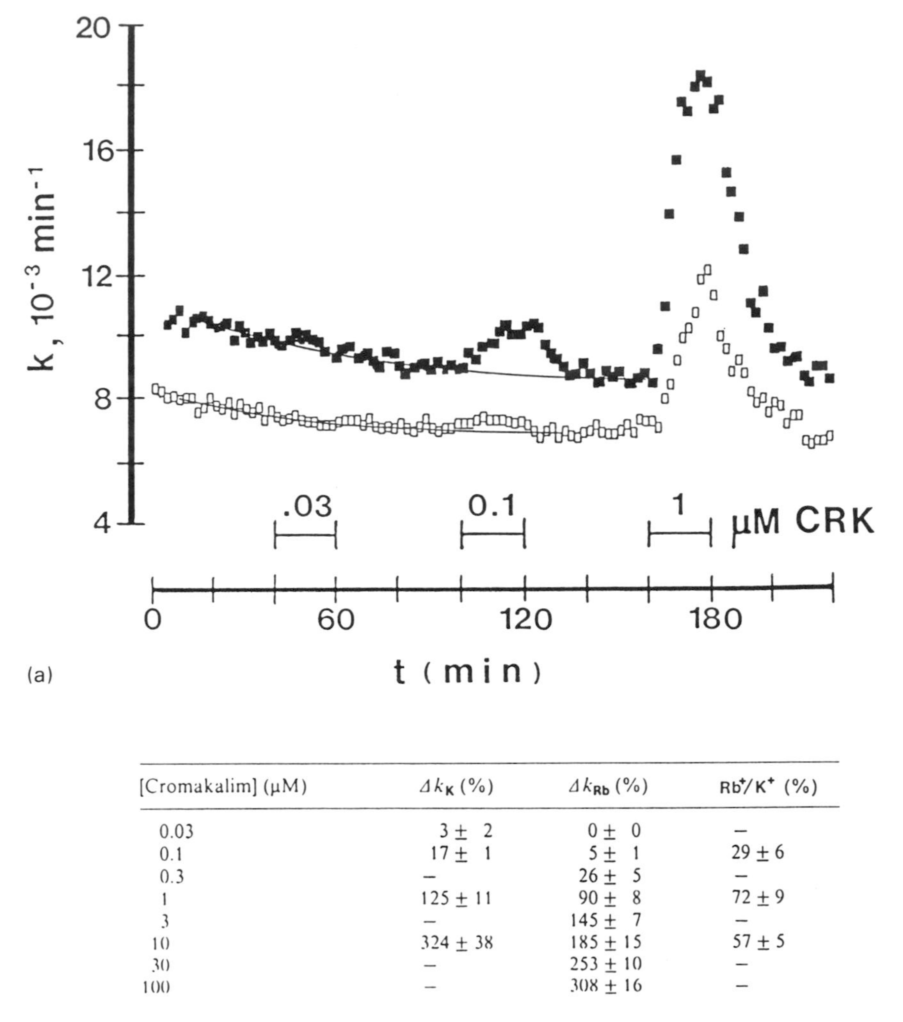

[Cromakalim] (μM)	Δk_K (%)	Δk_{Rb} (%)	Rb^+/K^+ (%)
0.03	3 ± 2	0 ± 0	–
0.1	17 ± 1	5 ± 1	29 ± 6
0.3	–	26 ± 5	–
1	125 ± 11	90 ± 8	72 ± 9
3	–	145 ± 7	–
10	324 ± 38	185 ± 15	57 ± 5
30	–	253 ± 10	–
100	–	308 ± 16	–

(b)

Fig. 26.3 Stimulation by cromakalim of $^{42}K^+$ and $^{86}Rb^+$ efflux from rat aorta. (a) Single traces of the rate of tracer efflux (*k*) from rat aortic strips preloaded with 5 μCi/ml $^{42}K^+$ (■) or $^{86}Rb^+$ (□). (b) Peak increases in *k* (in Δ%) of $^{42}K^+$ and $^{86}Rb^+$ efflux and the ratio of these values (in %) (mean values *x* ± SEM, *n* = 8). After Quast and Baumlin (1988).

concentrations of cromakalim in these preparations. Although the flux response was quantitatively underestimated by using $^{86}Rb^+$, the changes in $^{86}Rb^+$ efflux would seem to qualitatively reflect the cromakalim-induced increase in P_k in these vascular tissues (Quast and Baumlin, 1988). However, this appears not to hold for all KCOs (cf. minoxidil sulphate, which increases the efflux of $^{42}K^+$ (Bray *et al.*, 1988c; Meisheri *et al.*, 1988) but not of $^{86}Rb^+$ (Newgreen *et al.*, 1989)).

26.2.2 Vasodilator effects of cromakalim

The vasodilator profile of cromakalim (and other KCOs) shows characteristic features which distinguish it from most other vasodilators. Most importantly, cromakalim is able to relax contractions to low (<25 μM) but not high (>30 μM) KCl in various vascular preparations (Hamilton *et al.*, 1986; Weir and Weston, 1986b; Clapham and Wilson, 1987, here: see Fig. 26.4), and the presence of elevated KCl abolishes the vasorelaxant effect of cromakalim against other stimuli (e.g. noradrenaline) (Cook *et al.*, 1988b). This is a behaviour expected for an agent acting solely by opening K^+ channels, since in the presence of high KCl concentrations the actual membrane potential approaches the Nernst equilibrium potential for K^+ (see, for example, Häusler, 1983), thus reducing the achievable degree of hyperpolarization by opening of K^+ channels. Furthermore, since the Nernst potential under these conditions will be less negative than the threshold potential needed to increase the intracellular Ca^{2+} concentration (e.g. by opening voltage-operated Ca^{2+} channels), the agent will be ineffective.

Cromakalim inhibits the spontaneous activity of rat or guinea-pig portal vein (IC_{50} = 20–40 nM) by decreasing the duration of the contractile spikes and the frequency of the contractions without much effect on their amplitude (Quast, 1987; Hof *et al.*, 1988; Quast and Baumlin, 1988; here: see Fig. 26.2). An increase in tracer efflux could, however, be detected only at concentrations where mechanical inhibition was essentially complete (using $^{86}Rb^+$, Quast, 1987) or near the IC_{50} value ($^{42}K^+$; Quast and Baumlin, 1988). A similar situation exists in rat uterus where relaxation is seen with cromakalim, but no change in $^{86}Rb^+$ or $^{42}K^+$ efflux (Hollingsworth *et al.*, 1987, 1988). The currently favoured theory to explain the relaxation/flux discrepancy is that these drugs preferentially act on K^+ channels in pacemaker cells from where the myogenic contractions initiate. Hence the spontaneous activity of the tissue would be inhibited, but (due to the sparsity of these cells) changes in the whole tissue $^{86}Rb^+$ efflux or membrane potential

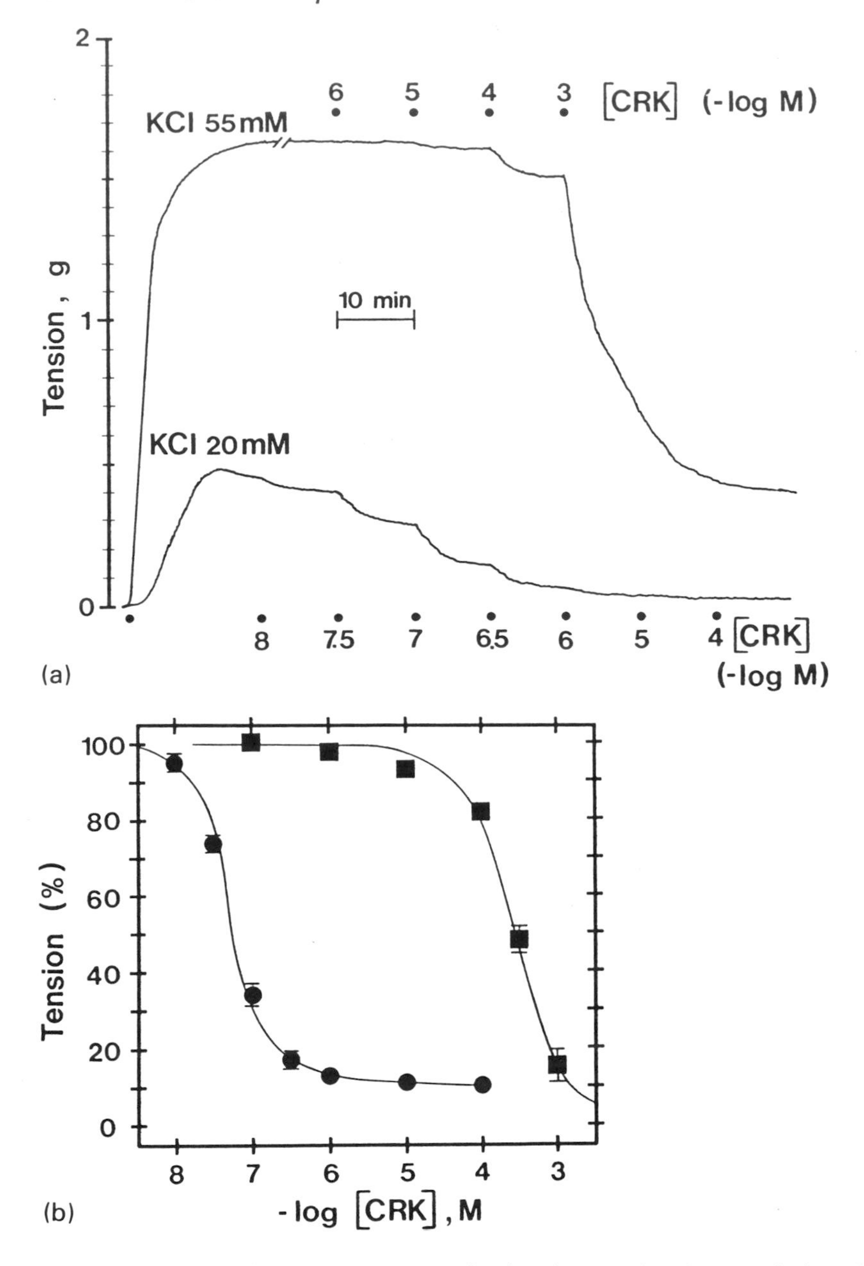

Fig. 26.4 Effect of cromakalim (CRK) against KCl-induced contractions in rat aortic rings. (a) Typical traces comparing the vasorelaxant effect of CRK against contractions induced by 20 and 55 mM KCl (experiments in a modified Krebs-Henseleit solution containing 5 mM Hepes, at pH 7.4 and 37 °C; KCl was increased at the expense of NaCl). (b) Concentration dependence of these effects (●; 20 mM, ■, 55 mM KCl; $x \pm$ SEM, $n = 6$). Cromakalim is about 6000 times weaker against 55 mM KCl than against 20 mM.

would be undetectable (Hamilton *et al.*, 1986; Quast, 1987; Hollingsworth, 1987, 1988).

A mechanistically interesting effect of cromakalim is seen against the tonic contraction of the rabbit aorta to noradrenaline. Here noradrenaline does not depolarize the membrane potential (Cauvin *et al.*, 1984; Bray *et al.*, 1988a) and consequently dihydropyridine Ca^{2+} antagonists are without effect (see, for example, Cook *et al.*, 1988b). Cromakalim, however, reduces this contraction by about 80% with midpoint at 0.3 μM (Cook *et al.*, 1988b). These data indicate that the cromakalim affects smooth muscle contraction by other mechanisms in addition to indirectly reducing opening of dihydropyridine-sensitive Ca^{2+} channels and evidence for an interference with refilling of intracellular Ca^{2+} stores has been presented (Bray *et al.*, 1988b; Cowlrick *et al.*, 1988; see below, section 26.4).

26.2.3 Blockers

The effects of cromakalim on P_k and contraction of smooth muscle can be inhibited in parallel by K^+ channel blockers. Quaternary ammonium ions (in particular tetraalkylammonium (TEA)), aminopyridines and procaine are effective in the millimolar range (vascular tissue: Coldwell and Howlett, 1987; Quast, 1987; Wilson *et al.*, 1988b; guinea-pig trachea: Allen *et al.*, 1986); at the same time they have a profound effect on basal tension of smooth muscle. Comparing the potency of externally applied symmetrical tetraalkylammonium ions as blockers of the cromakalim-induced $^{86}Rb^+$ efflux, Quast and Webster (1989) found that potency increased with increasing chain length up to the *n*-pentyl compound ($pK_{50} = 6.9$). This indicated that the tetraalkylammonium site of the channel accessible from the outside has a strong preference for bulky (but not too bulky) lipophilic cations in contrast to the majority of K^+ channels where this site is rather specific for TEA (see Cook and Quast, 1990, for a recent review).

The sulphonylurea glibenclamide (Fig. 26.1), a blocker with nanomolar potency of ATP-sensitive K^+ channels in the pancreatic ß-cell or related cell lines (Schmidt-Antomarchi *et al.*, 1987; Sturgess *et al.*, 1988; Zünkler *et al.*, 1988), inhibits the effects of cromakalim on P_k and/or contraction at 0.1 μM (Quast and Cook, 1989a; for tension see also Cavero *et al.*, 1989; Eltze, 1989; Wilson, 1989; Winquist *et al.*, 1989) without an effect on basal tension. In the limited concentration range up to 3 μM, the inhibition by glibenclamide of the effects of cromakalim on rat portal vein spontaneous activity and $^{86}Rb^+$ efflux were compatible with competitive antagonism, giving pA_2 values

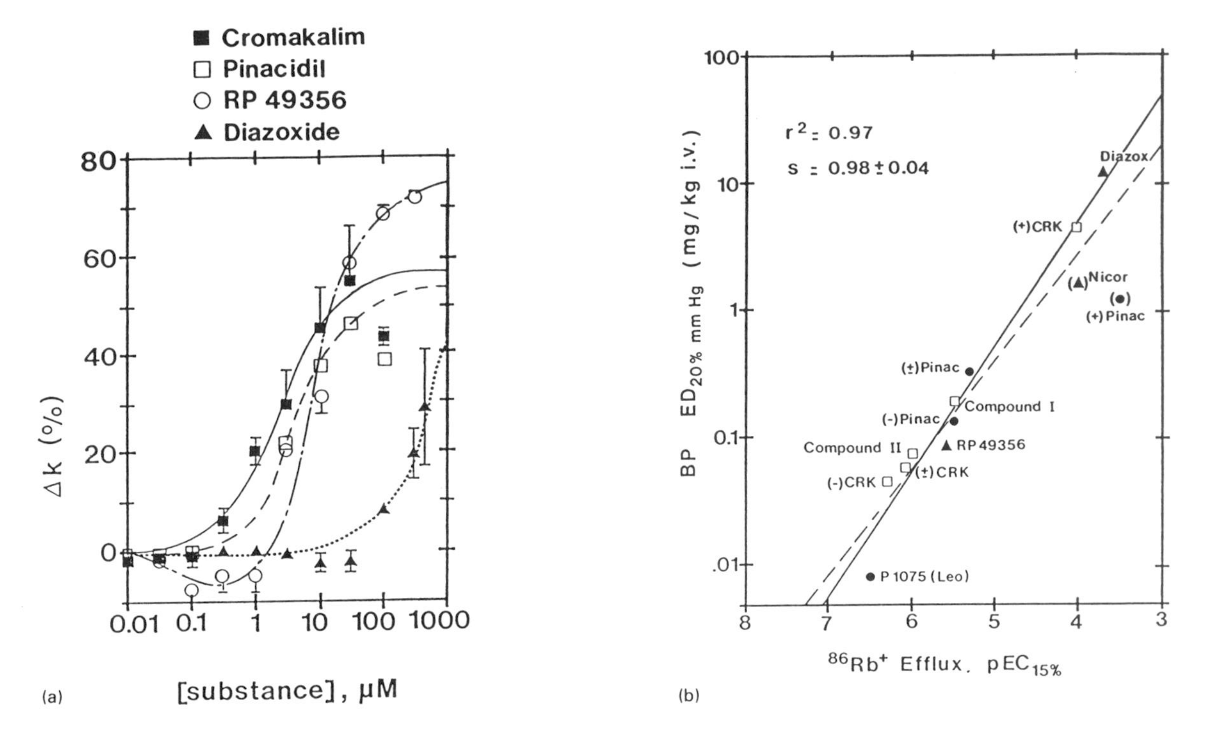

Fig. 26.5 Comparison of KCOs. (a) Effect on the rate of $^{86}Rb^+$ efflux (Δk, %) from rat portal vein. ($x \pm$ SEM, n = 4–6). (b) Correlation between blood pressure lowering activity of KCOs in anaesthetized rats and their potency at stimulating $^{86}Rb^+$ efflux from the rat portal vein. Blood pressure lowering was determined as described in Cook *et al*. (1988a) and is quantified by the intravenous dose necessary to lower blood pressure by 20%. $^{86}Rb^+$ efflux measurements were performed as described by Quast (1987) and are quantified by the negative logarithm of the concentration (M) which increases efflux by 15%. The broken line is the correlation obtained when all compounds are included (n = 13, r^2 = 0.9, slope = 0.84 ± 0.08); the solid line is obtained after exclusion of (+)-pinacidil and nicorandil (n = 11, r^2 = 0.97, s = 0.98 ± 0.04). Chemical names of the compounds not appearing in Fig. 26.1: P 1075 (Leo) = *N*-cyano-*N'*-(1,1-dimethylpropyl)-*N''*-3-pyridinylguanidine; compound I = (±)-3,4-dihydro-3-hydro-2,2-dimethyl-4-(2-oxo-1-piperazinyl)-2*H*-1-benzopyran-6-carbonitrile; compound II = (±)-*N*-(6-cyano-3,4-dihyro-3-hydro-2,2-dimethyl-2*H*-1-benzopyran-4-yl)-2-furancarboxamide. After Cook and Quast (1990).

of 7 (Quast and Cook, 1989a). Glibenclamide (although at high doses) also inhibits the blood pressure lowering effect of cromakalim in the rat (Cavero *et al.*, 1989; Quast and Cook, 1989a). Other sulphonylureas or related K_{ATP} channel blockers that lack the sulphonylurea group (e.g. AZ-DF 265, see Fig. 26.1) are also able to block cromakalim-stimulated $^{86}Rb^+$ efflux, their relative potencies suggesting that the sulphonylurea binding site of this channel resembles that of K_{ATP} channels in the pancreatic ß-cell (Quast and Webster, 1989).

26.2.4 Stereoselectivity

Cromakalim is a racemate where the substituents in the 3 and 4 position are in the *trans* configuration (see Fig. 26.1). The effects of cromakalim on vascular P_k and contractility in vitro and in vivo (blood pressure lowering) are (about 100-fold) stereoselective with the activity residing in the (−)-(3*S*, 4*R*) enantiomer (Buckingham *et al.*, 1986; Hof *et al.*, 1988; here: see Fig. 26.2).

26.2.5 Other K^+ channel openers

Vasodilators chemically quite unrelated to cromakalim (see Fig. 26.1 for structures) have recently been shown also to stimulate $^{86}Rb^+$ efflux from smooth muscle (Fig. 26.5(a); for details see Cook and Quast, 1990). Fig. 26.5(b) shows that an excellent correlation exists between the potencies of various KCOs to increase $^{86}Rb^+$ efflux from the rat portal vein and to lower blood pressure in the anaesthetized rat. This provides strong evidence for the hypothesis that the antihypertensive effect of these compounds (with the exception of pinacidil; see Cook *et al.*, 1989; Cook and Quast, 1990) relies on their ability to increase P_k in vascular smooth muscle.

26.3 NATURE OF THE K^+ CHANNEL OPENED

Effects of the KCOs have been observed in various smooth muscle preparations (blood vessels, trachea, intestine, bladder) and, generally at higher concentrations, also in heart, pancreas and neuronal tissue (see Cook and Quast, 1990, for a recent review). Evidence is emerging that this group of drugs may be itself heterogeneous (e.g. minoxidil sulphate opposed to the others; for a discussion see Cook and Quast, 1990) and that these drugs act on different K^+ channels according to the tissue chosen and, in one particular tissue, according to the concentration used. Examples for the latter include in portal vein a preferential action on pacemaker cells (Fig. 26.2) or the decrease in selectivity for K^+ over Rb^+ with increasing concentrations of cromakalim (Fig. 26.3) and pinacidil in several vessels. In the following we discuss a few points pertinent to the nature of the K^+ channels opened by these drugs, with focus on the results obtained in vascular smooth muscle.

26.3.1 Is the channel Ca^{2+} dependent?

Several electrophysiological studies have appeared which show that cromakalim can affect several Ca^{2+}-dependent K^+ channels in vascular smooth muscle (see Cook and Quast, 1990, for a review), in particular the large-conductance Ca^{2+}-activated K^+ channel (BK_{Ca}) (Kusano *et al.*, 1987; Trieschmann *et al.*, 1988; Gelband *et al.*, 1989). The conductance of the BK_{Ca} channel was not modified but the duration of the closed periods was reduced in the presence of cromakalim, such that the channel open probability was increased up to 60% (Gelband *et al.*, 1989; Trieschmann *et al.*, 1988). On the other hand it has been reported that charybdotoxin, a toxin purified from the venom of the scorpion *Leiurus quinquestriatus* and a potent (nanomolar) blocker of BK_{Ca} (Miller *et al.*, 1985; Smith *et al.*, 1986), does not block effects of cromakalim on $^{86}Rb^+$ efflux or vascular tone (Strong *et al.*, 1989; Winquist *et al.*, 1989). In addition, tetraethylammonium (which blocks BK_{Ca} channels at submillimolar concentrations) inhibits effects of the K^+ channel openers only in the millimolar range, which again argues against an action of these drugs on the BK_{Ca}. Apamin, a toxin from bee venom (Habermann, 1984) which selectively blocks a Ca^{2+}- dependent K^+ channel of small conductance (Romey *et al.*, 1984; Cook and Haylett, 1985; Blatz and Magleby, 1986), has been shown not to affect cromakalim activity in vitro or in vivo (Weir and Weston, 1986b; Cook and Hof, 1988).

Many studies show neither the vasorelaxant effects of the KCOs nor the increase in P_k they elicit to be modified by (selective) Ca^{2+} antagonists (Quast, 1987; Cook *et al.*, 1988b; Quast and Baumlin, 1988; Post *et al.*, 1989). Ca^{2+} removal increases the effect of cromakalim on $^{86}Rb^+$ efflux and prevents the rapid decline in this response seen in the presence of Ca^{2+} (Coldwell and Howlett, 1988; Post *et al.*, 1989).

Taken together these data suggest that although clear electrophysiological evidence shows that the KCOs can open K_{Ca} (in particular

BK_{Ca}), this appears not to be the way these drugs exert their pharmacological effects on the vasculature at low concentrations.

26.3.2 Is the channel ATP sensitive?

This question arose when it was shown that glibenclamide, the most potent (nanomolar) inhibitor of ATP-sensitive K^+ channels (K_{ATP}) in the pancreas, was also a potent, apparently competitive inhibitor of the effects of cromakalim on smooth muscle (pA_2 = 7, see section 26.2.3 for details). The similar inhibition of the KCOs (with the exception of minoxidil sulphate) suggests they act on the same K^+ channel to elicit their effects. As outlined above (see section 26.2.3), other sulphonylureas or compounds derived therefrom like AZ-DF 265 also inhibit cromakalim-stimulated $^{86}Rb^+$ efflux with potencies suggesting that the K^+ channel opened by cromakalim in vascular smooth muscle has indeed a binding site which resembles the sulphonylurea site of K_{ATP} in the pancreatic ß-cell (Quast and Webster, 1989).

A second intriguing fact is that diazoxide, an opener of K_{ATP} in the pancreas, is also a KCO in smooth muscle (Quast and Cook, 1989a). The comparison of cromakalim with diazoxide shows that the K^+ channel in smooth muscle must have a different pharmacology from K_{ATP} in pancreatic ß-cells, since cromakalim does not modify insulin secretion in vitro and in vivo or plasma glucose levels (Cook *et al.*, 1988a; Wilson *et al.*, 1988a., Quast and Cook, 1989a). Cromakalim thus shows selectivity for the K^+ channel in smooth muscle over K_{ATP} in ß-cells, whereas diazoxide opens both channels with a similar potency. An additional point of interest to emerge from these studies is that glibenclamide does not significantly increase tone in vitro or in vivo at doses blocking effects of cromakalim (Quast and Cook, 1989a). Hence, the channel opened by these drugs would appear not to be open under resting conditions, as originally suggested by Hamilton *et al.* (1986).

The third intriguing fact is that (at least part of) the cardiac effects of the KCOs are mediated via the opening of K_{ATP} in the heart. Sanguinetti *et al.* (1988) and Mestre *et al.* (1988) showed that the shortening of the cardiac action potential by cromakalim was inhibited by glibenclamide. In experiments on excised cardiac myocyte membranes, 300 μM cromakalim was also shown to increase (single-channel) currents through K_{ATP} (Escande *et al.*, 1988). RP 49356, another KCO, opens these channels by decreasing their affinity for ATP, thus relieving the blockade of K_{ATP} by physiological levels of ATP in the cell (Thuringer and Escande, 1989).

Very recently, evidence has been obtained at the single-channel level that K_{ATP} channels exist in freshly prepared cells from rat or rabbit mesenteric arteries (Standen *et al.*, 1989). In inside-out patches, channel activity was abolished by the presence of 1 mM ATP. Cromakalim (1 μM) restored the channel activity, and this could be inhibited by 20 μM glibenclamide. In line with these observations, the relaxation of the noradrenaline-induced tonic contractions of the rabbit mesenteric artery by cromakalim, diazoxide and pinacidil was reversed by glibenclamide (<1 μM), tolbutamide (500 μM) and Ba^{2+} (30–100 μM) but not by tetraethylammonium (< 1 mM) or charybdotoxin (50 nM). This direct electrophysiological demonstration of K_{ATP} channels in vascular smooth muscle, and the effect of cromakalim thereon, greatly strengthens the argument that these channels are indeed the primary target of the K^+ channel openers (but see Beech and Bolton, 1989, for different results), and that their opening underlies the mechano-inhibitory effect of these drugs. The pharmacological consequences of this finding have been discussed recently (Quast and Cook, 1989b). The K_{ATP} channels in smooth muscle may constitute a cellular defence mechanism in response to ischaemia or hypoxia. The changes in the metabolic state of the cell like a drop in intracellular ATP and pH associated with this condition might favour opening of these channels by K^+ channel openers (Quast and Cook, 1989b).

26.4 CONSEQUENCES OF K_+ CHANNEL OPENING UPON SMOOTH MUSCLE TONE

Since mechano-inhibitory effects of cromakalim seem invariably to be abolished in depolarized vessels, it is reasonable to assume (at least until experimental evidence proves otherwise) that all of its effects are coupled to its ability to open K^+ channels, thereby hyperpolarizing the cell membrane. Its ability to inhibit the 'electrically silent', dihydropyridine-insensitive tonic contractions of the rabbit aorta to noradrenaline (Cook *et al.*, 1988c; Bray *et al.*, 1988a, 1988b) requires a modification of the original postulate, that the relaxant effect of the K^+ channel openers was due to their hyperpolarizing action indirectly closing voltage-sensitive (slow) Ca^{2+} channels (Hamilton *et al.*, 1986). One must therefore ask the question, aside from dihydropyridine-sensitive Ca^{2+} channels, which other mechanisms affecting smooth muscle tone are likely to be modified by an increase in P_k and subsequent hyperpolarization of the membrane? The following possibilities might be worth considering.

(a) Dihydropyridine-insensitive voltage-sensitive Ca^{2+} channels have recently been identified in vascular smooth muscle (Bean *et al.*, 1986; Friedman

et al., 1986). By hyperpolarizing the membrane, cromakalim might prevent the opening of such channels.

(b) Receptor-operated cation channels might be modified by changing the cell membrane potential. At the moment, this possibility is merely speculative since the only such channel that has been characterized electrophysiologically (an ATP-operated cation channel in isolated cells from the rabbit ear artery) is relatively insensitive to changes in membrane voltage (Benham and Tsien, 1987). There may, however, exist many more types of receptor-operated cation channels.

(c) Na^+/Ca^{2+} exchange: due to the electrogenic nature of this ion transport system, hyperpolarization of the cell membrane is expected to favour Ca^{2+} extrusion (Mullins, 1979). The relevance of this effect in smooth muscle is difficult to test since specific inhibitors of this transport system are lacking (Kaczorowski *et al.*, 1988).

(d) Superficially (membrane) bound Ca^{2+}: the group of van Bremen has discussed the existence of a 'superficial' (membrane-bound) Ca^{2+} pool in smooth muscle (see, for example, Lodge and van Breeman, 1985; Lukeman and van Breeman, 1985) By hyperpolarizing the cell membrane cromakalim might interfere with the release of this pool from the membrane, which would also reduce contractions evoked by agonists.

(e) Refilling of intracellular Ca^{2+} stores: recently, evidence has been presented that cromakalim is able to prevent the refilling of intracellular Ca^{2+} stores by an as yet unknown mechanism (Bray *et al.*, 1988b; Cowlrick *et al.*, 1988). Inhibition of Ca^{2+} entry through a special 'refilling' channel or through one or other of the pathways of Ca^{2+} entry discussed above could account for this effect of cromakalim, with the consequence that contractions dependent primarily on the release of intracellular Ca^{2+} would be reduced in the presence of these drugs (Cook *et al.*, 1988b).

(f) Ca^{2+} release from intracellular stores: caffeine is known to release Ca^{2+} from the sarcoplasmatic reticulum (see Bolton, 1985, for a review). Recent studies show that cromakalim, pinacidil, RP 49356 and diazoxide inhibit caffeine-induced contractions in a glibenclamide-sensitive manner (Wilson, 1988; Bray *et al.*, 1989; Wilson and Hicks, 1989), suggesting that the KCOs somehow interfere with the ability of caffeine to release Ca^{2+} from the stores. In the perfused mesenteric bed of the rat, contractions to pulses of noradrenaline in Ca^2-free medium were inhibited concentration-dependently by cromakalim (Quast, 1989; Quast and Baumlin, 1990). This apparent inhibition of Ca^{2+} release from intracellular stores by cromakalim was stereoselective, inhibited by glibenclamide and abolished in depolarized

tissues, suggesting that it was associated with K^+ channel opening. However, this effect of cromakalim has not been observed in larger vessels (Wilson, 1988; Chiu *et al.*, 1987), indicating that this may vary according to the preparation (e.g. large/small vessel) being studied.

26.5 CONCLUSION

The field of KCOs in smooth muscle has developed very rapidy after the original description of cromakalim (Hamilton *et al.*, 1986). Numerous other KCOs (and blockers) of quite varying chemical structure have been described. Do they bind to the same channel, do they bind to different sites on the same channel or do they bind to different channels altogether? The availability of radioligands and more detailed electrophysiological data at pharmacologically relevant concentrations are of paramount importance in resolving these issues. The recent finding that cromakalim opens K_{ATP} in smooth muscle (Standen *et al.*, 1989) provides an explanation for its preferential dilatation of vessels in ischaemic tissue which was first reported by Angersbach and Nicholson (1988). Other targets of the KCOs may still be found (Beech and Bolton, 1989) and there is hope to find more specific drugs for the diverse indications where the KCOs have shown (or may be expected to show) efficacy, be it hypertension, angina pectoris, peripheral artery disease, cerebral ischaemia, asthma, irritable bladder, disorders of other smooth muscle and even alopecia areata (male pattern baldness).

REFERENCES

Allen, S.L., Boyle, J.P., Cortijo, J. *et al.* (1986) *Br. J. Pharmacol.*, **89**, 395–405.

Angersbach, D. and Nicholson, C.D. (1988) *Naunyn-Schmiedebergs Arch. Pharmacol.*, **337**, 341–6.

Bean, B.P., Sturek, M., Puga, A. and Hermsmeyer, K. (1986) *Circ. Res.*, **59**, 229–35.

Beech, D.J. and Bolton, T.B. (1989) *Br. J. Pharmacol.*, **98**, 851–64.

Benham, C.D. and Tsien, R.W. (1987) *Nature*, **328**, 275–8.

Blatz, A.L. and Magleby, K.L. (1986) *Nature*, **323**, 718–20.

Bolton, T.B. (1985) in *Control and Manipulation of Calcium Movement* (ed. J.R. Parratt), Raven Press, New York.

Bray, K.M., Weston, A.H., McHarg, A.D. *et al.* (1988a) *Br. J. Pharmacol.*, **93**, 205P.

Bray, K.M., Weston, A.H., McHarg, A.D. *et al.* (1988b) *Pflügers Arch.*, **411**, R202.

Bray, K.M., Brown, B.S., Duty, S. *et al.* (1988c) *Br. J. Pharmacol.*, **95**, 733P.

Bray, K.M., Longmore, J., Newgreen, D.T. and Weston, A.H. (1989) *Br. J. Pharmacol.*, **96**, 733P.

Buckingham, R.E., Clapham, J.C., Coldwell, M.C. *et al.* (1986) *Br. J. Pharmacol.*, **87**, 78P.
Cauvin, C., Lukeman, S., Cameron, J. *et al.* (1984) *J. Cardiovasc. Pharmacol.*, **6**, S630–S638.
Cavero,I., Mondot, S., Mestre, M. and Escande, D. (1988) *Br. J. Pharmacol.*, **95**, 643P.
Cavero, I., Mondot, S. and Mestre, M. (1989) *J. Pharmacol. Exp. Ther.*, **248**, 1261–8.
Chiu, P.J.S., Tetzloff, G., Ahn, H. and Sybertz, E. (1988) *Eur. J, Pharmacol.*, **155**, 229–37.
Clapham, J.C. and Wilson, C. (1987) *J. Auton. Pharmacol.*, **7**, 233–42.
Coldwell, M.C. and Howlett, D.R. (1987) *Biochem. Pharmacol.*, **36**, 3663–9.
Coldwell, M.C. and Howlett, (1988) *Biochem. Pharmacol.*, **37**, 4105–10.
Cook, N.S. and Haylett, D.G. (1985) *J. Physiol.*, **385**, 373–94.
Cook, N.S. and Hof, R.P. (1988) *Br. J. Pharmacol.*, **93**, 121–31.
Cook, N.S. and Quast, U. (1990) in *Potassium Channels: Structure, Classification, Function and Therapeutic Potential* (eds N.S. Cook) Ellis Horwood, Chichester, in press.
Cook, N.S., Quast, U. and Weir, S.W. (1988a) *Pflügers Arch.*, **411**, R46 (no. 81).
Cook, N.S., Weir, S.W. and Danzeisen, M.C. (1988b) *Br. J. Pharmacol.*, **95**, 741–52.
Cook, N.S., Quast, U. and Manley, P. (1989) *Br. J. Pharmacol.*, **96**, 181P.
Cowlrick, I.S., Paciorek, P.M. and Waterfall, J.F. (1988) *Br. J. Pharmacol.*, **95**, 640P.
Eltze, M. (1989) *Eur. J. Pharmacol.*, **165**, 231–9.
Escande, D., Thuringer, D., Leguern, S. and Cavero, I. (1988) *Biochem. Biophys. Res. Commun.*, **154**, 620–5.
Friedman, M.E., Suarez-Kurtz, G., Kaczorowski, G.J. *et al.* (1986) *Am. J. Physiol.*, **250**, H699–H703.
Furukawa, K., Itoh, T., Kajiwara, M. *et al.* (1981) *J. Pharmacol. Exp. Ther.*, **218**, 248–59.
Gelband, C.H., Lodge, N.J. and Van Breemen, C. (1989) *Eur. J. Pharmacol.*, **167**, 201–10.
Habermann, E. (1984) *Pharmacol. Ther.*, **25**, 255–70.
Hamilton, T.C., Weir, S.W. and Weston, A.H. (1986) *Br. J. Pharmacol.*, **88**, 103–11.
Häusler, G. (1983) *Fed. Proc.*, **42**, 263–8.
Hof, R.P., Quast, U., Cook, N.S. and Blarer, S. (1988) *Circ. Res.*, **62**, 679–86.
Hollingsworth, M., Amédée, T., Edwards, D. *et al.* (1987) *Br. J. Pharmacol.*, **91**, 803–13.
Hollingsworth, M., Edwards, D., Rankin, J.R. and Weston, A.H. (1988) *Br. J. Pharmacol.*, **93**, 199P.
Holzmann, S. (1983) *J. Cardiovasc. Pharmacol.*, **5**, 364–70.
Kaczorowski, G.J., Garcia, M.L., King, V.F. and Slaughter, R.S. (1988) *Handb. Exp. Pharmacol.*, **83**, 163–83.
Kusano, K., Barros, F., Katz, G., Garcia, M., *et al.* (1987) *Biophys. J.*, **51**, 54a.
Lodge, N.J. and van Breemen, C. (1985) *Blood Vessels*, **22**, 234–43.

Lukeman, S. and van Breemen, C. (1985) in *Recent Aspects in Calcium Antagonism* (ed. P.R. Lichtlen), Schattauer, New York, pp. 49–63.
Meisheri, K.D., Cipkus, L.A. and Taylor, C.J. (1988) *J. Pharmacol. Exp. Ther.*, **245**, 751–60.
Mestre, M., Escande, D. and Cavero, I. (1988) *Br. J. Pharmacol.*, **95**, 571P.
Miller, C., Moczydlowski, E., Latorre, R. and Phillips, M. (1985) *Nature*, **313**, 316–18.
Mullins, L.J. (1979) *Am. J. Physiol.*, **236**, C103–110.
Newgreen, D.T., Longmore, J. and Weston, A.H. (1989) *Br. J. Pharmacol.*, **96**, 116P.
Post, J.M., Smith, J.M. and Jones, A.W. (1989) *J. Pharmacol. Exp. Ther.*, **250**, 591–7.
Quast, U. (1987) *Br. J. Pharmacol.*, **91**, 569–8.
Quast, U. (1989) *Br. J. Pharmacol.*, 114P.
Quast, U. and Baumlin, Y. (1988) *Naunyn-Schmeidebergs Arch. Pharmacol.*, **338**, 319–26.
Quast, U. and Baumlin, Y. (1990) *Br. J. Pharmacol.*, submitted.
Quast, U. and Cook, N.S. (1989a) *J. Pharmacol. Exp. Ther.*, **250**, 261–71.
Quast, U. and Cook, N.S. (1989b) *Trends Pharm. Sci.*, **10**, 431–5.
Quast, U. and Webster, C. (1989) *Naunyn Schmeidebergs Arch. Pharmacol.*, **339** (Suppl R64).
Romey, G., Hugues, M., Schmid-Antomarchi, H. and Lazdunski, M. (1984) *J. Physiol.* (Paris), **79**, 259–64.
Sanguinetti, M.C., Scott, A.L., Zingaro, G.J. and Siegl, P.K.S. (1988) *Proc. Natl Acad. Sci. USA*, **85**, 8360–4.
Schmid-Antomarchi, H., DeWille, J., Fossett, M. and Lazdunski, M. (1987) *J. Biol. Chem.*, **262**, 15840–4.
Smith, C., Phillips, M. and Miller, C. (1986) *J. Biol. Chem.*, **261**, 14607–13.
Standen, N.B., Quayle, J.M., Davies, N.W. *et al.* (1989) *Science*, **254**, 177–80.
Strong, P.N., Weir, S.W., Beech, D.J. *et al.* (1989) *Br. J. Pharmacol.*, **98**, 817–26.
Sturgess, N.C., Kozlowski, R.Z., Carrington, C.A. *et al.* (1988) *Br. J. Pharmacol.*, **95**, 83–94.
Thuringer, D. and Escande, D. (1989) *Pflügers Arch.* **414** (Suppl. 1), S175.
Trieschmann, U., Pichlmaier, M., Klöckner, U. and Isenberg, G. (1988) *Pflügers Arch.*, **411**, R199.
Weir, S.W. and Weston, A.H. (1986a) *Br. J. Pharmacol.*, **88**, 113–20.
Weir, S.W. and Weston, A.H. (1986b) *Br. J. Pharmacol.*, **93**, 18P.
Wilson, C. (1988) *Br. J. Pharmacol.*, **95**, 570P.
Wilson, C. (1989) *J. Auton. Pharmacol.*, **9**, 71–8.
Wilson, C. and Hicks, F. (1989) *Br. J. Pharmacol.*, **96**, 222P.
Wilson, C., Buckingham, R.E., Mootoo, S. *et al.* (1988a) *Br. J. Pharmacol.*, **93**, 126P.
Wilson, C., Coldwell, M.C., Howlett, D.R. *et al.* (1988b) *Eur. J. Pharmacol.*, **152**, 331–9.
Winquist, R.J., Heaney, L.A., Wallace, A.A. *et al.* (1989) *J. Pharmacol. Exp. Ther.*, **248**, 149–56.
Zünkler, B.K., Lenzen, S., Manner, K. *et al.* (1988) *Naunyn-Schmiedebergs Arch. Pharmacol.*, **337**, 225–30.

27 Molecular and cellular mechanisms involved in the high resistance of neonatal brain to anoxia

Z. DRAHOTA,[1] J. MOUREK,[2] H. RAUCHOVÁ[1] AND S. TROJAN[2]
[1]Institute of Physiology, Czechoslovak Academy of Sciences, and [2]Institute of Physiology, Faculty of Medicine, Charles University, Prague, Czechoslovakia

27.1 INTRODUCTION

The fascinating problem of high resistance of newborn mammals to oxygen deficiency has been known for a long time but has remained unsolved. This phenomenon is mainly based on the specific metabolic properties of the neonatal brain but differences in structural organization and functional activity should also be considered.

Elucidation of this problem could help in understanding many functional and behavioural disorders originating during the early postnatal period of life and also provide new ideas as to improving the functional activity of brain cells during ageing.

In his classical book summarizing experimental work in this field Himwich (1951) emphasized the importance of anaerobic glycolysis as a complementary source of energy for neonatal brain functions during hypoxia or anoxia. This idea was considered for a long time as the most plausible explanation of the high resistance of neonatal brain to anoxia. However, Adolph (1948) had demonstrated that during hypoxia newborn mammals decrease their body temperature, which clearly indicates that besides glycolysis other regulatory factors are involved. Lister (1953) described the role of fetal haemoglobin in the better utilization of oxygen, and in our laboratory we found that in newborn animals partial anoxia induces a much lower oxygen consumption (Mourek, 1959) and much higher decrease of brain functional activity (Verley and Mourek, 1966).

All these findings indicate that many factors besides anaerobic glycolysis are responsible for the high resistance of neonatal organisms to anoxia. Elucidation of this mechanism could help to evaluate changes in brain energy metabolism connected with various neurological and psychiatric disorders and find possibilities for improvement of adaptive compensatory mechanisms. Data about metabolic, structural and functional maturation could also help to improve knowledge of regenerative processes during ageing. Improvement of brain energy metabolism is extremely important for the maintainance of brain regulatory and integrative functions.

27.2 UTILIZATION OF ACETOACETATE BY BRAIN CELLS DURING DEVELOPMENT

Development of oxidative metabolic pathways in the brain indicates that brain cells in the neonatal period are fully equipped for ATP production coupled to respiratory chain oxidation (Erecinska and Silver, 1989). Because the respiratory quotient of the adult brain is close to one it has been concluded that the brain cells utilize primarily substrates of carbohydrate origin (McIlwain, 1955).

In newborn suckling mammals, however, lipids are the main source of energy for functional activities because milk can be considered as a high-fat diet (Hahn *et al.*, 1981). In the liver of suckling rats fatty acids are oxidized at a much higher rate than in adult animals and acetoacetate is formed by liver cells as a substrate of lipid origin and released into blood (Drahota *et al.*, 1964). The level of acetoacetate in the blood of neonatal animals is ten times higher than in adult animals and is utilized as the energy substrate of lipid origin in conditions when the supply of energy substrates of carbohydrate origin is deficient. In this developmental period acetoacetate is utilized as energy substrate by all extrahepatic tissues including brain (Drahota *et al.*, 1965; Mourek, 1965). The high level of acetoacetate in the blood of newborn children indicates that acetoacetate is an important energy substrate of lipid origin also in the human organism (Melichar *et al.*, 1965).

That acetoacetate is an important energy substrate for newborn animals is further supported by the findings that during hypoxia or starvation its level in blood is highly increased (Koudelová *et al.*, 1986). This clearly indicates that this substrate could play an important role as a protective factor in various situations of neonatal hypoxia.

The higher capacity for acetoacetate utilization by neonatal brain cells should be expressed by a higher activity of mitochondrial enzymes responsible for its oxidation. In fact we found that mitochondria isolated from 5-day-old rats have a much higher capacity for acetoacetate oxidation than mitochondria from old animals (Table 27.1). Besides acetoacetate, also the rate of acetyl carnitine and palmityl carnitine oxidation is higher in the brain mitochondria from neonatal animals.

Our data indicate that mitochondria from newborn and adult brain can oxidize fatty acids (Krasinskaya *et al.*, 1985). Commonly it is accepted that fatty acids are not used as energy substrate in brain cells, in spite

Table 27.1 Oxidation of various substrates by newborn and adult mitochondria

Substrate (mmol/l)	*5-day-old (A)*	*90-day-old (B)*	*A/B*
Succinate (5)	19.1 ± 0.9	71.4 ± 9.2	1.3
Acetoacetate (50)	15.1 ± 3.9	2.8 ± 0.3	5.3
Acetyl carnitine (5)	11.3 ± 3.5	4.7 ± 3.5	2.4
Palmityl carnitine (0.01)	13.6 ± 4.3	9.5 ± 1.0	1.4

All values represent an average from five measurements. For each measurement mitochondria were prepared from 25 newborn and five adult brains, respectively. Mitochondria were isolated on Ficoll gradient as described by Clark and Niclas (1970). Oxygen consumption was measured using Clark oxygen electrode in medium containing (mmol/l) 100 KCl, 10 Tris-HCl, 2 $MgCl_2$, 4KH_2PO_4, 1 EDTA, 2.5 ADP and substrate as indicated. In the case of acetoacetate, acetyl carnitine and palmityl carnitine 0.3 mmol/l of malate was added.

of the fact that Beattie and Basford (1965) demonstrated that brain mitochondria are equipped with enzymes required for fatty acid oxidation. However, determination of fatty acid oxidation by brain mitochondria requires modification of experimental conditions that are commonly used for mitochondria from other tissues. Malate, as a sparker for complete fatty acid oxidation by the Krebs cycle, is important also in the case of brain mitochondria but the maximum rate of fatty acid oxidation can be observed only at a ten times lower malate concentration. Under these experimental conditions palmityl carnitine is oxidized by the liver as well as the brain mitochondria. However, the rate of palmityl carnitine oxidation is much higher in liver mitochondria and differences between the rate of palmityl carnitine oxidation in neonatal and adult mitochondria are not so pronounced as in the case of acetoacetate and acetylcarnitine (Table 27.1). This could indicate that acetoacetate is used by brain cells not only as energy substrate but also as an important substance for synthesis of brain phospholipids. Anderson and Connor (1988) demonstrated that in neonatal brain intensive synthesis of phospholipids occurs and that only a small portion of fatty acids reaching brain cells are used for oxidation. Most of them are incorporated into phospholipids because these processes occur in neonatal brain at maximum rate (Kishimoto *et al.*, 1965; Sinclair and Crawsford, 1972).

We may thus conclude that both newborn and adult brain mitochondria can oxidize fatty acids. But for neonatal brain cells acetoacetate as substrate of lipid origin is the most important substrate. It can be used for energy formation as well as for synthesis of various membrane structures and biologically active substances. It will be necessary to obtain more information about the transport of acetoacetate to the brain cells because, besides biogenesis of specific mitochondrial enzymes involved in acetoacetate oxidation, transport processes may also be involved. It is quite evident that all these processes in adult or senescent brain could supply additional energy-generating substrate for brain functional activity and can improve metabolic processes participating in regeneration of membrane structures. It is therefore evident that acetoacetate represents a very important substrate in the metabolism of neonatal brain cells. It is probable that these metabolic pathways specific for the neonatal period participate in the phenomenon of higher resistance of the neonatal brain to anoxia, but more experimental work will be required to elucidate all these mechanisms.

27.3 VARIOUS MECHANISMS INVOLVED IN THE HIGHER RESISTANCE OF NEONATAL BRAIN TO HYPOXIA

Fetal as well as neonatal hypoxia may induce irreversible disturbances in the functional activity of brain cells. It is well established that survival in hypoxia or asphyxia depends on the degree of central nervous system maturation. As brain maturation proceeds during postnatal development resistance to various types of oxygen deficiency highly decreases (Trojan and Šťastný, 1988).

One of the important factors determining the greater resistance of neonatal brain to oxygen deficiency is the ability to lower functional activity (Verley and Mourek, 1966). Also metabolic activity is highly depressed, as demonstrated by the changes in oxygen consumption (Mourek, 1959). In newborn rats a 70–80% decrease in oxygen consumption during hypoxia was found, whereas in adult animals the respiratory rate is not depressed at all.

The importance of glycolytic ATP supply during neonatal hypoxia was stressed several times (Trojan and Šťastný, 1988), but evidently this is only one of many factors participating in this phenomenon. The above-mentioned data indicate that regulatory mechanisms which can depress the rate of functional and metabolic activity are extremely important because energy utilization is highly depressed.

Another important factor that must be studied in more detail is the ability of neonatal brain to overcome the negative effects of acidosis connected with lactate production. In our previous work we found that very many enzyme reactions in neonatal brain are highly resistent to anoxia (Mourek and Trojan, 1963). Evidently in neonatal brain there are auxiliary mechanisms that help to overcome problems connected with the high rate of lactate production. Our data indicate that both a lower increase in pH and higher resistance of various enzymes to pH changes participate in these protective mechanisms.

We found that the decrease of pH in brain slices incubated in hypoxic conditions is much smaller in newborn than in adult animals (Table 27.2). Because the rate of glycolysis is not depressed it indicates the existence of a hydrogen sink in neonatal brain cells. At least two such mechanisms operate in the neonatal brain. One may be represented by the glycerolphosphate shuttle, which can transfer hydrogen to proton acceptors of the mitochondrial respiratory chain. The other could represent increased activity of hydrogenation reactions. These reactions, however, require reduced nicotinamide adenine dinucleotide phosphate (NADPH) and transhydrogenation (NADH–NADPH) as an energy-dependent process. However, Cook and Spence (1974) demonstrated that for the synthesis of fatty acids in the neonatal brain NADH can also be used as hydrogen donor. Together with our previous findings indicating a high activity of enzymes participating in lipogenesis in neonatal brain we may propose that fatty acid synthesis could represent another reaction which participates in the regulation of pH changes (Mourek, 1989).

Table 27.2 The effect of anoxia on pH in brain cortex from young and adult rats

	5-day-old		*90-day-old*	
	O_2	N_2	O_2	N_2
Control	7.26 ± 0.06	7.20 ± 0.04	7.35 ± 0.04	7.31 ± 0.05
After 30 min incubation	6.98 ± 0.05	6.92 ± 0.05	6.80 ± 0.03	6.92 ± 0.05
ΔpH	0.28	0.28	0.55	0.73
ΔpH%	−4	−4	−7.5	−10

All pH values represent an average from eight measurements. Samples of 10% brain cortex homogenate were incubated in Sörenson's medium at 37 °C in the presence of oxygen or under nitrogen.

In addition to reactions that help to reoxidize NADH and minimize lactate production it was found that in the neonatal brain there is a higher resistance of membrane-bound enzymes to pH changes induced by hypoxia, because during or after ischaemia or hypoxia the pH in brain cells rapidly decreases (Siesjö, 1978). We found that the activity of respiratory enzymes in immature brain cells is more resistant to acidification. Similarly, the activity of Na^+–K^+-ATPase in neonatal brain was more resistant to lower pH than the activity of adult brain (Table 27.3).

We may thus conclude that our data support the idea that higher resistance of the neonatal brain to anoxia or hypoxia cannot be fully

Table 27.3 The effect of pH on Na^+-K^+-ATPase activity in the brain cortex of young and adult rats

pH	*5-day-old (%)*	*90-day-old (%)*
7.4	0.175 ± 0.016 (100)	0.370 ± 0.070 (100)
7.0	0.157 ± 0.007 (90)	0.281 ± 0.035 (76)
6.4	0.171 ± 0.019 (99)	0.050 ± 0.022 (13)

All values represent an average from six measurements. The ouabain-sensitive Na^+-K^+-ATPasae (EC 3.6.1.3) was determined in a medium containing 30 mmol/l Tris-HCl, 6 mmol/l $MgCl_2$, 20 mmol/l KCl, 100 mmol/l NaCl. Ouabain where added was 1 mmol/l. Samples wre incubated for 16 min at 37 °C, pH was 7.4. Enzyme activity is expressed as μmol P min^{-1} mg $protein^{-1}$.

explained by a higher rate of ATP formation through glycolysis. Besides this, other protective mechanisms may be important. Evidently, the hydrogen sink reactions and the higher resistance of the most important metabolic reactions to pH changes are involved. However, more experimental data will be necessary before we can provide clear answers to the very important questions of how to help the neonatal brain in struggling with hypoxia and anoxia, through which mechanisms metabolic defects induced by anoxia and hypoxia can be repaired and how to use all this knowledge in protecting brain cells against degenerative processes occuring during ageing.

REFERENCES

Adolph, E.F. (1948) *Am. J. Physiol.*, **155**, 366–77.

Anderson, G.J. and Connor, W.E. (1988) *Lipids*, **23**, 286–90.

Beattie, D.S. and Basford, R.E. (1965) *J. Neurochem.*, **12**, 103–17.

Clark, J.B. and Nicklas, W.J. (1970) *J. Biol. Chem.*, **245**, 4724–31.

Cook, H. and Spence, M.W. (1974) *Biochim. Biophys. Acta*, **369**, 129–41.

Drahota, Z., Kleinzeller, A. and Kostolanská, A. (1964) *Biochem. J.*, **93**, 61–7.

Drahota, Z., Hahn, P., Mourek, J. and Trojanová, M. (1965) *Physiol. Bohemoslov.*, **14**, 134–6.

Erecinska, M. and Silver, I.A. (1989) *J. Cereb. Blood Flow Metab.*, **9**, 2–19.

Hahn, P., Seccombe, D.W. and Towell, M.E. (1981) in *Physiological and Biochemical Basis for Perinatale Medicine* (eds M. Monset-Couchard and A. Minkowski), Karger, Basel.

Himwich, H.E. (1951) *Brain Metabolism and Central Disorders*, Williams and Williams, Baltimore.

Kashimoto, J., Davies, W.E. and Radin, N.S. (1965) *J. Lipid Res.*, **6**, 532–6.

Koudelová, J., Mourek, J. and Drahota, Z. (1986) *Physiol. Bohemoslov.*, **35**, 414–19.

Krasinskaya, I.P., Mourek, J., Drahota, Z. *et al.* (1985) *Physiol. Bohemoslov.*, **34**, 121–5.

Lister, G. (1953) *Pediatr. Res.*, **18**, 172–7.

McIlwain, A. (1955) *Biochemistry and the Central Nervous System*, Churchill, London.

Melichar, V., Drahota, Z. and Hahn, P. (1965) *Biol. Neonat.*, **8**, 348–52.

Mourek, J. (1959) *Physiol. Bohemoslov.*, **8**, 106–11.

Mourek, J. (1965) *Physiol. Bohemoslov.*, **14**, 502–6.

Mourek, J. (1989) *Physiol. Bohemoslov.*, **38**, 223–30.

Mourek, J. and Trojan, S. (1963) *Physiol. Bohemoslov.*, **12**, 372–6.

Siesjö, Bo. K. (1978) *Brain Energy Metabolism*, Wiley, New York.

Sinclair, A.J. and Crawford, M.A. (1972) *FEBS Lett.*, **26**, 127–9.

Trojan, S. and Šťastný, F. (1988) in *Handbook of Human Growth and Developmental Biology*, Vol. I, Part C (eds E. Meisami and P.S. Timiras), CRC Press, Boca Raton, Florida, pp. 101–22.

Verley, R. and Mourek, J. (1966) *Physiol. Bohemoslov.*, **15**, 122–31.

28 Metabolic activation and mechanisms of MPTP toxicity

D. DI MONTE, K.P. SCOTCHER AND E.Y. WU
California Parkinson's Foundation, San Jose, California, USA

28.1 INTRODUCTION

In 1982 the abrupt onset of a severe Parkinsonian syndrome in young drug addicts led to the discovery of the neurotoxicant 1-methyl-4-phenyl-1,2,3,6-tetrahydropyridine (MPTP) (Langston *et al.*, 1983). The great interest prompted by this discovery within the scientific community as well as the general public is quite justified if the following two considerations are made. First, MPTP-induced Parkinsonism in humans is virtually indistinguishable from the idiopathic disease. Patients who injected themselves with a 'synthetic heroin' contaminated by MPTP exhibited all of the motor features of Parkinsons disease (PD), including the classic triad of tremor, bradykinesia and rigidity (Langston *et al.*, 1983). Secondly, administration of MPTP to primates has provided us with the closest animal model for PD. For example, exposure of monkeys to this compound causes a severe degeneration of dopaminergic neurones in the zona compacta of the substantia nigra (Burns *et al.*, 1983; Langston *et al.*, 1984a); this, of course, represents the major pathological feature of PD as well.

Besides the primate model, other in vivo and in vitro experimental systems have been utilized in order to clarify the molecular mechanisms of MPTP neurotoxicity; a major goal of these studies is to provide us with information on events possibly involved in the pathogenesis of PD. Shortly after its discovery, MPTP was found to be metabolized to a fully oxidized pyridinium species, the 1-methyl-4-phenylpyridinium ion (MPP^+), by monoamine oxidase type B (MAO B) (Chiba *et al.*, 1984). MPP^+ possesses most of the toxic properties attributed to its parent compound (Bradbury *et al.*, 1986; Di Monte *et al.*, 1987), raising the possibility that it may ultimately be responsible for the damaging effects of MPTP. Thus, a process of 'metabolic activation', as shown in Fig. 28.1, may be critically involved in MPTP neurotoxicity.

If MPTP conversion to MPP^+ is necessary in order to induce tissue injury, then the scheme represented in Fig. 28.1 allows us to identify two different levels at which this sequence of toxic events may be blocked. It should be possible to prevent toxicity by: (a) inhibiting the formation of the toxic metabolite, or (b) counteracting the biochemical changes which follow MPP^+ production and actually lead to cell death. These two alternative strategies of protection against MPTP toxicity will be discussed in the following paragraphs. Some of the information already available will be summarized and interesting areas for future investigations will be pointed out.

28.2 INHIBITION OF TOXIC METABOLITE FORMATION

Experiments performed with different animal models and, most convincingly, with primates have shown that inhibitors of MAO B dramatically counteract the neurotoxic effects of MPTP. In the mouse model, these inhibitors protect against the depletion of striatal dopamine caused by MPTP (Heikkila *et al.*, 1984) and, in monkeys, they prevent the neurochemical, behavioural and pathological changes due to treatment with this neurotoxicant (Langston *et al.*, 1984b). These data clearly confirm the toxic role played by MAO B in the bioactivation of MPTP.

As already mentioned, the MPTP model is expected to provide clues on the molecular mechanisms involved in the pathogenesis of PD. Hence the interaction between MAO B and MPTP and the protective effects of MAO B inhibitors against MPTP toxicity have prompted scientists to investigate the role of this enzyme in PD. On the basis of the results with the MPTP model, a clinical trial was undertaken three years ago in which the specific MAO B inhibitor, deprenyl, was administered to Parkinsonian patients at an early stage of their disease. Data from this study have been recently published (Tetrud and Langston, 1989) and reveal that deprenyl significantly slows down the progression of the disease, delaying the need for *L*-dopa therapy. More detailed information on the mechanism of action of deprenyl in Parkinsonian patients is needed in order to relate the effects of this drug to specific events involved in the pathogenesis of PD. It is likely, however, that MAO B plays a role in these events which might include a process of metabolic activation of MPTP-like toxic agent(s).

Other interesting considerations may arise from the knowledge of the MAO B-mediated bioactivation of MPTP. For example, the neurotoxic action of MPTP has been found to be age dependent, being significantly more pronounced in older than in younger mice (Langston *et al.*, 1987).

This effect has been related to the levels of MPP^+ generated in the brain after MPTP administration: because of an increased activity of MAO B, older animals produce greater levels of this toxic metabolite. An increase in MAO B activity in the brain as well as in other tissues seems to be a consistent feature of ageing and has been found in different animal species and in humans (Strolin-Benedetti and Dostert, 1989). Thus, the possibility of this enzyme being involved in the activation of toxic compounds may result in an age-related increased susceptibility to toxic injury. If neurodegenerative disorders such as PD are proven to be caused by environmental or/and endogenous toxins, the age-dependent increase in MAO B activity in the brain might contribute to make people more susceptible to these diseases.

More recently, detailed studies have been performed in order to characterize the chemical and biochemical properties responsible for the neurotoxicity of MPTP-like compounds (Youngster *et al.*, 1989; Sayre, 1989). The following factors seem to be required; (a) the MPTP analogues must be substrates for oxidation by MAO; (b) the corresponding MPP^+ analogues have to be concentrated within dopaminergic neurones by the dopamine reuptake pump: and (c) these metabolites must be able to inhibit mitochondrial respiration (this last factor will be discussed in detail in the next paragraph). Although all these properties contribute to determine the neurotoxic effects of MPTP analogues, the oxidation by MAO is likely to be the initial limiting step. The MPP^+ analogues need to be generated from the corresponding lipophilic parent compounds within the brain, or else their charged structures would limit their access to the central nervous system through the blood–brain barrier.

28.3 PROTECTION AGAINST BIOCHEMICAL CHANGES

Once the toxic metabolite is generated as a consequence of a process of metabolic activation (see Fig. 28.1), tissue damage may still be prevented by counteracting the biochemical changes that lead to cytotoxicity. In order to achieve protection at this level after MPTP exposure, it was first necessary to identify the biochemical mechanism of toxicity of MPP^+. The initial hypothesis on the mechanism of MPP^+-induced cell damage involved the generation of oxygen radicals. Although some indirect evidence has been reported in favour of this hypothesis, most experimental data speak against a role of oxidative stress in MPP^+ toxicity. Furthermore, it remains unclear how oxygen radicals would be generated in the presence of MPP^+. The proposed mechanism of ‘redox cycling’ with molecular oxygen would require the one-electron reduction of MPP^+; this reaction is very unlikely to occur in a

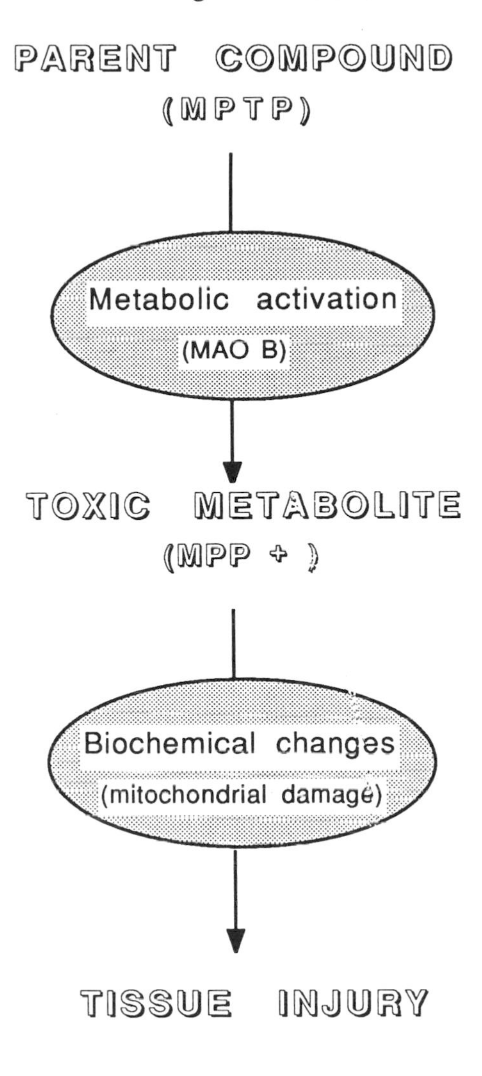

Fig. 28.1

biological system because of the low reduction potential of MPP^+ (−1.2 V) (Sayre, 1989). A comparison between the toxic properties of MPP^+ and those of compounds (such as the herbicide paraquat) able to generate oxygen radicals via redox cycling supports these conclusions (Di Monte *et al.*, 1986b).

An alternative explanation for the cytotoxic effects of MPP^+ stemmed from studies revealing that this metabolite of MPTP inhibits reduced nicotinamide adenine dinucleotide (NADH)-linked substrate oxidation in mitochondria (Nicklas *et al.*, 1985). The relatively high concentrations of MPP^+ necessary to achieve this inhibition can be reached within mitochondria since MPP^+ is actively concentrated across the inner membrane via an energy-dependent mechanism (Ramsay and Singer, 1986). As a consequence of the impaired mitochondrial function, cells would be unable to perform most of their activities, which require energy substrates in the form of ATP. A clear correlation between depletion of ATP and cytoxicity induced by MPP^+ has been demonstrated using different in vitro cellular systems (Di Monte *et al.*, 1986a; Denton and Howard, 1987; Singh *et al.*, 1988). ATP depletion has also been observed in synaptosomal preparations exposed to either MPTP or MPP^+, suggesting that mitochondrial damage may occur in nerve terminals following the generation of MPP^+ in the brain in vivo (Scotcher *et al.*, 1990).

Additional studies are needed to conclude definitively that MPTP-induced Parkinsonism is the result of a 'mitochondrial disorder'. Most of all, data obtained with different in vitro experimental systems need to be confirmed using in vivo animal models, including the primate one. In fact, as already mentioned, behavioural, neurochemical and pathological features of PD can only be reproduced when MPTP is administered to monkeys. Nevertheless, currently available data have prompted scientists to investigate the possibility that an impairment of energy production by mitochondria may be involved in the pathogenesis of neurodegenerative disorders; some very interesting results have already been produced by testing this relatively new working hypothesis.

In a preliminary communication, researchers have recently reported that the activities of mitochondrial NADH cytochrome c reductase (rotenone sensitive) and NADH coenzyme Q (CoQ) reductase are significantly reduced in autopsy specimens of the substantia nigra from patients affected by PD as compared to matched controls (Schapira *et al.*, 1989). Similarly, immunoblotting studies using antisera against complexes I, II and IV have revealed that the 30, 25, and 24 kDa subunits of complex I were moderately to markedly decreased in four out of the five Parkinsonian patients tested (Mizuno *et al.*, 1989). Finally, the activity of NADH–ubiquinone oxidoreductase has been reported to be approximately 50% lower than controls in mitochondria isolated from platelets of subjects with PD (Parker *et al.*, 1989), and a decrease in the activities of complexes I, II and IV has been measured in the skeletal muscles of Parkinsonian patients (Bindoff *et al.*, 1989). A correct interpretation of these data will be possible only when information from more extensive studies becomes available, allowing us to determine how consistent and specific these abnormalities are in PD. Nevertheless, some considerations on the possible role of mitochondrial injury in the pathogenesis of PD may be suitable at the present time. For example, the fact that inhibition of mitochondrial respiration by MPP^+ occurs at the level of complex I, the activity of which is also reduced in Parkinsonian patients, may not be coincidental and should not be overlooked.

If the mechanism of toxicity of MPTP/MPP^+ involves an impairment of mitochondrial production of energy supplies in the form of ATP, how could we possibly counteract cell damage induced by these compounds? In vitro studies have pointed out some possible strategies. An effective 'antidote' against cell death caused by either MPTP or MPP^+ has been found in compounds able to bypass the mitochondrial inhibition and to stimulate ATP production in the cytosol. Good substrates for glycolysis, such as fructose (in liver cells; Di Monte *et al.*, 1988) or glucose (in PC12 cells; Denton and Howard, 1987), protect against the ATP depletion induced by MPP^+ and, consequently, prevent cytoxicity. The design of similar protective strategies against the neurotoxic effects of MPTP exposure in vivo represents a very important area for future investigations, as it may lead to new approaches for the treatment of neurodegenerative disorders.

In conclusion, the mechanism of MPTP toxicity seems to be adequately represented by the scheme of Fig. 28.1, indicating that: (a) MPTP-induced tissue damage is a consequence of a process of metabolic activation; (b) generation of MPP^+ is the critical step leading to cytotoxicity, probably via an impairment of mitochondrial function; (c) protective strategies can be designed to prevent MPTP toxicity before and after the metabolic production of MPP^+; and (d) some of these strategies may have implications for the treatment of PD, thus emphasizing the relevance of the MPTP model of Parkinsonism.

REFERENCES

Bindoff, L.A., Birch-Machin, M., Cartlidge, N.E.F. *et al.* (1989) *Lancet*, **ii**, 49.

Bradbury, A.J., Costall, B., Domeney, A.M. *et al.* (1986) *Nature*, **319**, 56–7.

Burns, R.S., Chiueh, C.C., Markey, S.P. *et al.* (1983) *Proc. Natl. Acad. Sci. USA*, **80**, 4546–50.

Chiba, K., Trevor, A.J. and Castagnoli, N., Jr (1984) *Biochem. Biophys. Res. Commun.*, **120**, 574–8.

Denton, T. and Howard, B.D. (1987) *J. Neurochem.*, **49**, 622–30.

Di Monte, D., Jewell, S.A., Ekstrom, G. *et al.* (1986a) *Biochem. Biophys. Res. Commun.*, **137**, 310–15.

Di Monte, D., Sandy, M.S., Ekstrom, G. and Smith, M.T. (1986b) *Biochem. Biophys. Res. Commun.*, **137**, 303–9.

Di Monte, D., Ekstrom, G., Shinka, T. *et al.* (1987) *Chem. Biol. Interact.*, **62**, 105–16.

Di Monte, D., Sandy, M.S., Blank, L. and Smith, M.T. (1988) *Biochem.*

Biophys. Res. Commun., **153**, 734–40.
Heikkila, R.E., Manzino, L., Cabbat, F.S. and Duvoisin, R.C. (1984) *Nature*, **311**, 467–9.
Langston, J.W., Ballard, P., Tetrud, J.W. and Irwin, I. (1983) *Science*, **219**, 979–80.
Langston, J.W., Forno, L.S., Rebert, C.S. and Irwin, I. (1984a) *Brain Res.*, **292**, 390–4.
Langston, J.W., Irwin, I., Langston, E.B. and Forno, L.S. (1984b) *Science*, **225**, 1480–2.
Langston, J.W., Irwin, I. and DeLanney, L.E. (1987) *Life Sci.*, **40**, 749–54.
Mizuno, Y., Ohta, S., Tanaka, M. *et al.* (1989) *Biochem. Biophys. Res. Commun.*, **163**, 1450–5.
Nicklas, W.J., Vyas, I. and Heikkila, R.E. (1985) *Life Sci.*, **36**, 2503–8.
Parker, W.D., Jr, Boyson, S.J. and Parks, J.K. (1989) *Ann. Neurol.*, **26**, 719–23.
Ramsay, R.R. and Singer, T.P. (1986) *J. Biol. Chem.*, **261**, 7585–7.
Sayre, L.M. (1989) *Toxicol. Lett.*, **48**, 121–49.
Schapira, A.H.V., Cooper, J.M., Dexter, D. *et al.* (1989) *Lancet*, **i**, 1269.
Scotcher, K.P., Irwin, I., DeLanney, L.E. *et al.* (1990) *J. Neurochem.*, **54**, 1295–301.
Singh, Y., Swanson, E., Sokoloski, E. *et al.* (1988) *Toxicol. Appl. Pharmacol.*, **96**, 347–59.
Strolin-Benedetti, M. and Dostert, P. (1989) *Biochem. Pharmacol.*, **38**, 555–61.
Tetrud, J.W. and Langston, J.W. (1989) *Science*, **245**, 519–22.
Youngster, S.K., Sonsalla, P.K., Sieber, B. and Heikkila, R.E. (1989) *J. Pharmacol. Exp. Ther.*, **249**, 820–8.

29 Neurochemical changes produced by perinatal exposure to lead

E. DANIELE,[1] M.D. LOGRANO,[1] C. LOPEZ,[2] F. BATTAINI,[3] M. TRABUCCHI[3] AND S. GOVONI[1]
[1]Department of Pharmacobiology, University of Bari, Italy; [2]Institute of Pharmacology, University of Milan, Italy; [3]Chair of Toxicology, 2nd University of Rome, Italy

29.1 INTRODUCTION

Lead poisoning has been a medically recognized entity since ancient times. This fact and the wide industrial use of this metal have generated an enormous number of studies on the toxic effects of lead and on the underlying mechanisms of toxicity.

For this reason the investigations on lead may well represent a useful paradigm to deepen our understanding of the health significance of dangerous substances in general and to identify the necessary steps for their removal from the human environment. The recognition and description of signs and symptoms of lead intoxication was already correct at the end of the eighteenth century (Pleck, 1799), when it was considered as a disease affecting mostly miners, smelters and painters.

Childhood lead poisoning was first identified at the turn of this century in the United States, as a disease almost exclusively of poor minorities living in deteriorated houses painted with lead-containing paints. After 1940 the process became a public health concern following the introduction of organic lead derivatives in gasoline (Bryce, 1971); in particular, the interest in lead intoxication has grown greatly in recent decades because of the increased use and diffusion of this metal in the environment.

At present the major environmental sources of lead include household dust, interior paint removal, contaminated soil, industrial sources and improper removal of external paint. Dust and soil are contaminated by motor-car exhaust and deterioration of old paint. Lead is adsorbed into the organism largely through ingestion. Paint chips, dust and soil are frequently ingested by young children. Bellinger *et al.* (1986a) have studied the relationship between blood lead level and some sets of variables (environmental lead sources, mouthing activity and sociodemographic characteristics) in 249 middle- to upper-middle class children at 2 years of age in Boston. The results of this study indicated that two variables are particularly important in determining blood lead levels: environmental lead and mouthing activity. Lead is also adsorbed by inhalation: motor-car exhaust and dust particles smaller than 5μm are retained by the lungs.

Some factors may influence the amount of lead which will be adsorbed; the major is nutritional deficiency. In fact both animal and human studies have recently shown that deficiencies in iron, calcium and zinc will result in greater gastrointestinal adsorption of lead (Mahaffey, 1981). An iron deficiency without anaemia can multiply the gastrointestinal absorption of lead several times.

In spite of the concern about the diffused presence of lead in the environment, occupational exposure is still present in small low-technology factories.

It is well known that high level exposure induces toxic effects in multiple organs, including cardiovascular and nervous systems. In particular, at central nervous system (CNS) level neuropsychological effects of lead, including reduction in intelligence and behaviour changes, have been described. Of great interest are the neuropsychological effects at blood lead levels well below those associated with obvious toxicity symptoms. Along this line during the past ten years a number of studies have focused on low-level exposure in young children.

29.2 LEAD EXPOSURE AND INTELLECTUAL PERFORMANCE

In 1972 an English research team (David *et al.*, 1972) described the features of lead-induced mental deterioration in children. The importance of this paper was to stimulate the scientific community to extend the clinical and epidemiological observations about the effects induced by low doses of lead.

It is now recognized that lead may act as a neurotoxin at very low doses, so that its presence in the environment is a serious threat to the health of large portions of the human population. Moreover clinical and experiemental studies indicate that developing and young organisms are more susceptible to lead neurotoxicity: the signs are generally subtle, requiring the evaluation of complex behavioural and psychological tests to reveal mental impairment. In a recent review of this matter, Needleman (1983) proposed some directions to avoid methodological problems and showed that high lead subjects are deficient in IQ, auditory processing and attention, and display more disordered classroom behaviour. Their electroencephalographic pattern is different from low-lead counterparts and shows an increase of slow delta waves (Winneke *et al.*, 1983). High-lead subjects, three years after the initial study, spend more time in distracted, off-task activity. On the contrary, European studies performed by Winneke *et al.* (1983) have indicated that lead-exposed children present an impairment in visual motor integration and reaction performance, without deficits of general intelligence when the data are corrected for social background variables. The discrepancy between the US and the European studies may result from the assessment of IQ and the criteria chosen for discriminating between high and low exposure. On the other hand, the results of these studies agree that lead is an important neurotoxin and that the population at greatest risk for adverse health effects from environmental exposure to lead is children. The maximum safe blood lead concentration for an individual child is regarded to be about 25 μg/dl by the Center for Disease Control (1985). This concentration is presumed to allow an adequate margin of safety against serious health effects. However, recent epidemiological data have reported adverse intellectual effects resulting from environmental exposure at lower doses (Bellinger *et al.*, 1984, 1986ab, 1987). To stress the importance of the problem, it must be noted that three to four million children in the US have a blood lead over 25 μg/dl and 50 000 of them over 39 μg/dl. Lead is widely present in drinking water, particularly in the UK, where about 45% of the 18 million households in England, Wales and Scotland have lead in their water distribution systems (Department of the Environment, 1981). The relation between blood lead and water lead follows a cube root function and Moore (1983) has shown that, in adults, raised water lead is associated with serious health effects such as ischaemic heart disease, renal insufficiency, gout and hypertension. The same group has shown a negative correlation between maternal blood lead and gestational age at term, and positive correlation between high blood lead levels during infancy and mental retardation (Moore, 1983; Moore *et al.*, 1977). In these studies more subtle neurological indices were not examined, but biochemical effects at blood lead levels below 25 μg/dl were demonstrated: (a) inhibition of haem synthesis with measurable increases in the concentration of δ-aminolaevulinic acid (Meredith *et al.*, 1978); (b) inhibition of dihydrobiopterin reductase, which is an essential coenzyme in neurotransmitter synthesis (Blair *et al.*, 1982); (c) supranormal plasma biopterins in children (Blair *et al.*, 1982).

Moreover, in the absence of changes of biological indicators of toxicity, physiological dysfunctions and disturbances in subjective feelings of well-being may be demonstrated (Arnvig *et al.*, 1980; Hanninen *et al.*, 1980).

Immature organisms have been shown to be more susceptible to the adverse neurological effects of chronic lead exposure. Lead is absorbed and retained more completely in prenatal life (Ziegler *et al.*, 1978);

moreover, the developing CNS is more sensitive to the toxic effect of lead (Konat, 1984; Shellenberger, 1984). Recent studies have investigated the developmental effects of prenatal exposure to lead, finding a relationship between early (prenatal) exposure to lead and delayed cognitive development (Bellinger *et al.*, 1984, 1987; Dietrich *et al.*, 1987). In these investigations cord blood lead level is considered as a direct indication of the amount of lead that the infant was exposed to during prenatal life, because high correlations have been shown to exist between maternal and fetal blood lead levels (Buchet *et al.*, 1978). To emphasize the extent of the problem concerning prenatal lead exposure, research carried out on 802 infants born in Buffalo between November 1987 and April 1988 clearly showed that there were a significant number of newborns prenatally exposed to lead: 60% of the infants had cord blood lead levels in the range of 4–20 μg/dl (Shucard *et al.*, 1988). This finding, together with those of other investigations, requires further research to determine the entity and permanence of lead following prenatal exposure and the possible role of other concurrent factors.

29.3 THE DEVELOPMENT OF ANIMAL MODELS

The concern about the behavioural teratology of lead as well as the CNS effects of low-level lead exposure in children and adults have stimulated the development of animal models of lead intoxication which provide a useful tool to understand the biochemical and behavioural modification induced by lead.

Exposure of rats during development to high doses of lead produces encephalopathic lesions comparable to those observed in man (Winder *et al.*, 1983). Behavioural effects of lead may precede and be manifest in the absence of morphological damage of CNS. The effect of lead on behaviour is discussed and many discrepancies are present in the literature due to the tests performed, animals, strains and protocols of lead exposure. Many researches show lead-induced alterations of mechanisms of attention and hyperactivity (Silbergeld and Goldberg, 1973, 1975; Golter and Michaelson, 1975; Maker and Lehter, 1975; Grant *et al.*, 1975), while others do not find these modifications (Gray and Reiter, 1977; Modak *et al.*, 1975). McCarren and Eccles (1983), testing the effect of lead at three different doses present in the drinking water of lactating dams, observed an increase in activity only at a single dose and at a specific age. Other investigations indicate that the effect of lead is complex, resulting in an increase of behavioural reactivity and in a disrupted performance in complex discrimination tasks than exclusively in hyperkinesis (Winneke *et al.*, 1977, 1982). Moreover, in studies using a battery of sensorimotor tests and pharmacological manipulations, Fox (1979) has shown that lead-exposed rats exhibit delayed maturation, altered developmental patterns and long-term CNS disturbances. Lead-exposed rats also exhibit an altered drug responsivity: in particular they present a paradoxical or a reduced stimulatory response to amphetamine and fenfluramine. Zenick and Goldsmith (1981) have suggested that the development of hyposensitivity occurs as a consequence of the exposure of the animal during neonatal life, since the alteration was present also in animals which were not exposed to lead after weaning. Some behavioural studies have been carried out on primates (cynomolgus monkeys). Lead (daily ingestion in milk for one year after birth, producing levels of 30–100 μg/dl) alters the social development of infant monkey (Bushnell and Bowman, 1979). Another study, designed to assess spatial learning and memory in adult monkeys exposed to lead from birth (blood lead concentrations from 3 to 25 μg/dl), demonstrated an impairment in their ability to learn the alternation task. The most pronounced effects were manifested during acquisition of performance and during the longer delay values (Rice and Karpinski, 1988).

The behavioural effects of lead have stimulated research on the underlying neurochemical changes.

The general biochemical effects of lead on the CNS include modifications in protein or DNA concentrations and decrease in respiratory activity; but generally these changes occur at high levels of exposure. Other biochemical modifications are changes in lipid composition and competition with calcium effects on enzyme systems (Winder and Kitchen, 1984).

29.4 NEUROCHEMICAL CORRELATES OF LEAD EXPOSURE

The neurochemical effects of lead have been demonstrated to involve almost all the neurotransmitters that have been so far investigated (Winder and Kitchen, 1984). Many studies have been carried out on the effect of lead on acetylcholine, catecholamine, γ-aminobutyric acid (GABA), serotonin and peptide transmitters.

However, it is not known whether lead-induced changes in neurotransmitter functions are primary effects of lead or occur as the result of modifications induced by the metal on systems particularly susceptible to the toxic action that then drive compensatory changes in other neurotransmitters. A peculiar observation is that the alterations in neurotransmitter function are area specific.

Dubas and Hrdina (1978) found that serotonin levels of lead-exposed rats at 8 weeks of age were decreased in cortex and hypothalamus and unmodified in striatum and midbrain. Another study (Govoni *et al.*, 1979) showed that dopamine turnover is decreased in striatum and increased in nucleus accumbens of lead-exposed rats. In the same animals, GABA-binding sites were found decreased in striatum, increased in cerebellum and unchanged in cortex and hypothalamus (Govoni *et al.*, 1980; Memo *et al.*, 1980). The selective action of lead may reflect regional accumulation and distribution of the metal or differences in the mechanisms regulating synthesis, release and uptake of the involved neurotransmitter in the various brain areas.

Experimental data support both the hypotheses. Collins *et al.* (1984) have demonstrated that hippocampus accumulates lead, confirming previous findings of selective lead accumulation in this brain region (Fjerdingstad *et al.*, 1974); however, the reason for this preferential accumulation is unknown.

On the other hand, the second possibility is supported by several lines of evidence derived from in vivo and in vitro studies. An interesting example of this is given by the interaction of lead with dopaminergic transmission in different brain areas such as striatum and nucleus accumbens. The biochemical data indicate that lead induces an impairment of dopaminergic transmission both at presynaptic (dopamine and DOPAC (dihydroxy phenyl acetic acid) content, dopamine turnover) and post-synaptic level (D1 and D2 receptors, dopamine-stimulated and dopamine inhibited adenylate cyclase). This is supported by behavioural data and motor abnormalities induced by lead in laboratory rodents (Silbergeld and Goldberg, 1973; Zenick and Goldsmith, 1981; Moresco *et al.*, 1988).

Table 29.1 Area specific lead-induced changes in dopaminergic transmission

Parameter	*Striatum*	*N. Accumbens*	*Hypothalamus*	*Pituitary*
- DA content	-	-	U	-
- DOPAC cont.	D	I	D	-
- DA turnover	D	I	-	-
- DA uptake	D	I	-	-
- Sodium dependent cocaine binding	D	-	-	-
- Sodium independent cocaine binding	U	U	-	-
- SCH 23390 binding (D1 receptors	U	U	-	-
- Sulpiride displaceable spiperone binding (D2 receptors)	I	D	-	D
- DA stimulated adenylate cyclase	U	U	-	-
- DA inhibited adenylate cyclase	U	D	-	D

DA = Dopamine; DOPAC = Dihydroxyphenylacetic acid.
D = decreased, I = increased, U = unmodified.
Reference to the original data are in Govoni *et al.*, 1979, 1980; Lucchi *et al.*, 1986; Memo *et al.*, 1980; Missale *et al.*, 1984; Moresco *et al.*, 1988.

29.5 EXAMPLES OF SELECTIVE NEUROCHEMICAL EFFECTS OF LEAD

29.5.1 Dopaminergic transmission

Results from in vivo lead exposure (summarized in Table 29.1) have shown that lead intoxication depresses dopamine turnover in the striatum, enchances it in the nucleus accumbens and the frontal cortex and does not induce changes in the substantia nigra (Govoni *et al.*, 1979). The uptake of dopamine is decreased in striatum and increased in nucleus accumbens (Govoni *et al.*, 1986).

These presynaptic modifications are followed at the postsynaptic level by adaptive changes of the dopamine receptors and of the adenylate cyclase system.

The functional state of adenylate cyclase, basal or dopamine stimulated, is not affected in any of the areas examined after lead intoxication (Govoni *et al.*, 1980). On the contrary, dopamine-inhibited adenylate cyclase is unchanged in striatum and decreased in nucleus accumbens (Govoni *et al.*, 1986). Dopamine receptors present a complex pattern of variations after in vivo lead exposure which seem to represent an adaptive response of the postsynaptic membrane to the altered activity of the presynapsis. Along this line, the effects of doses of lead not causing gross toxicity were investigated both on biochemical and behavioural markers of D1 and D2 receptors; [^{3}H]SCH 23390 binding parameters are unmodified by the treatment both in striatum and in nucleus accumbens.

In contrast (−) sulpiride-displaccable [^{3}H]spiroperidol binding is increased in striatum and decreased in nucleus accumbens. The behavioural data are consistent with the neurochemical alterations. The exploratory activity was similar in both groups, whereas basal motility values were significantly higher in lead-treated rats, possibly reflecting the higher dopamine turnover in mesolimbic areas such as in nucleus accumbens. The D1 agonist SKF 38393 did not modify locomotor activity either in control or in lead-treated rats. Apomorphine induced hypomotility in both groups, but this response was attenuated in the lead-treated group.

The attenuation of apomorphine-induced hypomobility may occur as a consequence of the subsensitivity of D2 receptors in nucleus accumbens. No behavioural expression of D2 receptor increase was found in the striatum: the lack of correlation between behavioural and biochemical data in the case of striatal D2 receptor supersensitivity may be explained by considering that the final effect of apomorphine is the result of the stimulation of D2 receptors in different brain areas differently modulated by lead.

The effects of lead on dopaminergic transmission are not limited to the striatum and nucleus accumbens but involve brain areas particularly important for neuroendocrine control. The tubero-infundibular dopamine neurones, controlling prolactin release, seem to be highly sensitive to this metal. A decrease in dopamine turnover at the hypothalamic level in lead-exposed rats has been shown (Govoni *et al.*, 1978).

Lead also induced changes in dopamine recognition sites, characterized by a decreased [^{3}H]sulpiride binding in the pituitary (Govoni *et al.*, 1984) (Table 29.1). In this area a reduced ability of bromocriptine to inhibit adenylate cyclase activity is observed, as well as a reduced dopamine inhibition of vasoactive intestinal peptide-stimulated cAMP formation (Lucchi *et al.*, 1986).

All these changes may be explained on the basis of a decreased number of D2 receptors. It is difficult to establish the sequence of events leading to changes of dopamine metabolism and receptors. Pituitary dopamine receptors are in fact subsensitive in spite of the decreased dopaminergic tone in the hypothalamus. It is possible that the decreased sulpiride binding and the consequent reduced ability of dopamine to inhibit cAMP formation are due to a direct effect of lead on the receptor protein.

The lead-induced dysfunction of the hypothalamic dopaminergic system is in agreement with the increase in serum prolactin concentrations observed in lead-treated rats (Govoni *et al.*, 1978). The interference of lead with central neuroendocrine processes may contribute to the changes in sexual development and reproductive ability observed during lead intoxication in rodents (Jacquet *et al.*, 1977). These observations have stimulated clinical investigations in order to establish whether lead-exposed workers have abnormal plasma prolactin concentrations.

Plasma prolactin, zinc protoporphyrin (Zpp) and blood lead were measured in the blood of workers of pewter factories (Govoni *et al.*, 1987). Although serum prolactin was in the normal range (1–10 ng/ml) in all subjects, dividing the population under test into groups having blood lead and Zpp below and above the normal safe level (40 μg/dl), the mean value for prolactin concentration of the subgroup having Zpp and blood lead higher than 40 μg/dl was significantly higher (+ 47%) than that observed in the group of workers having Zpp and blood lead less than 40 μg/dl. Similar findings, i.e. a correlation between blood lead and plasma prolactin levels, were found in a recent study carried out on occupationally lead-exposed workers in Argentina by Roses *et al.*, (1989).

In conclusion, lead may potentially alter the secretion of pituitary hormones and this may have a role in the disturbances of subjective feelings of well-being reported by exposed workers. On the other hand, a change in plasma prolactin concentrations may represent an early biological marker indicating a central effect of lead for threshold levels of exposure, without relevant clinical symptoms.

29.5.2 Calcium metabolism

Part of the neurotoxic effect of lead may be attributed to the ability of this ion to interact with other positively charged physiological ions (Silbergeld, 1982) or their carriers (Hexum, 1984); on this line the role of calcium has been particularly investigated. Lead alters calcium metabolism, cellular calcium homeostasis (Pounds, 1984; Audesirk, 1985) and movements at presynaptic level (Silbergeld, 1977; Winder, 1983; Cooper and Manalis, 1983); the metal added in vitro to striatal synaptosomes is able to promote the influx of ^{45}Ca into the terminals (Silbergeld, 1977).

Further investigations about the mechanisms by which lead interacts with synaptic calcium movements have shown an increase in [^{3}H]nitrendipine (a specific probe for L-type calcium channels) binding to striatal membranes (Rius *et al.*, 1984). However, when the membranes are washed with chelating agents, the number of binding sites of lead-exposed rats returns to control values, indicating that the increase is due to the presence of lead in the membranes. Moreover, the data of recent research (Rius *et al.*, 1988) indicate that this effect displays regional selectivity. In fact, no changes in [^{3}H]nitrendipine binding have been observed in hippocampus of treated rats, even though this area is able to accumulate lead (Collins *et al.*, 1984).

In addition, in vitro exposure to lead stimulated nitrendipine binding maximally in striatum, while in hippocampus almost no stimulation occurs. This selective sensitivity to lead of [^{3}H]nitrendipine binding sites is in agreement with previous observations indicating that lead has different effects on neurotransmitters according to the brain area and may indeed represent one of the fundamental mechanisms mediating the selective neurotoxicity of lead. The action of lead in in vitro preparations and the observation that membranes prepared from in vivo treated animals presented binding characteristics similar to those of controls, after in vitro treatment with chelating agents, suggests a direct action of lead on voltage-sensitive calcium channels.

Electrophysiological studies on neuromuscular junctions (Cooper and Manalis, 1983) confirm this hypothesis, showing that lead first inhibits evoked release and then stimulates the rate of spontaneous release of the neurotransmitter. This has been explained by hypothesizing that lead blocks calcium channel on or very near the external surface of the presynaptic membrane and then enters the intracellular components, where it induces an increase of free calcium concentration by altering the Ca^{2+} sequestration by mitochondria and other organelles.

Although there are many inconsistencies about this hypothesis, there is agreement that mitochondria are key structures in the toxic action of lead. Recent investigations indicate that interference between lead and calcium may occur also at the calmodulin-sensitive processes in peripheral models such as erythrocytes (Goldstein and Ar, 1983) and in brain (Chao *et al.*, 1984).

29.6 CONCLUDING REMARKS

At present the molecular mechanisms of lead toxicity are not fully elucidated; however, the available experimental and clinical data strongly indicate the necessity to reduce exposure to this toxin.

The complex pattern of lead-induced behavioural dysfunctions is paralleled by complex neurochemical changes that occur at blood lead levels well below those associated with obvious symptoms of toxicity. In particular, lead seems to act as a specific neurotoxin able to interfere with complex molecular mechanisms, rather than as a non-specific protein denaturating agent.

In this view an interference of lead with other factors (pharmacological treatment, health status, ageing, environmental agents) affecting neuronal systems modified by the metal should be expected.

As an example, since it has been shown that lead affects learning performance, what will be the combined effect of ageing and lead exposure on cognitive functions? The answer to questions such this will require a further step in the evolution of neurotoxicology.

Toxicological studies have indeed proceeded from the investigation of gross toxicity to the measurement of subtle neurobehavioural and neurochemical processes. Now the time has come to study the action of subliminal exposure to neurotoxic substances, not only on the early development but also in respect to the long-term consequences during adulthood and senescence when age-related physiological changes may unmask unforeseen toxicity.

REFERENCES

Arnvig, E., Grandjean, P. and Backman, J. (1980) *Toxicol. Lett.*, **5**, 399–404.
Audesirk, G. (1985) *Prog. Neurobiol.*, **24**, 199–231.
Bellinger, D.C., Needleman, H.L., Leviton, A. *et al.* (1984) *Neurobehav. Toxicol. Teratol.*, **6**, 387–402.
Bellinger, D., Leviton, A. and Rabinowitz, M. (1986a) *Paediatrics*, **77**, 826–33.
Bellinger, D., Leviton, A., Needleman, H. *et al.* (1986b) *Neurobehav. Toxicol. Teratol.*, **8**, 151–61.
Bellinger, D., Leviton, A., Waternaux, C. *et al.* (1987) *N. Engl. J. Med.*, **316**, 1037–43.
Blair, J., Hilburn, M., Leerning, R. *et al.* (1982) *Lancet*, **i**, 964–68.
Bryce, S.D. (1971) *Chem. Br.*, **7**, 54–60.
Buchet, J.P., Roels, H., Hubermont, G. and Lauwerys, R. (1978) *Environ. Res.*, **15**, 494–503.

Burchfiel, J., Duffy, F., Bartels, P.H. and Needleman, H.L. (1980) *Low Lead Exposure* (ed H.L. Needleman), Raven Press, New York, p. 75.
Bushnell, P.J. and Bowman, R.R. (1979) *Neurobehav. Toxicol.*, **1**, 207–19.
Center for Disease Control (1985) *Morbidity and Mortality Weekly Report*, United States, **34**, 66–73.
Chao, S.H., Suzuki, Y., Zysk, J.R. and Cheung, W.Y. (1984) *Mol. Pharmacol.*, **26**, 75–82.
Collins, M.F., Whittle, E. and Singhal, R.L. (1984) *Can. J. Physiol. Pharmacol.*, **62**, 430–5.
Cooper, G.P. and Manalis, R.S. (1983) *Biochem. Neurotoxicol.*, **4**, 69–84.
David, O., Clark, J. and Voeller, K. (1972) *Lancet*, **ii**, 900–3.
Department of the Environment and Welsh Office (1981) *Lead in the Environment*, (DOE Circular no. 22/82; WO Circular no. 31/82). HM Staionery Office, London.
Dietrich, K.N., Kraff, K.M., Bornschein, R.L. *et al.* (1987) *Paediatrics*, **80**, 721–30.
Dubas, T.C. and Hrdina, P.D. (1978) *J. Environ. Pathol. Toxicol.*, **2**, 473–7.
Fjerdingstad, E.J., Danscher, G. and Fjerdingstad, E. (1974) *Brain Res.*, **80**, 350–55.
Fox, D.A. (1979) *Neurobehav. Toxicol.*, **1**, 193–206.
Goldstein, G.W. and Ar, D. (1983) *Life Sci.*, **33**, 1001–6
Golter, M. and Michaelson, I.A. (1975) *Science*, **187**, 359–61.
Govoni, S., Montefusco, O., Spano, P.F. and Trabucchi, M. (1978) *Toxicol. Lett.*, **2**, 333–7.
Govoni, S., Memo, M., Spano, P.F. and Trabucchi, M. (1979) *Toxicology*, **12**, 343–9.
Govoni, S., Memo, M., Lucchi, L. *et al.* (1980) *Pharmacol. Res. Commun.*, **12**, 447–60.
Govoni, S., Lucchi, L., Battaini, F. *et al.* (1984) *Toxicol. Lett.*, **20**, 237–41.
Govoni, S., Lucchi, L., Missale, C. *et al.* (1986) *Brain Res.*, **381**, 138–42.
Govoni, S., Battaini, F., Castelletti, L., *et al.* (1987) *J. Environ. Pathol. Toxicol. Oncol.*, **7**, 13–15.
Grant, L.D., Howard, J.L., Alexander, S. and Krigman, M.R. (1975) *Environ. Health Perspect.*, **10**, 267–9.
Gray, L.E. and Reiter, L.W. (1977) *Toxicol. Appl. Pharmacol.*, **41**, 140–9.
Hanninen, H., Mantere, P., Hernberg, S. *et al.* (1980) *Neurotoxicology*, **1**, 333–47.
Hexum, T.D. (1984) *Biochem. Pharmacol.*, **23**, 3441–7.
Jacquet, P., Gerber, G.B., Leonard, A. and Maef, J. (1977) *Experentia*, **33**, 1375–7.
Konat, G. (1984) *Neurotoxicology*, **5**, 87–96.
Lucchi, L., Govoni, S., Memo, M. *et al.* (1986) *Toxicol. Lett.*, **32**, 255–60.
Mahaffey, K. (1981) *Nutr. Rev.*, **39**, 353–62.
Maker, H.S., Lehrer, G.M. and Silides, D.M. (1975) *Environ. Res.*, **10**, 76–9.
McCarren, M. and Eccles, C.V. (1983) *Neurobehav. Toxicol. Teratol.*, **5**, 527–32.
Memo, M., Lucchi, L., Spano, P.F. and Trabucchi, M. (1980) *Toxicol. Lett.*, **6**, 427–32.
Meredith, P., Moore, M., Campbell, B. *et al.* (1978) *Toxicology*, **9**, 1–9.
Missale, C., Battaini, F., Govoni, S. *et al.* (1984) *Toxicology*, **33**, 81–90.
Modak, A.T., Weintraub, S.T. and Stavinoha W.B. (1975) *Toxicol. Appl. Pharmacol.*, **34**, 340–5.

Moore, M. (1983) in *Lead Versus Health: Sources and Effects of Low Level Lead Exposure.* (eds. M. Rutter, and R. Russell Jones), Wiley, Chichester.
Moore, M., Meredith, P. and Goldberg, A. (1977) *Lancet*, **i**, 712–19.
Moresco, R.M., Dall'Olio, R., Gandolfi, O. *et al.* (1988) *Toxicology*, **53**, 315–22.
Needleman, H.L. (1983) *Neurotoxicology*, **4**, 121–8.
Pleck, G.J. (1799) *Tossicologia ossia dottrina intorno i veleni ed i loro antodoti*, Venezia.
Pounds, J.C. (1984) *Neurotoxicology*, **5**, 295–332.
Rice, D.C. and Karpinski, K.F. (1988) *Neurotoxicol. Teratol.*, **10**, 207–14.
Rius, R.A., Lucchi, L., Govoni, S. and Trabucchi, M. (1984) *Brain Res.*, **332**, 180–3.
Rius, R.A., Govoni, S., Bergamaschi, S. *et al* (1988) *The Science of the Total Environment*, **71**, 441–8.
Roses, O.E., Alvarez, S., Conti, M.I. *et al.* (1989) *Bull. Environ. Contam. Toxicol.*, **42**, 438–42.
Shellenberger, M.K. (1984) *Neurotoxicology*, **5**, 177–212.
Shucard, J.L., Shucard, D.W., Patterson, R. and Guthrie, R. (1988) *Neurotoxicology*, **9**, 317–26.
Silbergeld, E.K. (1977) *Life Sci.*, **20**, 309–18.
Silbergeld, E.K. (1982) *Mechanisms of Actions of Neurotoxic Substances*, Raven Press, New York, pp. 1–23.
Silbergeld, E.K. and Goldberg, A.M. (1973) *Life Sci.*, **13**, 1275–83.
Silbergeld, E.K. and Goldberg, A.M. (1975) *Neuropharmacology*, **14**, 431–44.
Winder, C. (1983) *Biochem. Pharmacol.*, **31**, 3717–21.
Winder, C. and Kitchen, I. (1984) *Prog. Neurobiol.*, **22**, 59–87.
Winder, C., Garten, L.L. and Lewis, P.D. (1983) *Neuropathol. Appl. Neurobiol.*, **9**, 87–108.
Winneke, G., Brockhaus, A. and Baltissen, R. (1977) *Arch. Toxicol.*, **37**, 247–63.
Winneke, G., Lilienthal, H. and Werner, W. (1982) *Arch. Toxicol.*, **5**, 84–93.
Winneke, G., Kraemer, U., Brockhaus, A. *et al.* (1983) *Int. Arch. Occup. Environ. Health*, **51**, 231–3.
Zenick, H. and Goldsmith, M. (1981) *Science*, **212**, 569–71.
Ziegler, E., Edwards, B., Jensen, R. *et al.* (1978) *Paediatr. Res.*, **12**, 29–34.

30 Electrophoretic studies of the interaction of neuroleptic drugs with brain tissue

J.W. GORROD, M.S. FELDMAN, A. KLER AND A. ROSEN
Chelsea Department of Pharmacy , King's College London (University of London), Manresa Road, London SW3 6LX, UK

30.1 INTRODUCTION

The most obvious side effects of neuroleptic drugs are those on the function of the central nervous system (CNS), the endocrine system, pigmentation in the eye, photosensitization of the skin, and impairment of liver and gastrointestinal function.

Side effects in the CNS include Parkinson's syndrome (akinesia, rigidity and tremor), akathaesia (inability to sit still, intolerance of inactivity, restless movement), tardive dyskinesia (symptoms include smacking and licking of lips, sucking and chewing movements, rolling and protrusion of the tongue, grotesque grimaces, and spastical facial contortions) and dystonic reactions (Burki, 1979; Christensen *et al.*, 1970; Maickel *et al*, 1974; Zirkle and Kaiser, 1974). Whereas most of these therapeutically undesirable effects are reversible and disappear after discontinuation of the drug, or lowering of drug dosage, tardive dyskinesia, which occurs later in the course of treatment with neuroleptic drugs, usually persist for many years and may be permanent.

Not all neuroleptic drugs induce these side effects to the same extent (Table 30.1). Thioridazine (TRZ) and clozapine (CLZ) exhibit a lower

Table 30.1 Comparative extrapyramidal dysfunction of five neuroleptic drugs.

Drug	*Extrapyramidal dysfunction symptoms*[a]*
Fluphenazine	Considerable
Haloperidol	Considerable
Chlorpromazine	Moderate
Thioridazine	Slight
Clozapine	Slight

[a]Data obtained from the British National Formulary

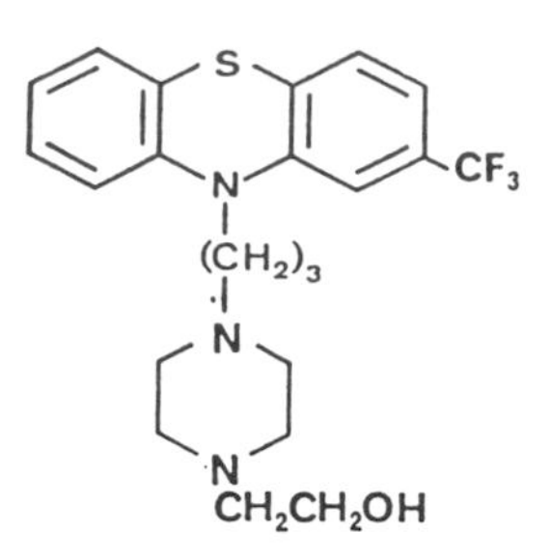

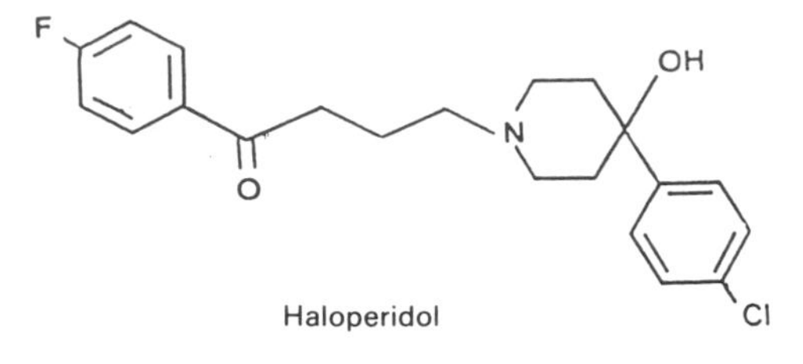

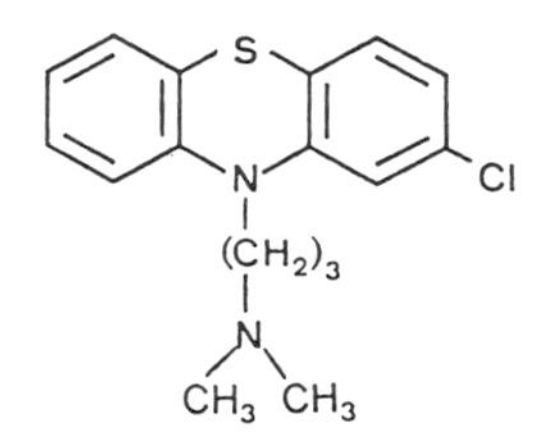

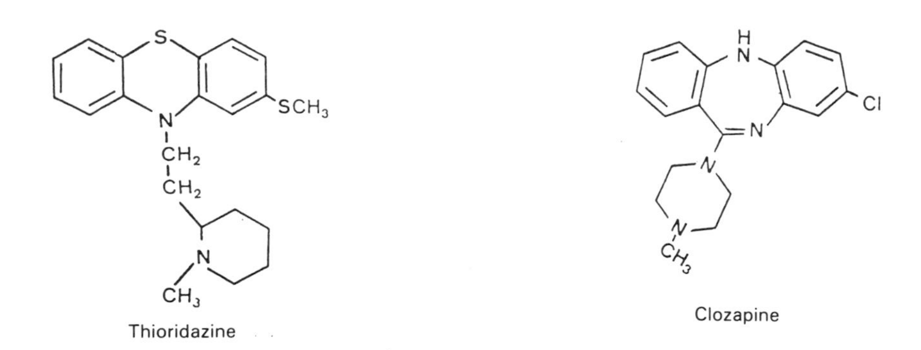

Fig. 30.1 Structures of neuroleptic drugs used in this study.

extrapyramidal liability, whilst some phenothiazines, such as chlorpromazine (CPZ) and fluphenazine (FPZ) and butyrophenones, i.e. haloperidol (HPD), have higher neurotoxic incidence (Zirkle and Kaiser, 1974). These extrapyramidal symptoms may be related to their action on the basal ganglia, particularly on the neostriatum (Zirkle and Kaiser, 1974). Lesions which impair function of the dopaminergic pathway from the substantia nigra to the striatum have been associated with Parkinsonism.

We have been concerned with the interaction of a number of drugs (Fig. 30.1) with neuromacromolecules in various regions of the brain in the presence of an active drug-metabolizing system as a possible mechanism for the neurotoxicity of these drugs and we report the following findings obtained using electrophoretic techniques.

30.2 MATERIALS AND METHODS

All laboratory animals were obtained from the Animal Unit, King's College (University of London), London, UK. General laboratory reagents were from the usual commercial sources; sodium dodecyl sulphate (SDS) was from Sigma Chemical Co. UK, glucose-6-phosphate (disodium salt) (G6P), glucose-6-phosphate dehydrogenase (from yeast) (G6PD) and nicotinamide adenine dinucleotide phosphate (NADP) (disodium salt) were obtained from Boehringer Ltd, UK. Scintillation chemicals were obtained from BDH Chemicals Ltd, UK.

CPZ was donated by May and Baker Ltd, UK (5a,6,7,8,9,9a-^{14}C). CPZ was purchased from Amersham Interntional p.l.c., Feltham, UK. TRZ, CLZ and generally tritiated analogues were donated by Sandoz Products Ltd, Feltham, UK. HPD and generally tritiated HPD were donated by Janssen Pharmaceuticals Ltd, Marlow, UK. FPZ and its tritiated analogue were obtained from E.R. Squibb and Sons Ltd, Hounslow, UK.

30.2.1 Preparations of tissue fractions

Laboratory animals were sacrificed by cervical dislocation, exsanguinated, the brain immediately removed by dissection and placed in ice-cold isotonic Tris/KCl buffer (pH7.4). All subsequent operations were performed at 0–4°C. Tissue was blotted dry and weighed (wet weight) and homogenized in two volumes of fresh cold Tris /KCl buffer. The brains were dissected into seven major regions (amygdala, cerebellum, corpus striatum, cortex, hypothalamus, medulla and midbrain). The excised brains were individually placed on a glass dissecting tray chilled by ice. The cerebellum (Figure 30.21) was gently teased away from the brain stem to remove the medulla, including the pons, a cut at 30° to the vertical (line 2) was made. The remainder of the brain was placed on its dorsal surface and a vertical slice between the optic chiasma and the hypothalmus (line 3) separated the front section, containing most of the corpus striatum, from the rest of the brain. The two corpus striatum regions could easily be seen and all excess tissue was dissected away. The remainder of the brain was orientated on the surface exposed by slice (3). A vertical slice was made (4) separating the ventral third of the brain from the rest. This contained the two amygdala regions which

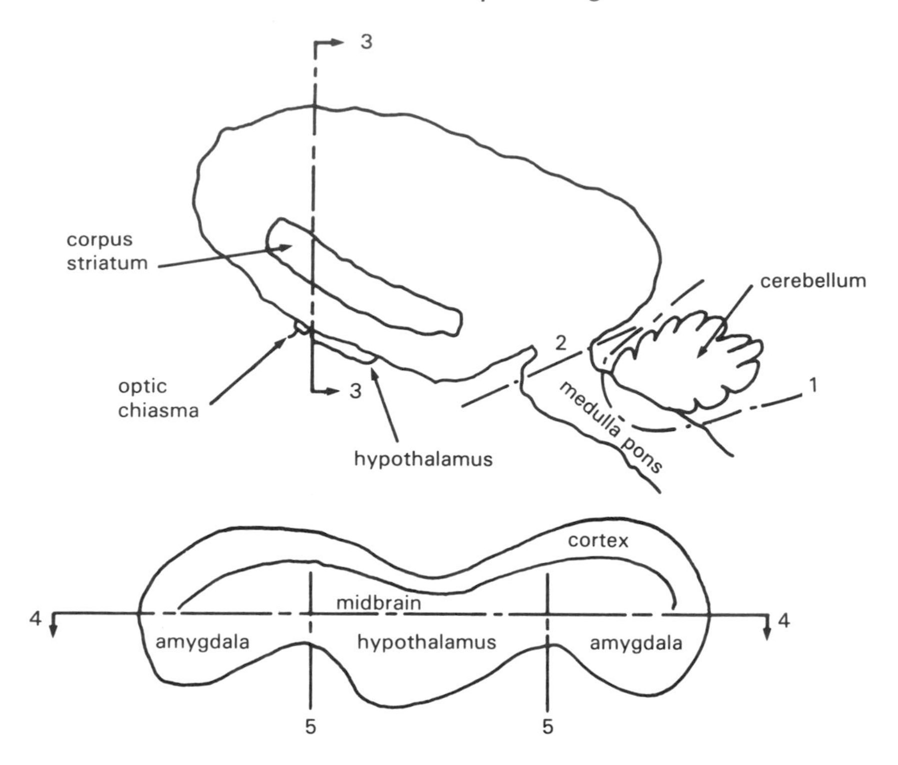

Fig. 30.2 Sites of dissection of rat brain to obtain discrete brain regions.

lie lateral to the hypothalamus. After removing the amygdala regions, the excess tissue surrounding the hypothalamus was dissected away. The cortex contained in the last section, was gently eased away from the midbrain.

All dissected tissue was inspected and any large blood vessels or blood clots removed. All subsequent operations were carried out at 0–4°C. Tissue was blotted dry and weighed (wet weight) and homogenized in two volumes of fresh ice cold isotonic Tris/KCl using an Ultra Turrax tissue homogenizer. This homogenate was transferred to plastic tubes and centrifuged at 9000 rev/min for 20 min. The supernatant, or '9000 × *g* fraction', was centrifuged at 40 000 rev/min (140 000 × *g*) for 70 min. The resulting supernatant (soluble) fraction was removed and the microsomal pellet resuspended in fresh buffer and further purified by recentrifugation at 140 000 × *g* for 70 min. The 'washed microsomes' were resuspended in fresh buffer (1–2 ml/g original wet weight).

30.2.2 Standard incubations

Incubations were carried out in 12 ml Steralin blood sample tubes at 37 °C using a shaking water bath. A typical microsomal incubation mixture contained substrate dissolved in 0.05 ml ethanol, 2 μmol NADP, 10 μmol G6P, 3 units G6PD, 20 μmol $MgCl_2$ in 2.0 ml 0.2 M phosphate buffer (pH 7.4) and 1.0 ml tissue. Prior to incubation, the cofactors were pre-incubated for 5 min to allow the generation of NADPH. The tissue and substrate were then added.

30.2.3 Electrophoretic examination of drug/protein interaction in several brain regions

Rat brains were removed and dissected into seven major regions (amygdala, cerebellum, corpus striatum, cortex, hypothalamus, medulla and midbrain), and the microsomal and soluble fractions prepared as described. Protein concentrations were determined using the method of Read and Northcote (1981). The interaction of the neuroleptic drugs with brain fractions was examined as shown in the flow diagram (Fig. 30.3). In order to investigate the influence of microsomal metabolism on any

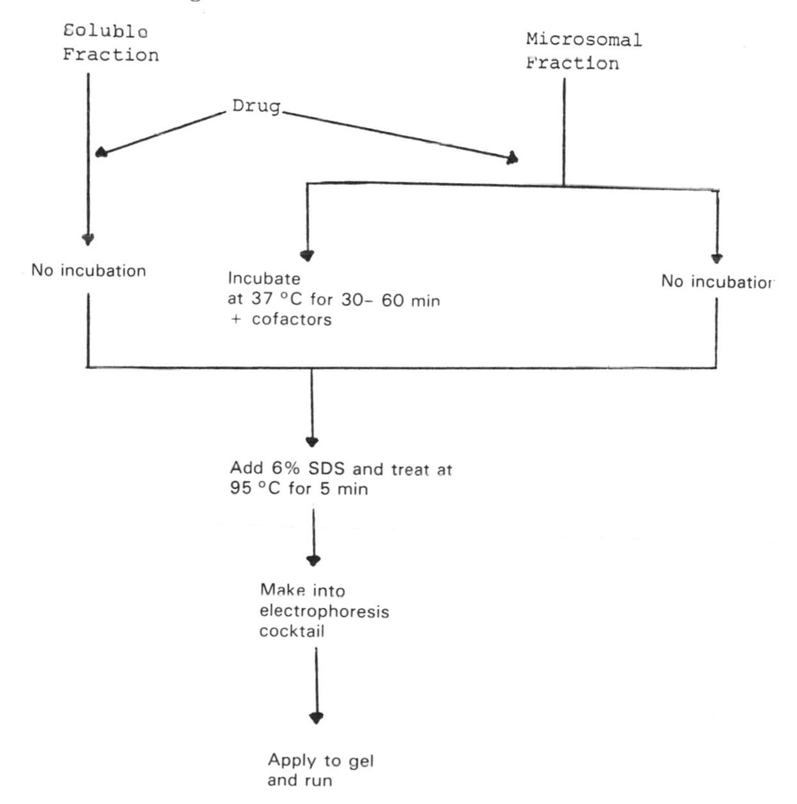

Fig. 30.3 Flow diagram for the preparation of samples for electrophoretic examination.

interactions, labelled drug was incubated for 30 min with microsomal preparations. In all cases a 2/1 molar ratio of drug to protein, assuming a protein average molecular weight of 60 000, was used. Incubations were terminated by addition of 1.0 ml 6.0% SDS and held at 95 °C for 5 min. An aliquot (0.3 ml) of the dissolved protein sample was treated with buffer to obtain a final composition of 2.0% SDS, 10.0% glycerol, 0.001% bromophenol blue, 0.0625 M Tris/HCl, pH 6.8, 1 mM ethylenediamine-tetraacetic acid (EDTA), 2-mercaptoethanol 10%, and 30.0% protein. The cocktail was then held for 5 min to 95°C to ensure the sample was completely dissolved. Cocktails were subjected to electrophoresis on gels ranging from 9 to 12% polyacrylamide (usually 10.5%). Resolving gels (12.0 cm, 0.75 mm thick) were buffered in 3.0 M Tris/HCl, pH 8.8, and 25% polyacrylamide plugs sealed the bottom of the plates. After polymerizing the resolving gel with 1.0% (w/v) *N,N,N′N′*-tetramethyl-methanediamine and 2.0% (w/v) ammonium persulphate, a 5.0% stacking gel (3.0 cm, 0.75 mm thick) was cast on top. Aliquots (50–150 μl) of the electrophoresis cocktails were placed in pre-formed wells in the stacking gel. Gels were developed at 35 mA/gel, constant current (Shandon Vokam 400-100 power pack), in running buffer (0.025 M Tris, 0.192 M glycine, 0.1% SDS, pH 8.3) until the tracking dye migrated to within 0.5 cm of the bottom of the gel, essentially as described by Kler and Rosen (1985). Gels to be photographed were stained and fixed with 7.0% acetic acid, with 35.0% methanol, 0.1% Coomassie Brilliant Blue dye plus 3.0% formaldehyde. For determination of radioactivity in gel electrophoresis samples, 1 cm wide tracks were cut from gel and each divided into 30 segments (0.5 cm). Each segment was placed in 4.0 ml of cocktail/Scintrain tissue solubilizer (10 : 1), shaken and left in the dark for 2 h prior to counting. Radioactivity measurements were carried out using a Packard Tricarb Scintillation Counter Model 314EX.

30.3 RESULTS AND DISCUSSION

Treatment of biological samples with SDS at 95 °C and subsequent electrophoresis is clearly a drastic technique. It might well have been supposed that all drugs would be stripped from the protein during denaturation and run at the front of the electrophoretogram. In practice, incubation of the brain region preparations with the various neuroleptic drugs gave levels of radioactivity distributed along each electrophoresis track which afforded drug/protein binding profiles. The general, non-specific drug/protein interaction was determined by comparison of the level of radioactivity for each tissue sample plus the drug with that of the drug run on a separate track (Fig. 30.4). Affinity of drug for a certain protein was assessed by the detection of a peak of radioactivity defined by at least four points, above background.

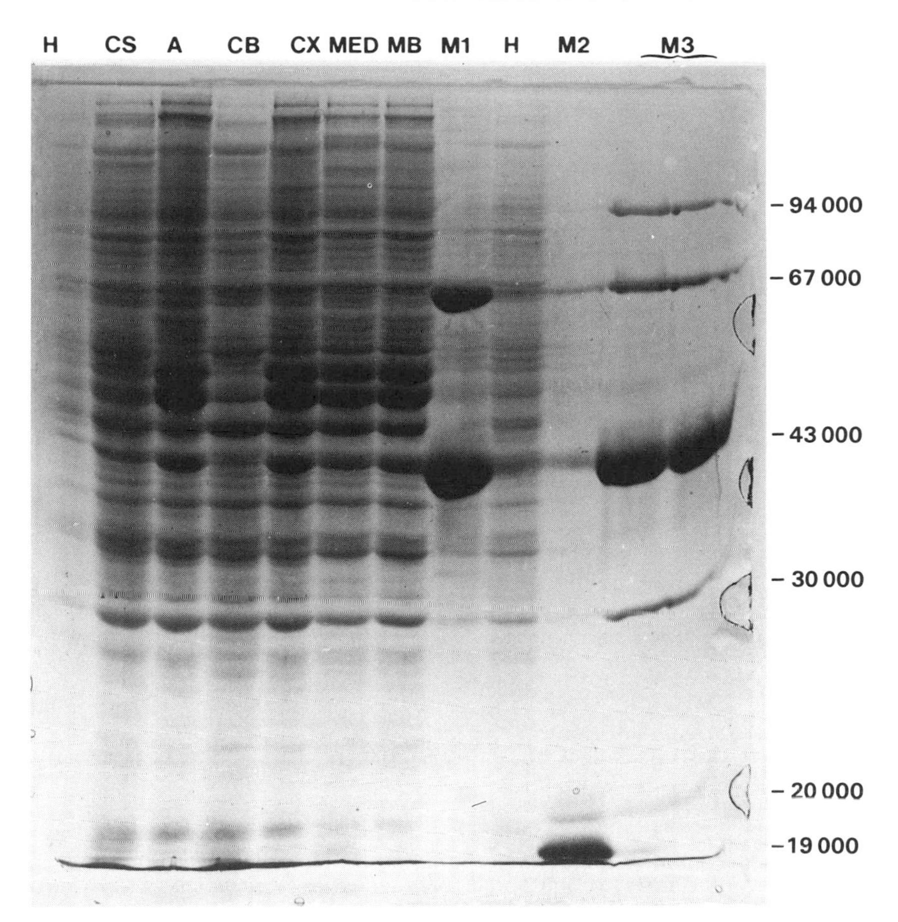

Fig. 30.4 SDS-PAGE gels of rat brain soluble proteins after staining with Coomassie Blue. A, amygdala; CB, cerebellum; CS, corpus striatum; CX, cortex; H, hypothalamus; ME, medulla; MB, midbrain; M1, albumin and ovalbumin; M2, RNAase; M3, marker mixture.

CPZ, FPZ and HPD exhibit protein interaction which is not observed with either TRZ or CLZ (Fig. 30.5). FPZ and HPD, in general, show higher levels of non-specific binding than CPZ. Although one might expect this to be an artefact of the relative metabolic instability of the tritium label, the fact that much lower levels of binding were found with tritiated TRZ and CLZ does not support this idea. Assuming these phenomena to be related to biological causes, such binding profiles suggest that those drugs which produce more potent extrapyramidal side effects have higher non-specific binding profiles. It seems that binding is higher in microsomal proteins, particularly in the region of 45–55 kDa, and binding in the microsomal fraction is reduced, in most cases, after

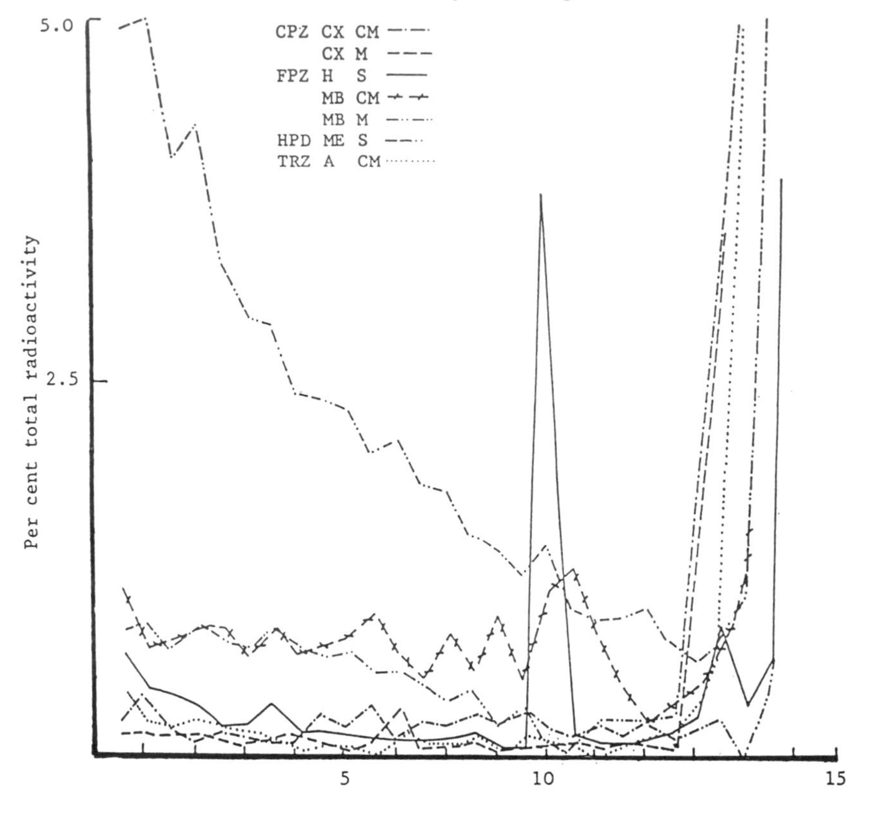

Fig. 30.5 Representative drug/protein binding profiles after electrophoresis. CX, cortex; H, hypothalamus; MB, midbrain; ME, medulla; A, amygdala; CM, control microsomes; M, microsomes; S, soluble.

incubation has allowed any metabolism to occur. The presence of discrete peaks also indicate a higher affinity for microsomal proteins in the same molecular weight region (Table 30.2).

From a biochemical viewpoint, the microsomal preparations are composed of RNA, phospholipids, ribosomal and vesicular proteins (Testa and Jenner, 1976). From a functional aspect, microsomes contain various enzyme systems, such as NADPH-linked cytochrome P-450, NADH-linked cytochrome b_5, various phosphatases, glucuronyl tranferases, cytochrome c reductases and flavoproteins. In addition to the structural ribosomal and vesicular proteins, many other proteins resulting from protein synthesis at ribosomes might also be associated with microsomes.

The soluble fraction contains a number of enzyme systems, including alcohol dehydrogenase, NADH-linked nitroreductase, aldehyde reductase, other oxidoreductases and various conjugating enzyme systems.

Table 30.2 Molecular weights of proteins exhibiting possible higher affinity for neuroleptic drugs in seven regions of rat brain. Numbers represent approximate molecular weights of peak centres along electophoresis track. A, amygdala; CB, cerebellum, CS, corpus striatum; CX, cortex; H, hypothalamus; ME, medulla; MB, midbrain; CM, control microsomes; MI, microsomal incubate; S, soluble. (For drug abbreviations see text.)

Drug	*Protein*	*A*	*CB*	*CS*	*CX*	*H*	*ME*	*MB*
CPZ	CM	39 000		100 000 51 000 37 500			94 000 55 000 48 000 40 000	
	MI		50 000 43 000		48 000	50 000	38 000	
	S						94 000 48 000 35 000	
FPZ	CM			67 000 48 000 25 000				50 000 36 500
	MI	50 000	53 500 43 000 32 500		53 500 25 000		47 500	
	S					35 000 24 000		
HPD	CM	94 000 60 000 30 000	94 000 52 000	55 000	94 000 55 000 46 500 32 500			94 000 47 500
	MI	94 000 60 000 30 000	94 000 52 000	55 000 46 000	55 000 35 000	75 000 53 500 37 500	High non-specific	
	S						High non-specific	
TRZ	CM					94 000		
	MI							
	S							
CLZ	CM							
	MI							
	S							

Since most of the binding of neuroleptic drugs occurs with microsomal protein, particularly in the 45 000–55 000 region, these drugs appear to have a greater affinity for structural and functional microsomal proteins. Since cytochrome P-450s have molecular weights of approximately 50 000, it is possible that a drug may bind with cytochrome P-450 prior to the initiation of metabolism. Following metabolism, the metabolite is released in a form which has a lower affinity for protein.

This is in agreement with the observed reduction in radioactivity following incubation of drugs with the active microsomal metabolizing system. If dopamine receptor proteins are associated with the microsomal fraction, as is likely, this could explain the affinities observed for microsomal proteins at other molecular weights.

Finally, neuroleptic drugs have also shown affinity for other proteins, such as soluble proteins with molecular weights of 24 000 and 48 000 and microsomal proteins of 55 000 and 75 000. These approximate molecular weights correspond to certain brain proteins which have been previously identified: S-100 (24 000) and 14-3-2 (50 000) soluble proteins (Grasso *et al.*, 1977; Perez *et al.*, 1970) and neurofilaments and glial filaments (55 000 and 69 000) (Runge *et al.*, 1981).

These observations generally indicate that a relationship may exist between drugs which induce extrapyramidal disorder and binding to neuromacromolecules. However, these preliminary observations will need to be investigated further to distinguish between any interaction relating to the pharmacology of the drug and those responsible for undesirable side effects.

REFERENCES

Burki, H.R. (1979) Extrapyramidal side effects. *Pharmacol. Ther.*, **5**, 525–34.

Christensen, E., Moller, J.E. and Faurbye, A. (1970) Neuropathological investigation from patients with diskinesia. *Acta Psychiatr. Scand.*, 4614–23.

Grasso, A., Roda, G., Hogue-Angeletti, R.A. *et al.* (1977) Preparation and properties of the brain specific protein 14-3-2. *Brain Res.*, **124**, 497–507.

Kler, A. and Rosen, A. (1985) Difference in the pattern of soluble proteins from rat brain regions. *J. Neurochem.*, **44**, 1333–9.

Maickel, R.P., Fedynskyj, N.M., Potter, W.Z., and Manian, A.A. (1974) Tissue localization of 7 and 8-hydroxychloropromazines. *Toxicol. Appl. Pharmacol.*, **28**, 8–17.

Perez, V.J., Olney, J.W., Cicero, T.J. *et al.* (1970) Wallerian degeneration in rabbit optic nerve: cellular localization in central nervous system of the S-100 and 14-3-2 proteins. *J. Neurochem.*, **17**, 511–19.

Read, S.M. and Northcote, D.H. (1981) Minimization of variation in response to different proteins of the coomassie blue G dye-binding assay for protein. *Anal. Biochem.*, **166**, 53–64.

Runge, M.S., Schlaepfer, W.W. and Williams R.C., Jr (1981) Isolation and characterization of neurofilaments from mammalian brain. *Biochemistry*, **20**, 170–175.

Testa, B. and Jenner, P. (1976) *Drug Metabolism: Chemical and Biochemical Aspects.* Marcel Dekker, New York.

Zirkle, C.L. and Kaiser, C. in *Psychopharmacological Agents*, Vol. 3 (ed. M. Gordon), Academic Press, New York, pp. 39–128.

Index

Page numbers in **bold** refer to figures, and those in *italic* to tables.